Sediments:
Chemistry and Toxicity of In-Place Pollutants

Edited by
Renato Baudo
John P. Giesy
Herbert Muntau

LEWIS PUBLISHERS, INC.
Ann Arbor Boca Raton Boston

Library of Congress Cataloging-in-Publication Data

Sediments: chemistry and toxicity of in-place pollutants/ edited by Renato Baudo, John P. Giesy, Herbert Muntau.
   p. cm.

Conclusions of the workshop sponsored by the Italian Hydrobiological Institute in Verbania-Pallanza, Novara, Italy, in 1989.
   Includes bibliographical references and index.
   ISBN 0-87371-252-8
   1. Soil pollution — Congresses. 2. Lake sediments — Congresses.
3. Water — Pollution — Congresses. I. Baudo, R. II. Giesy, John P.
III. Muntau, Herbert. IV. Istituto italiano di idrobiologia (Pallanza, Italy)

TD878.S43   1990
628.1'68'091692 — dc20   90-6123

LEWIS PUBLISHERS, INC.
121 South Main Street, Chelsea, Michigan 48118

PRINTED IN THE UNITED STATES OF AMERICA

Foreword

Any elements or compounds can be considered to be pollutants when they exist in excess concentrations in the wrong place. Both naturally occurring and synthetic substances can be pollutants. Human activities often mobilize and redistribute natural substances such that they can cause adverse effects. Due to chemical and physical transport processes these chemicals often become associated with the bottom sediments of aquatic systems. Once there, these pollutants can be transformed, become inactivated and buried, or reenter the water column. Also, they can cause toxic effects to benthic organisms.

The above observations seem obvious and, while not simple, it is possible to measure the concentrations of chemicals in sediments. However, to assess the importance of these in-place pollutants one must know more than how much of each chemical exists in the sediment. It is necessary to know the forms in which the chemicals exist and how available they are to benthic organisms or to be transported to the water column. It is also necessary to know the rate of transfer between the sediments and the water.

Sediments are a complex and dynamic matrix, which changes rapidly and often. Naturally occurring sediments can be inhospitable to benthic organisms due to small concentrations of dissolved oxygen and great concentrations of reduced gases and ammonia. Sediments can often be polluted with a number of contaminants, ranging from inorganic compounds such as nutrients or toxic heavy metals to complex synthetic organic chemicals. The ecotoxicologist is confronted with questions such as: How toxic are the sediments and what is causing the toxicity? How long will the toxicity last, and how much would sediment need to be diluted or cleaned before it would no longer be toxic? Questions of this type are difficult to answer because of a lack of understanding of the processes which control the movement and toxicity of in-place pollutants and methods to measure the concentrations and toxicity of these in-place pollutants.

Our experience in Lake Orta, which is contaminated with copper and ammonia and is acidified, has led us to realize that while we had a good understanding of the processes operating in the water column, to delve into the sediments would require new and better techniques. For that reason, we along with Michigan State University initiated the Pallanza Workshop series. The goal was to bring together recognized experts from around the world to discuss the current state of understanding of in-place pollutants and produce a published volume which reviews the current state of the art and identifies future needs and possible approaches.

The results of that effort are presented here in a comprehensive compendium of the study of elemental fluxes to and from sediments, bioconversions, bioavailability, and toxicity assessment. We at the Italian Institute of Hydrobiology are satisfied with the start we have made, but much remains to be learned.

Dr. R. de Bernardi, Director
Istituto Italiano di Idrobiologia
Pallanza, Italy, 1989

Preface

Lake Orta (Novara, Italy) first became contaminated with copper and ammonia in 1926. Continued pollution eventually killed all of the fish and most the phytoplankton, zooplankton and benthic invertebrates in this subalpine lake. Oxidation of ammonium ion (NH_4^+) has resulted in acidification of the lake. In 1989 the pH of the water column of the lake was approximately 4.0. The copper concentration in the water column was approximately 30 $\mu g/L$, and the lake could be described as being almost completely devoid of life. Even bacterial processes of some types were greatly depressed. In 1989 the loadings of pollutants had been essentially stopped, and remedial action in the form of liming had begun.

Limnologists have long studied Lake Orta and the contamination and its effects on the organisms in the water column documented. However, when the remedial actions were begun in 1989, scientists realized that while predictions about the response of the chemistry and biology of the water could be made, it was extremely difficult to predict what would happen in the sediments. It was concluded that scientists had too little information about the chemistry and toxicology of pollutants in sediments and assessment methods to predict what would happen in the sediments as a result of liming the water column.

The Italian Hydrobiological Institute has sponsored the workshop on "Fates and Effects of In-Place Pollutants in Aquatic Ecosystems" in Verbania-Pallanza, Novara, Italy to discuss processes which control the chemistry, dynamics, bioavailability, and toxicity of pollutants in lake sediments and methods available to monitor and assess the dynamics and toxicity of contaminants in lake sediments. While the workshop was focused on contaminants in subalpine lakes and on remedial action, the reviews and insights on the processes controlling the chemistry and movement of pollutants in sediments and the methods of studying them will be of interest and have practical applicability to limnologists, aquatic toxicologists, engineers, and lake managers.

The conclusions of the workshop attendees were that the cycling and effects of in-place pollutants are complex and improved methods will be needed to be able to understand the underlying processes. Particularly, methods to study processes in the sediment-water interface microlayer under field conditions are needed. Existing methods, which make use of benthic chambers, sedimentation traps, and box corers, should be improved, since the results currently cannot be extrapolated to the natural conditions of whole ecosystems. Also, laboratory studies of processes, while necessary, can be misleading. Studies of rates of flux under field conditions are difficult and subject to artifacts, but "peepers" seem to offer a technology which can be exploited to good advantage.

The chemistry and dynamics of in-place pollutants are poorly known and predictive power is quite poor. Long-term investigation on whole-lake releases of contaminants from sediments under controlled conditions, as well as the results of remedial actions, demonstrate the inadequacy of predictions based on empirical or theoretical models alone.

This bleak picture of understanding of processes involving pollutants in sediments would seem to indicate that one would have no chance to understand or predict the toxic effects of these pollutants. However, it was demonstrated at the workshop that significant progress is being made to provide the tools to measure and map the distributions of concentrations of pollutants and their effects in sediments. Methods which use batteries of simple bioassays and chemical and physical manipulations allow for the classification of the toxicity of complex mixtures of pollutants and determination of the causes of toxic effects.

This volume represents the first in a series of books which will document the most recent understanding of the processes which affect in-place pollutants and methods to assess the sources, causes, and intensity of adverse effects of toxic pollutants in sediments and point the way toward more effective methods of monitoring, predicting, and controlling their effects in aquatic systems.

R. Baudo, J. Giesy, and H. Muntau
Pallanza, Italy, 1989

Acknowledgments

This volume resulted from the first Pallanza workshop, which was held August 28–30, 1989 in Verbania-Pallanza, Novara, Italy. That workshop and preparation of this volume were supported by many institutions. The primary financial and logistical support for the workshop were provided by the Italian Institute of Hydrobiology, Pallanza (National Research Council of Italy). Financial and logistical support was also provided by the Department of Fisheries and Wildlife, Pesticide Research Center and Agricultural Experiment Station of Michigan State University, East Lansing, Michigan, U.S.A. The workshop was cosponsored by the Society of Environmental Toxicology-Europe, The Hague, The Netherlands.

As with all endeavors of this type, many individuals made the workshop and subsequent volume possible. First, Dr. R. de Bernardi, Director of the Italian Institute of Hydrobiology, provided the finances, direction, and leadership for the workshop. Dr. R. Mosello assisted with local arrangements, provided information about Lake Orta, and led the excursion to Lake Orta. Local arrangements were organized by P. Panzani, C. Corbella, L. Nobili, P. Poletti, and V. Baudo. Graphic design and artwork were coordinated and produced by P. Poletti. Visual aids and photographic services were provided by J. Toriati. J. Thompson, P. Jones, and R. Hoke of Michigan State University provided logistic and communications support.

The following persons assured the quality of the contributions to this volume by serving as reviewers and referees:

- P. F. Landrum, Great Lakes Environmental Research Laboratory, NOAA, Ann Arbor, Michigan, U.S.A.
- D. L. King, Department of Fisheries and Wildlife, Michigan State University, East Lansing, Michigan, U.S.A.
- F. M. D'Itri, Institute of Water Research, Michigan State University, East Lansing, Michigan, U.S.A.
- G. A. Burton, Department of Biology, Wright State University, Dayton, Ohio, U.S.A.
- D. Long, Department of Geology, Michigan State University, East Lansing, Michigan, U.S.A.

Dr. Renato Baudo is Senior Scientist at the Italian Institute of Hydrobiology, National Research Council. He graduated from the High School of Chemical Engineering in 1969, adding a degree in Biological Sciences from the University of Milan in 1975. Since 1970 he has worked at the Italian Institute of Hydrobiology, National Research Council; his main intereste are in the fields of cycling and ecotoxicological effects of trace elements and organic micropollutants in aquatic environments.

He has been Consultant to the Italian Government and to the Commission of the European Communities for the protection of freshwater and marine life; he is also a member of the Council of the Society of Environmental Toxicology and Chemistry — Europe.

John P. Giesy is Professor of Fisheries and Wildlife at Michigan State University in East Lansing, Michigan, where he is also affiliated with the Pesticide Research Center, Center for Environmental Toxicology and Ecology, and Evolutionary Biology Program. He attended Alma College in Alma, Michigan where he obtained a Bachelors of Science Degree *summa cum laude* with honors in Biology in 1970. Prof. Giesy obtained Masters and Doctor of Philosophy Degrees in Limnology from Michigan State University in 1971 and 1974, respectively. From 1974 until 1981 he was affiliated with the Savannah River Ecology Laboratory and a faculty member in the Department of Zoology at the University of Georgia. Prof. Giesy considers himself an aquatic toxicologist with interests in many aspects of this field, including both the fates and effects of potentially toxic compounds and elements. He has conducted research into the movement, bioaccumulation, and effects of toxic substances at different levels of biological organization, ranging from biochemical to ecosystem. Prof. Giesy has done extensive research in the areas of metal speciation, multispecies toxicity testing, biochemical indicators of stress in aquatic organisms, fate and effects of polycyclic aromatic hydrocarbons, and photo-enhanced toxicity of organic compounds. Currently Prof. Giesy and his research group are actively studying the toxicity and reproductive effects of organic compounds on fish and fish-eating birds and mammals with a special interest in raptors such as hawks and eagles. Prof. Giesy has been active in the development and application of methods for the assessment of the toxicity of contaminated sediments, especially in the North American Great Lakes. He has received more than $6,000,000 from many local, state, federal, and international agencies and organizations to conduct his research, which has resulted in the publication of 121 peer-reviewed publications and hundreds of lectures, world-wide. Prof. Giesy works frequently in Europe with many universities, research establishments, and government agencies. He is the recipient of the Sigma Xi Meritorious Research Award and the Ciba-Geigy Agricultural Recognition Award. Prof. Giesy served as President of the Society of Environmental Toxicology and Chemistry in 1990–1991.

Herbert Muntau is Senior Scientist at the Environment Institute Joint Research Centre of the Commission of the European Communities, Ispra, Italy. He holds degrees in Chemical Engineering and Chemistry, dedicating most of his 25-year professional career to the biogeochemical cycling of trace metals in aquatic ecosystems and the analytical aspects of such studies, including speciation. In the frame of these activities much attention was paid to the general problem of analytical reliability leading to the development of some one hundred environmental reference materials. In recent years the problems of in-place pollutants and pollutant non-point source identification and quantification have occupied major space in his research activities.

Contents

Chapter 4
Spatial and Temporal Variations in Microbial Processes in Aquatic
Sediments: Implications for the Nutrient Status of Lakes

Richard G. Carlton and Michael J. Klug

Chapter 5
Mechanisms Controlling Fluxes of Nutrients Across the Sediment/
Water Interface in a Eutrophic Lake

René Gächter and Joseph S. Meyer

Chapter 9
Freshwater Sediment Quality Criteria: Toxicity Bioassessment 265
John P. Giesy and Robert A. Hoke

CHAPTER 1

Lesser Known In-Place Pollutants and Diffuse Source Problems

Renato Baudo and Herbert Muntau

1.0 INTRODUCTION

The depletion of groundwater reserves is a growing concern in all member countries of the European Community. Surface water from rivers and lakes is already used intensively, and this will increase as the basis for drinking water supply in European countries in the future. Thus, it is becoming more and more imperative that water supplies be protected from contamination. Presently, national and European directives regulate the release of contaminants into the waterways by wastewater discharge. The correct implementation of these directives is certainly more of an administrative than a scientific problem. As point sources of pollution are more effectively controlled through monitoring, regulations, and technology, the attention of environmental scientists is shifting to the more difficult task of controlling non-point sources of pollution. Such contaminants reach the aquatic systems by a number of uncontrolled pathways, or non-point sources. However, a great deal of semantic confusion exists in the literature with respect to "point sources" and "non-point sources", suggesting the need for clearer definition in the future.

2.0 SEDIMENTS AS DIFFUSE SOURCES

Bottom sediments play an important role both as a sink where contaminants can be stored and as a source of these contaminants to the overlying water and to biota. Sediments have only recently been considered as non-point sources of environmental contaminants. In fact, contaminated sediments have not been included in the comprehensive reviews of non-point sources published regularly by the *Journal of the Water Pollution Control*

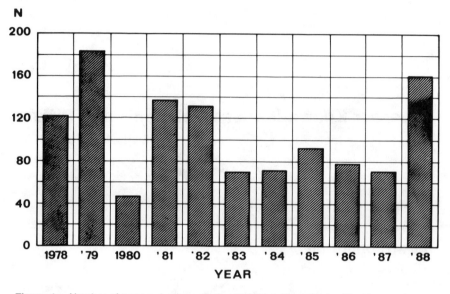

Figure 1. Number of papers dealing with non-point sources reviewed in 10 years by the *Journal of the Water Pollution Control Federation.*

Federation since 1978 (Figure 1). In these reviews (Browne 1978, Browne & Grizzard 1979, Browne 1980, 1981, Browne *et al.* 1982, Younos & Smolen 1983, Younos *et al.* 1984, Newell *et al.* 1985, Kenimer *et al.* 1986, Younos *et al.* 1987, Wyatt *et al.* 1988), typically, the non-point sources considered have been the various forms of land drainage and atmospheric deposition. Runoff from land can be further subdivided. One category is agricultural runoff, which also includes forestry management, such as clear cutting of forests. A second is urban runoff, including road runoff, sewer system infiltrations, highway runoff, and deicing. A third category is solid waste leachate, sludge disposal on land. Another is runoff from coal piles, surface storage of oil shale, waste rock dumps undergoing pyritic oxidations, and leaching from surface or subsurface mining. Atmospheric loading, a major non-point source of contaminants, includes chemicals in rain, dry deposition, and snowpack runoff, and all of these may reflect long-range transport of contaminants from many human activities.

The loadings from all of these sources may include the sediment burden (particulate matter which causes turbidity), bacteria and viruses, and adsorbed and dissolved contaminants such as metals or organo-xenobiotics. Most of the published papers deal with two major groupings. One is nutrients or pesticides, in relation to the agricultural runoff. The second is oil and grease, hydrocarbons, or metals, in urban, suburban, and rural runoff.

Several models have been developed to estimate the export rates of the various elements and compounds as a function of land use. An example is given in Table 1. The contributions of metals and organic micropollutants from agricultural activities are usually not significant compared to urban runoff (Ihnat 1982, cited in Younos & Smolen 1983). From 85 to 99% of the lead entering the Great Lakes comes from non-point sources (International Refererence Group on Great Lakes Pollution from Land Use Activities 1978, cited in Browne & Grizzard 1979). "Amongst the numerous non-point sources of contaminants – caused by human activities or 'natural' – to aquatic systems, urban runoff accounted for 71% of the higher molecular weight polycyclic aromatic hydrocarbons (PAHs) and 36% of the total PAHs entering Narragansett Bay" (Hoffman et al. 1984, cited in Newell et al. 1985). In Rhode Island, runoff from roads "could be the source of over 50% of the solids, polycyclic aromatic hydrocarbons, lead, and zinc loads that enter the Pawtucket River" (Hoffmann et al. 1985, cited in Kenimer et al. 1986). However, the point sources in urban areas can have significant impacts on surface waters because "accidental or deliberate discharges from point sources under wet weather conditions were often the primary cause of wet weather impacts" (Heaney et al. 1981, cited in Browne et al. 1982).

3.0 LESSER KNOWN POLLUTANTS

3.1 External Non-Point Sources

Currently, most researchers concentrate on a small number of elements, including Cu, Zn, Pb, Cd, Cr, Ni, Fe, Mn, and Hg; a large data set on these elements has been accumulated. However, few authors have considered rarer trace elements except in relation to specific sources. Examples of trace elements emanating from non-point sources include:

1) molybdenum from soil polluted by industrial waste products (Grigoryan & Galstyan 1981, cited in Browne et al. 1982)
2) boron and fluorine leaching from surface storage of oil shale (Bates et al. 1981, cited in Browne et al. 1982)
3) selenium produced by the mining and smelting of copper-nickel ores at Sudbury (Nriagu & Wong 1983, cited in Younos et al. 1984)
4) silver from an acid mine drainage (Jones 1986, cited in Younos et al. 1987)
5) arsenic from urban retention and recharge basins (Nightingale 1987, cited in Wyatt et al. 1988)
6) selenium and tellurium produced by large-scale combustion of fossil fuels and accumulated in snowfields (Chiou & Manuel 1988) and water bodies (Dumont et al. 1980)
7) Ag, As, Be, Co, Mo, Se, Sn, Te, Tl, V, and Zn, emitted along with Cd, Cu,

Table 1. Metal Export Rates (kg ha^{-1} year^{-1})

Land Use	Zn	Pb	Cu
Large lot single family	0.07–0.24	0.06–0.22	—
Single family	0.10–0.20	0.12–0.23	—
Multifamily	0.31–0.39	0.68–0.84	—
High–rise residential	1.03–1.15	1.72–1.91	—
Shopping center	3.15–3.24	2.72–2.81	—
Central business district	2.90–2.99	6.66–6.84	—
Agriculture	0.005–0.300	0.002–0.080	0.002–0.090
Cropland	0.026–0.083	0.005–0.006	0.014–0.064
Improved pasture	0.019–0.172	0.004–0.015	0.021–0.038
Forest	0.01–0.03	0.01–0.03	0.02–0.03
General urban	0.3–0.6	0.14–0.50	0.05–0.13
Residential	0.02	0.06	0.03
Commercial	0.25–0.43	0.17–1.10	0.07–0.13
Industrial	3.50–12.0	2.2–7.0	0.29–1.30

Source: After Brown & Grizzard (1979).

Hg, Mn, Ni, and Pb during fossil fuel burning and ore smelting (Campbell *et al.* 1985)

8) arsenic, beryllium, and vanadium, from fossil fuel burning power plants (Sharma *et al.* 1980; Fishbein 1981; Vaccarino *et al.* 1983)

9) tin from fossil fuel combustion, cement production, and extractive metallurgy (Morgan *et al.* 1977)

Occasionally, rare elements were routinely monitored in nationwide sampling networks for air or water pollution. This was the case for V and Ba along with Pb, Zn, Cd, Ni, Mn, Cu, Cr, and Fe in air samples collected at 15 stations in Belgium from 1972 to 1977 (Kretzschmar *et al.* 1980). Long-term trends in the concentrations of As have been monitored in water from over 300 locations in major U.S. rivers (Smith *et al.* 1987, cited in Wyatt *et al.* 1988).

However, both runoff from land and atmospheric fallout are external non-point sources to aquatic ecosystems, whereas contaminated sediments represent a potential internal diffuse source. As pointed out by Salomons *et al.* (1987), "Sediments are the ultimate sink for pollutants. However, before these sediments become part of the sedimentary record (e.g., deeply buried) they are able to influence the composition of surface waters."

3.2 Internal Diffuse Sources

In terms of residence times, the situations in rivers and lakes are very different. Following the abatement of chemical pollution into the rivers, elevated concentrations of pollutants continue for a period, depending on a number of physical factors such as discharge rate, stream velocity, and the morphology of the waterway. Also, decreasing amounts of contaminants in the dissolved and/or particulate phase, which had been deposited along the river in earlier times, continue to be released from the sediments due to leaching and resuspension. Provided that inputs of contaminants are stopped definitively, sediments in rivers eventually "cleanse" themselves. Sometimes this washout can be fairly rapid. For instance, when the source of Hg to the St. Clair—Detroit River system was stopped in the early 1970s, the concentrations of mercury in all components of the ecosystem decreased rapidly.

The situation for lakes, especially deep lakes, is different. A number of processes cause contaminants to migrate to the sediments and have relatively long residence times. Contaminants which enter as dissolved atoms or molecules are removed from the water column through such chemical processes as precipitation and coprecipitation and such biological processes as uptake by biota. The resulting particles are then mixed with those already in place to create a larger spectrum of differing types and sizes. These particles then undergo a series of transport, settling, resuspension, and deposition processes. All of them are accompanied by a series of chemical reactions which change both the particle matrix and the particle-associated contaminants.

Adsorption-desorption processes, degradation of organic matter, transformation of iron and manganese oxyhydrates to sulfides and vice versa, and biotransformation of contaminants ingested by organisms produce a steady change in the absolute and relative concentrations and forms of contaminants. The concentrations in the sediments are generally much greater than in overlying water. In such cases, the sediments may be a source of contaminants to the water column. The rates and forms of contaminants released are generally unknown, but biological factors, such as bioturbation, tend to enhance the net rates of release.

Bioturbation can affect the movement of contaminants to the water column by: 1) pumping interstitial water, usually rich in contaminants, out of the sediment; 2) transporting particulate matter from deeper sediment layers to the sediment-water interface; 3) depositing fecal pellets on the sediment surface; and 4) creating horizontal and vertical disturbance of sediment layering (Petr 1977).

In river sediments the movement of contaminants associated with sediments can be very dynamic, especially due to advection during certain seasons, and the movement of contaminants to and from lake sediments can be very dynamic as well. For a correct understanding of the respective role of external and internal pollutant loadings, it is necessary to quantify this source of contaminants to the water column, not only under "natural", undisturbed conditions, but also in situations where the aquatic system and its equilibria have been drastically disturbed. For example, construction of a dam or lock, dredging, or alterations of water flow may alter the circumstances. Under certain conditions, some in-place pollutants are likely to be an important source for the water body (Baudo 1987, Thomas 1987). This is often the case for nutrients such as nitrogen and phosphorus, and also for trace elements and organic micropollutants. For instance, from the calculated mass balance for a harbor in Michigan, PCB contamination was concluded to be due to a non-point source, most likely in-place pollutants in sediments (Richardson *et al.* 1985). In the Wabigoon/English River system, "Hg in contaminated surface sediments is almost certainly the primary source of the mercury now entering the water and biota in this contaminated watercourse" (Parks & Hamilton 1987). In the river Elbe, which is "heavily polluted with non-degradable or almost non-degradable substances such as heavy metals and organochlorine compounds, . . ., it is apparent that sediment and organism accumulation of pollutants is the 'critical path' for ecological effects in the Port of Hamburg" (Tent 1987).

While the internal release of nutrients from sediments in lakes has been studied extensively, organic micropollutants and trace metals have been less often considered, probably due to the inherent analytical difficulties associated with their determination at small concentrations in the water column.

4.0 ENVIRONMENTAL HAZARDS

The potential risk of many of these contaminants has not yet been completely evaluated. As a result, many authors have proposed that environmental hazards be estimated by comparing the concentrations of contaminants in the top layer of sediments with that in deeper sediments (Håkanson 1980). This would allow for the calculation of relative enrichment factors but not for the calculation of hazard, which can only be done with some type of laboratory or field bioassay.

Not only are sediments a source of contaminants to the overlying water column, but toxicants associated with sediments can have direct, adverse effects on organisms that live in or near the sediments. Also, once they are contaminated, benthic organisms can act as a source of contaminants to pelagic organisms. Until now most studies on the accumulation by and effects of contaminants on benthic invertebrates have been done for only a few predominant metals. For instance, classifications of sediment pollution have been proposed for only eleven metals or metalloids (Table 2, after Gambrell *et al.* 1983, and Thomas 1987; see also Chapter 9 in this volume). However, other elements such as Ag, B, Be, Mo, Pt, Se, Sn, Tl, and V, which are known to be biologically active, might be as dangerous as the more commonly studied ones and could represent a problem in the future

Table 2. Toxicity Classifications of Sediments in Relation to Selected Element Concentrations (mg element kg^{-1} dry weight)

Element	Ontario (MOE)	EPA (1977)			Screening Level
		Non-polluted	Moderately Polluted	Heavily Polluted	
Total Hg	0.3	<1.0		>1.0	1.0
Pb	50	<90	90–200	>200	50
Zn	100	<90	90–200	>200	75
Fe	10,000	<17,000	17–25,000	>25,000	20,000
Cr	25	<25	25–75	>75	100
Cu	25	<25	25–50	>50	50
As	8	<3	3–8	>8	
Cd	1			>6	2
Ni	25	<20	20–50	>50	50
Mn		<300	300–500	>500	500
Ba		<20	20–60	>60	

Sources: Gambrell et al. (1983), Thomas (1987).

due to their increased use. Even the rarest elements, despite their scarcity, are mined in considerable quantity every year because of their use in several human activities. Then they are released from point and non-point sources into the environment at an increased rate with respect to geological weathering. During this phase they enter the hydrographic network both in particulate and dissolved forms. On a worldwide basis most metals are estimated to be carried preferentially by the suspended particles (Table 3, Martin & Meybeck 1979). These estimates were made with the Dissolved Transport Index (DTI), the ratio of dissolved metal concentration to its total transport in rivers. These calculations suggest that a substantial part of the total load eventually becomes associated with sediments.

Following another approach, Dall'Aglio et al. (1986) measured the "geochemical mobility" of various elements. This is defined as the tendency of an element to move in natural water and to remain there in a stable, dissolved form relative to the mean content of the crust and the hydrosphere. Elements like Ag, As, Mo, and Sb were found to be highly mobile (Table 4).

When information about toxicity and availability are combined, many other elements join the three currently on the "black list" of the most

Table 3. Dissolved Transport Index in Rivers. Percentage of the Concentration of Dissolved Element over its Total Transport

%	Elements
90–50	Br, I, S, Cl, Ca, Na, Sr
50–10	Li, N, Sb, As, Mg, B, Mo, F, Cu, Zn, Ba, K
10–1	P, Ni, Si, Rb, U, Co, Mn, Cr, Th, Pb, V, Cs
1–0.1	Ga, Tm, Lu, Gd, Ti, Er, Nd, Ho, La, Sm, Tb, Yb, Fe, Eu, Ce, Pr, Al

Source: Martin & Meybeck (1979).

Table 4. Metal Content of the Geo- and Hydrosphere in the Absence of Obvious Pollution, and Geochemical Mobility

Element	Mean Crust mg kg^{-1}	Mean Soil Range mg kg^{-1}	Freshwater Range μg L^{-1}	Seawater Range μg L^{-1}	Geochemical Mobility[a]
Ag	0.07	0.01–8	0.01–3.5	0.03–2.7	+ + +
As	1.5	0.1–40	0.2–230	0.5–3.7	+ + + +
Cd	0.11	0.01–2	0.01–3	<0.01–9.4	+
Cr (III)	100	5–1500	0.1–6	0.2–50	*
Cr (VI)					+ + + +
Cu	50	2–250	0.2–30	0.05–12	+
Fe	4.1%	4–55%	10–1500	0.03–70	+
Hg	0.05	0.01–0.5	0.0001–2.8	0.01–0.22	*
Mn	950	20–10000	0.02–130	0.03–21	+
Mo	1.5	0.1–40	0.03–10	4–10	+ + +
Ni	80?	2–750	0.02–27	0.13–43	+ + +
Pb	14	2–300	0.06–120	0.03–13	+
Sb	0.2	0.2–10	0.01–5	0.18–5.6	+ + +
Se	0.05	0.01–2	0.02–1	0.052–0.2	+ +
Zn	75	1–900	0.2–100	0.2–48	+ + +

Sources: Bowen (1979) and Dall'Aglio *et al.* (1986).
[a] + + + + very high, + + + high, + + moderate, + low, * very low.

dangerous pollutants (Cd, Pb, and Hg; Table 5, Winner *et al.* 1980). Also, when the rates of mining are compared to the rates of natural cycling, elements with ratios which exceed a factor of ten can be considered potentially hazardous (Bowen 1979). These elements are shown in Table 6.

When the chemical compositions of marine animals and plants were com-

Table 5. Classification of Elements According to Toxicity and Availability

Nontoxic	Very Toxic and Relatively Accessible	Toxic but Very Insoluble or Very Rare
Na K Mg Ca H O N	Be Co Ni Cu Zn Sn	Ti Hf Zr W Nb Ta Re
C P Fe S Cl Br	As Se Te Pd Ag Cd	Ga La Os Rh Ir Ru Ba
F Li Rb Sr Al Si	Au Hg Tl Pb Sb Bi	

Source: Winner *et al.* (1980).

Table 6. Classification of Elements According to the Ratio: Rate of Mining/Rate of Natural Cycling

Ratio <10	Ratio >10
As, B, Bi, Co	Ag, Au, C, Cd
Fe, Mo, Ni, P, U	Hg, Mn, N, Pb
	Sb, Sn, W, Zn

Source: Bowen (1979).

Table 7. Classification of Elements (Normalized over Si) as a Function of the Ratio: Concentration of Algae/Concentration of Pelagic Clay

Biophile Elements (Ratio 100–1000)	Biophobe Elements (Ratio 1)	Intermediate Behavior
Sn, Zn, P, Hg, As, Cd, Ag, Se, U, K, Ca, B, Mg, Na (and possibly Au, I, Re, Sr, S, Br)	Mn, Th, Co, Sc, Fe, Ti, Ni, Cr, Ba, Si, V, Cs, Mo, F	Pb, Cu, Ga, Sb, W, Rb, Li

Source: Li (1984).

pared with those of clay, seawater, and soil, it was found that living organisms apparently separate elements into biophile and biophobe categories (Li 1984). Normalizing the concentrations over Si, the biophile elements show an enrichment factor (concentration organism/concentration pelagic clay) of between 100 and 1,000 while the biophobe elements have values around 1 (Table 7). This does not necessarily mean that the biophile elements are essential for life, but very likely they are involved in the same physicochemical processes.

Some limits have been proposed on the concentrations of potentially hazardous but less often studied metals to protect aquatic organisms and in drinking water (IRPTC Legal File 1983, Tiravanti *et al.* 1987). Some of these regulations are summarized in Table 8.

Among the relatively few studies that have dealt with many of the elements that, for one reason or another, could be considered potentially dangerous for the aquatic environments, even fewer have considered in detail the possible tendency to accumulate in bottom sediments. However, in view of the previously reported information on similar elements, sedi-

Table 8. Proposed Limits (mg L^{-1}) for the Protection of Aquatic Life and Drinking Water

Element	Aquatic Life[a]	Country[b]	Drinking Water[a]	Country
Ag	0.001–0.00025	USA+, Italy+	—	—
As	0.05	EEC	0.05	EEC
Ba	0.5	Switzerland	0.1	Italy
Be	0.0002	USSR	0.0002	USSR
Cr	0.005–0.250*	Many	0.05	Many
Ni	0.05–0.25**	Many	0.05–0.25**	Many
Sb	0.05	USSR	0.01	EEC
Se	0.010	Canada, Australia	0.010	EEC, Australia
Sn Inorganic	1.0	Italy+	—	—
Organic	0.001			

Sources: IRPTC Legal File (1983), Tiravanti *et al.* (1987).
[a] *, depending on chemical species and water hardness; **, depending on water hardness.
[b] +, proposed.

ments could act both as a sink and as an internal non-point source of these as well.

5.0 RISK ASSESSEMENT

In the context of quantitative evaluations of contaminant releases from sediments, a number of questions arise. For instance, how much of the pollutant is present in the critical sediment layer of the system? Estimates of sediment-associated contaminant inventories are based on analyses of a small number of sediment samples, and estimates of the total sediment surface may not be very accurate. Consequently, if high resolution is required, a large number of samples will be needed to permit accurate mapping of sediments. Alternatively, inventory estimates may be based on sedimentation flux measurements. However, these are more difficult to make than static measurements.

Our approach has been to measure both the species as well as total concentrations of the more common and less common metals in surficial sediments at a high enough density in a well-planned sampling grid to allow the construction of distribution maps of the "active" sediment layers by kriging techniques (Baudo, this volume). A major component of the uncertainty of the level of reliability of such an approach is in how the sampling plan is developed.

Recently, heterogeneity of spatial distribution of metals in sediments has been studied in Monvallina Bay of Lake Maggiore (Muntau *et al.* 1986). This area has been exposed to multimetal pollution. Sediment samples have been taken in an extremely dense station network (Figure 2: 30 samples/km^2; 115 samples on 3.8 km^2). However, similar sampling station densities are generally not feasible for larger lakes due to the analytical work load involved. In our studies we have observed complex patterns of the distributions of concentrations of metals. The influence of lake bottom morphology on sediment sorting and deposition has been evident. The map of the distribution of V concentrations (Figure 3) indicates the likely sources for the element. From this the burden potentially available for the ecosystem can be calculated despite a lack of information with which to judge the potential environmental risk. The application of geostatistical methods to metal speciation data and the related visualization described above for total metal concentrations yields additional information on metal origin, pathways, and fate. Assessment of the mobility of the total mass of the contaminant deposited in the reactive sediment layer is another area of uncertainty; it will be treated in detail in several chapters of this volume.

Once contaminant inventories, distribution patterns, and relative mobili-

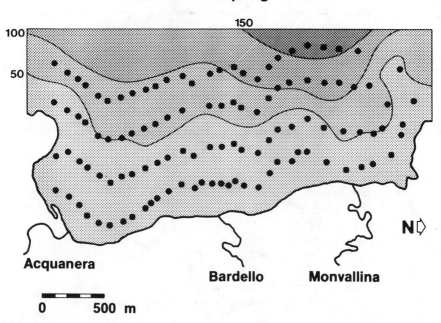

Figure 2. Sediment sampling stations in Monvallina Bay.

ties have been assessed, the next most critical steps are to determine the average net release rates and the potential harmful effects on the biota. A number of laboratory approaches have been devised, but their accuracy in field conditions is uncertain. Future studies aimed at describing the fates of better and lesser known pollutants must be encouraged.

In particular, the most accurate approach to predict, describe, and assess possible risks for the ecosystem should be more clearly identified. If possible, the whole procedure should be standardized, from planning the sampling to toxicity assessments. Then routine monitoring of threatened environments could readily provide a warning of the appearance of the first undesirable effects. Only at this moment is the great effort of identifying, checking, and controlling the pollutant, or pollutants, responsible for the contamination truly justified.

MONVALLINA BAY
VANADIUM   mg/kg dry w.t.

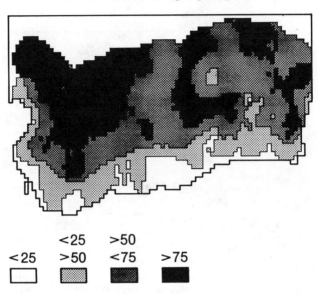

<25   >50
<25 >50 <75 >75

Figure 3. Vanadium distribution in sediments from Monvallina Bay.

REFERENCES

Baudo, R., 1987. Heavy metal pollution and ecosystem recovery. In O. Ravera (Ed.), *Ecological Assessment of Environmental Degradation, Pollution and Recovery.* Elsevier, Amsterdam, The Netherlands: 325–352.

Bowen, H.J.M., 1979. *Environmental Chemistry of the Elements.* Academic Press, New York: 333 pp.

Browne, F.X., 1978. Nonpoint sources. *J. Water Pollut. Control Fed.* 50: 1665–1674.

Browne, F.X., 1980. Nonpoint sources. *J. Water Pollut. Control Fed.* 52: 1506–1510.

Browne, F.X., 1981. Non-point sources. *J. Water Pollut. Control Fed.* 53: 901–908.

Browne, F.X. & T.J. Grizzard, 1979. Nonpoint sources. *J. Water Pollut. Control Fed.* 51: 1428–1444.

Browne, F.X., J.B. Orr, T.J. Grizzard & B.L. Weand, 1982. Non-point sources. *J. Water Pollut. Control Fed.* 54: 755–763.

Campbell, P.G.C., J.N. Galloway & P. Stokes, 1985. *Acid Deposition: Effects on Geochemical Cycling and Biological Availability of Trace Elements.* National Technical Information Service, Springfield, VA: PB85-176139/GAR.

Chiou, K.Y. & O.K. Manuel, 1988. Chalcogen elements in snow: Relation to emission source. *Environ. Sci. Technol.* 22: 453–456.

Dall'Aglio, M., R. Marchetti & E. Sabbioni, 1986. Distribuzione e circolazione dei metalli tossici in natura e loro effetti sulla biosfera. In M. Beccari, A.C. Di Pinto, D. Marani, M. Santori & G. Tiravanti (Eds.), *I metalli nelle acque: origine, distribuzione, metodi di rimozione. IRSA Quaderni* 71: 3–78.

Dumont, J.N., T.W. Schultz & S.R. Freeman, 1980. Uptake, depuration, and distribution of selenium in *Daphnia* and its effects on survival and ultrastructure. *Arch. Environ. Contam. Toxicol.* 9: 23–40.

Fishbein, L., 1981. Sources, transport and alterations of metal compounds: An overview. I. Arsenic, beryllium, cadmium, chromium, and nickel. *Environ. Health Perspect.* 40: 43–64.

Gambrell, R.P., C.N. Reddy & R.A. Khalid, 1983. Characterization of trace and toxic materials in sediments of a lake being restored. *J. Water Pollut. Control Fed.* 55: 1201–1210.

Håkanson, L., 1980. An ecological risk index for aquatic pollution control. A sedimentological approach. *Water Res.* 14: 975–1001.

IRPTC Legal Files, 1983. *International Register of Potentially Toxic Chemicals,* Vol. I and II. United Nations Environment Programme, Geneva, Switzerland.

Kenimer, A.L., T.M. Younos & S. Mostaghimi, 1986. Nonpoint sources. *J. Water Pollut. Control Fed.* 58: 603–606.

Kretzschmar, J.G., I. Delespaul & T. De Rijck, 1980. Heavy metal levels in Belgium: A five year survey. *Sci. Total Environ.* 14: 85–97.

Li, Y.-H., 1984. Why are the chemical compositions of living organisms so similar? *Schweiz. Z. Hydrol.* 46: 177–184.

Martin, J.M. & M. Meybeck, 1979. Elemental mass balance of material carried by major world rivers. *Mar. Chem.* 7: 173–206.

Morgan, J.J., W. Bach & E. Eriksson, 1977. Source functions: Fossil fuel combustion products, radionuclides, trace metals, and heat. Group report. In W. Stumm (Ed.), *Global Chemical Cycles and Their Alteration by Man.* Abakon Verlagsgesellschaft, Berlin: 291–311.

Muntau, H., M. Van Son, R. Baudo, P. Schramel, G. Marengo, A. Lattanzio & L. Amantini, 1986. Heavy metal variability in sediments and related sampling strategies. Paper presented at the EUCHEM Conference on Sampling Strategies and Techniques in Environmental Analysis, RIVM, Bilthoven, The Netherlands, January 20–24, 1986.

Newell, A.D., T.M. Younos, M.D. Smolen, S. Mostaghimi, T.A. Dillaha & R.P. Maas, 1985. Nonpoint sources. *J. Water Pollut. Control Fed.* 57: 630–634.

Parks, J. & A. Hamilton, 1987. Accelerating recovery of the mercury-contaminated Wabigoon/English River system. *Hydrobiologia* 149: 159–188.

Petr, T., 1977. Bioturbation and exchange of chemicals in the mud-water interface. In H.L. Golterman (Ed.), *Interactions Between Sediments and Freshwater.* W. Junk, The Hague, The Netherlands: 216–226.

Richardson, W.L., K.R. Rygwelski & R.P. Winfield, 1985. Mass balance of toxic substances in an IJC Class A area of concern. In *Programs and Abstracts of the 28th Conference on Great Lakes Research,* University of Wisconsin-Milwaukee, June 3–5, 1985: 61.

Salomons, W., N.M. de Rooij, H. Kerdijk & J. Bril, 1987. Sediment as a source for contaminants? *Hydrobiologia* 149: 13–30.

Sharma, R.P., R.D.R. Parker, S.G. Oberg, D.R. Bourcier & M.P. Verma, 1980. Toxicological aspects of vanadium: A by-product of western energy development. In *Health Implications for New Energy Technology*. Ann Arbor Science, Ann Arbor, MI: 645–652.

Tent, L., 1987. Contaminated sediments in the Elbe estuary: Ecological and economic problems for the Port of Hamburg. *Hydrobiologia* 149: 189–199.

Thomas, R.L., 1987. A protocol for the selection of process-oriented remedial options to control *in situ* sediment contaminants. *Hydrobiologia* 149: 247–258.

Tiravanti, G., C. Baldan, R. Baudo, R. Marchetti & M. Pettine, 1987. Gruppo 4: Metalli. In *Atti del Convegno Criteri e limiti per il controllo dell'inquinamento delle acque. Dieci anni di esperienze*. Roma, 26–27 Giugno 1986. Istituto di Ricerca sulle Acque. *Quaderni* 75: 195–336.

Vaccarino, C., G. Cimino, M.M. Tripodo, G. Lagana, L. Lo Giudice & R. Matarese, 1983. Leaf and fruit necroses associated with vanadium rich ash emitted from a power plant burning fossil fuel. *Agric. Ecosyst. Environ.* 10: 275–283.

Winner, R.W., M.W. Boesel & M.P. Farrell, 1980. Insect community structure as an index of heavy-metal pollution in lotic ecosystems. *Can. J. Fish. Aquat. Sci.* 37: 647–655.

Wyatt, L., J. Spooner, W. Berryhill, S.L. Brichford & A.L. Lanier, 1988. Nonpoint sources. *J. Water Pollut. Control Fed.* 60: 925–933.

Younos, T.M. & M.D. Smolen, 1983. Non-point sources. *J. Water Pollut. Control Fed.* 55: 748–752.

Younos, T.M., M.D. Smolen, C.A. Eiden, R.P. Maas, S.A. Dressing & T.A. Dillaha, 1984. Non-point sources. *J. Water Pollut. Control Fed.* 56: 689–692.

Younos, T.M., M.D. Smolen, S. Mostaghimi & J. Spooner, 1987. Nonpoint sources. *J. Water Pollut. Control Fed.* 59: 487–490.

CHAPTER 2

Sediment Sampling, Mapping, and Data Analysis

Renato Baudo

1.0 INTRODUCTION

Since the early studies of Zuellig (1956), freshwater and marine sediments have been often used to trace environmental contamination and to predict potential ecological modifications. However, until now no agreement has been reached regarding the standardization of the method of investigation, even though many authors claimed the need for such a standardization, with special reference to the procedures for sediment sampling, preservation, physical, chemical, and biological examination, and presentation of the results (Sly 1969, Foerstner & Wittmann 1979, Robbe 1981, de Groot *et al.* 1982).

This is not at all surprising, considering the complexity of the subject and the implications for the scientific, social, and economic applications of the results. In fact, according to Håkanson & Jansson (1983), as many as 12 different factors might influence the informative value of the sediment samples: type of water system (lentic or lotic), prevailing bottom dynamics, size of the water body, bottom roughness, anthropogenic factors, sediment chemical conditions, sediment physical and biological characteristics, number of samples, type of sampling net, sampling devices, sample handling, reliability of laboratory analysis; however, "no systematic study has yet been introduced which accounts for even half of these 12 factors."

Awaiting such a systematic study, this review will concentrate on the choice of the sampling tools, the selection of the sampling net, and the estimation of the required number of samples, whereas some of the other points will be covered in detail by the next chapters of this volume. Finally, some techniques largely used to produce sediment mapping will be compared—that is, to extrapolate a continuous distribution from the discrete one based on the actual observations.

2.0 SAMPLING DEVICES

As early as 1969, Sly pointed out that the choice of devices for sampling is of paramount importance to assure the representativeness of the true conditions. However, sampling is often governed by the availability of a given sampler or by the need to compromise between different requirements. Often, in fact, the same sample is intended to be used for physical, chemical, and biological investigations. Unfortunately, usually one kind of sampler may be appropriate for a specific task, but not for others.

Reviews of the different types of sampling devices have been published by several authors (Hopkins 1964, Wright *et al.* 1965, Bouma 1969, Sly 1969, Elliott & Tullett 1978, Robbe 1981, Elliott & Tullett 1983, Håkanson & Jansson 1983). Of those discussed, three main categories of sediment samplers can be recognized: dredges, grabs, and corers.

The first is essentially a vessel that can be dragged across and which digs into the sediment, from a boat (Elliott & Drake 1981b) or with a handle (Rofes 1980), to collect the surficial layer: this is mostly intended for collecting the benthic fauna, and it has the advantage of covering a large area of the bottom in a way which produces an "average" sample (Figure 1). However, it is usually difficult to establish the real surface scanned by the device and the depth of the sediment layer affected by the operation. In addition, the sediment material is disturbed by the mixing with the overlay-

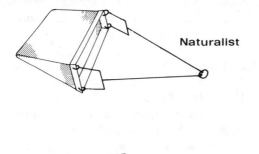

Naturalist

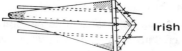

Irish

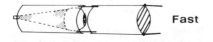

Fast

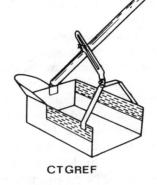

CTGREF

Figure 1.   Dredges commonly used to collect bottom fauna. (Modified from Elliott & Drake 1981b, Robbe 1981. With permission.)

ing water, so that no study on the porewater can be made (Robbe 1981). Since the dredge walls actually are made of fabric or have a net on the bottom, the dredges act as sieves, and only relatively coarse sediment material is retained, along with the bottom fauna.

The second type, the grab, is the favorite one for biologists (Dall 1981), since it produces large samples of known surface from the top layers of sediment, which are the ones that are more likely to be inhabited by the bottom flora and fauna. Grabs are usually composed of two metallic jaws that can be closed, after the grab reaches the bottom, either automatically or by sending a mechanical, acoustical, or electric signal from the surface. Examples of these grabs are shown in Figure 2. The drawbacks of these devices are the unpredictable penetration, which depends on the density of the sediment and on the weight and speed of the grab when it hits the bottom, and the more or less pronounced perturbations of the interface due to both the impact and the subsequent opening of the grab for removing the sample.

The third type of sediment sampler is the corer (Mackereth 1958, 1969, Richards & Keller 1961, Sly & Gardener 1970, Williams & Pashley 1979, Rofes & Savary 1981; Figure 3): in principle, this consists of a tube that can be inserted into the sediment to collect a cylindrical sample (but a square-barrel corer has been proposed by Kuehl et al. 1985).

The corers are usually designed to provide the maximum amount of sample with a minimum disturbance of the sediment itself. Since the first coring devices, the performance of corers has been improved to make them more practical and to provide a better core. Successive modifications are based either on intuitive judgment or on studies of the corer dynamics, even by analyzing the motion pattern (Zhong et al. 1986), to increase penetration and, at the same time, minimize disturbance and trouble in core recovery.

Many ways of deploying cores have been developed. Some corers may be operated by hand in shallow waters, others by scuba divers (Anima 1981, Bonem & Pershouse 1981, Martin & Miller 1982) to a depth of about 30 m, from a manned research submersible (Bothner & Valentine 1982), or from a boat, with or without the help of whinches (Anderson & Hess 1969, Hoyt & Demarest 1981, Irwin et al. 1983).

From a boat, many corers are operated simply by gravity. The coring device and its mechanical support are simply allowed to fall from the surface (Phleger 1951, Burns 1966, Brinkhurst et al. 1969, Hamilton et al. 1970, Kemp et al. 1971, Blakar 1978, Frithsen et al. 1983, Twinch & Ashton 1984). For these types, it has long been recognized that the size of the tubes affects the efficiency of recovery of the sediment (Emery & Dietz 1941, Emery & Hulsemann 1964, Wright et al. 1965, Patton & Griffin 1969, Hongve & Erlandsen 1979, Sly 1981). In addition, it has often been claimed that the corers introduce a noticeable disturbance and distortion of the

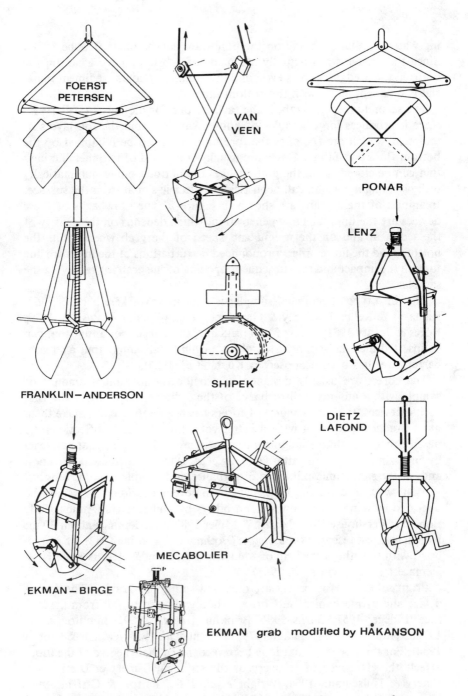

Figure 2. Examples of different types of grabs. (Modified from Robbe 1981, Håkanson & Jannson 1983, Blomqvist 1985. With permission.)

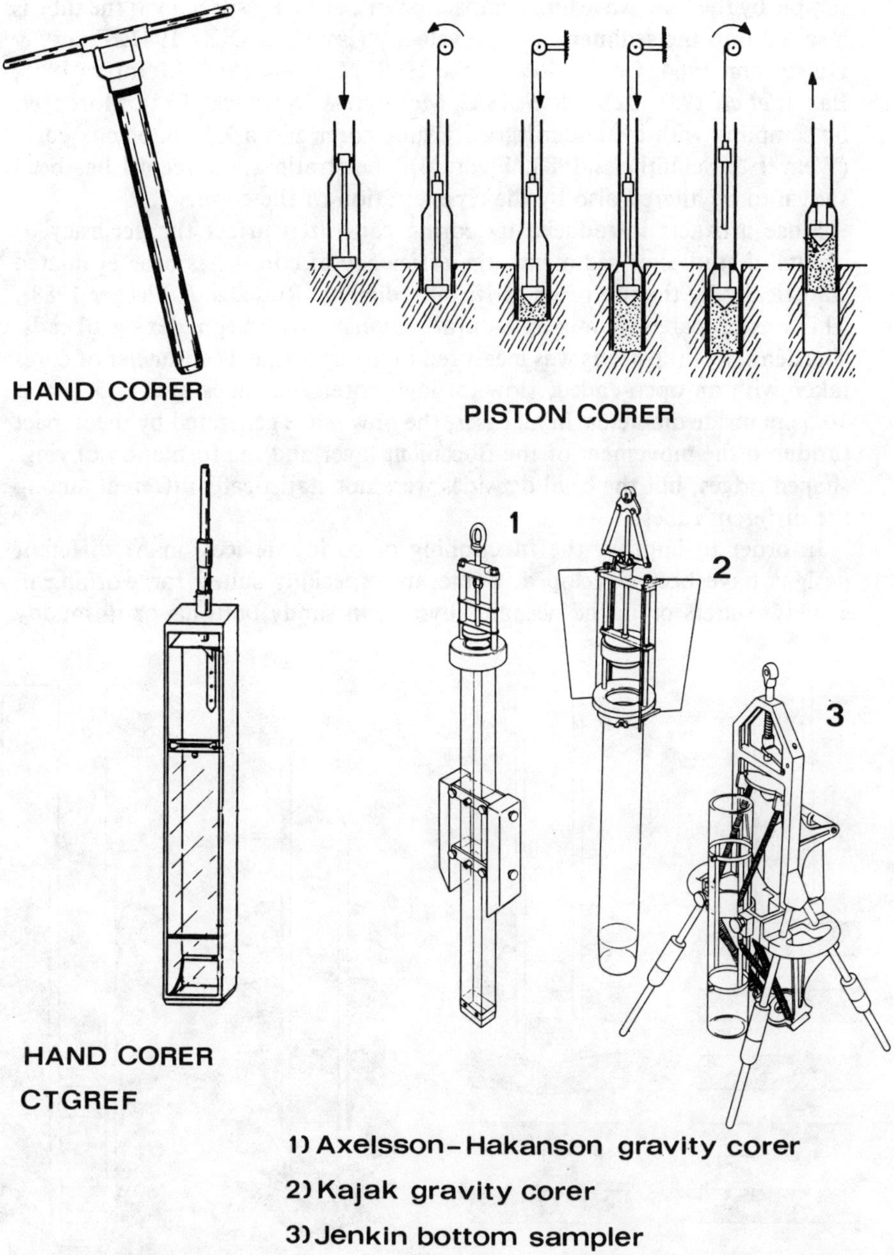

HAND CORER

PISTON CORER

HAND CORER

CTGREF

1) Axelsson–Hakanson gravity corer

2) Kajak gravity corer

3) Jenkin bottom sampler

Figure 3. Different types of corers. (Modified from Robbe 1981, Blomqvist 1985. With permission.)

sample by the bow-wave and compaction of the core, especially if the tube is inserted into the sediment at high velocity (Emery & Dietz 1941, Emery & Hulsemann 1964, Craib 1965, Burke 1968, Milbrink 1968, McIntyre 1971, Baxter *et al.* 1981, Lebel *et al.* 1982, McIntyre & Warwick 1984). Moreover, by sampling with a standard gravity tube corer and a Kastenlot box corer (Weaver & Schultheiss 1983, Figure 4), the stratigraphic record has been shown to be altered also by the repenetration of the corer.

Since artifacts introduced by coring can often affect the accuracy of results of studies of sediments, the efficiency of corers has been evaluated statistically by the use of an artificial sediment (Rutledge & Fleeger 1988). This was prepared by mixing natural sediments with Sephadex® gel beads. The density of the beads was measured in the uppermost centimeter of cores taken with an open-ended, flow-through corer and tubes of 2.6, 5.7, and 10.5 cm inside diameter. In all cases, the bow-wave generated by the impact produced the movement of the flocculent layer and the formation of ring-shaped ridges, but the bead densities were not statistically different among the different tubes.

In order to improve the functioning of coring devices, many different designs have been developed. Some are especially suited for working in shallow waters or in the oceanic abysses, in sandy bottoms or in muddy

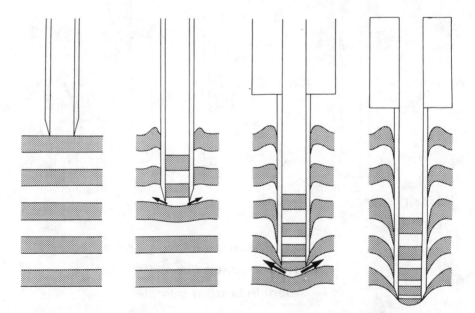

Figure 4.   Artifacts produced during core operations: plastic layers (grey bands) are short-ened more then stiff layers (white bands). (Modified from Weaver & Schultheiss 1983. With permission.)

sediments, hand operated or triggered from the boat. As a result, corers can be either very simple and inexpensive (Satake 1983, Ali 1984, Perillo *et al.* 1984) or rather complex and expensive. The device developed by Morris & Peters (1981), for instance, is a multipurpose stainless steel corer that can take many types of different samples: short cores from the sediment-water interface, subsurface cores (1.2 m long) in a square box barrel, archive cores 1 m long in 10-cm-diameter round liners, and long samples (3 m and over) in 16- and 10-cm-diameter liners.

The piston corer has been suggested as a standard coring device because it seems to be practical and to perform well. In the piston corer the support holds a mechanism that, after the device has reached the bottom, pushes the tube into the sediment by the action of springs, rubber bands, compressed air, or other suitable mechanic forces; this corer has been proposed as a standard, but Ross & Riedel (1967), Inderbitzen (1968), and McCoy (1980) criticized its use, especially because the free passage of water in the tube is prevented by the presence of the piston, and the latter, coming in contact with the bottom, tends to stir the fluid interface.

Other corers are particularly useful for sampling on hard substrates. For instance, McCoy & Selwyn (1983/1984) developed a special corer that uses a hydrostatic motor (to convert the difference in hydrostatic pressure into energy) and succeeded in driving a gravity corer into stiff marls where the conventional types were unsuccessful. A number of authors have described pneumatic vibration corers operated by commercial compressors, claiming they recover undisturbed samples from mud, peat, indurated calcareous sand, and coral (Lanesky *et al.* 1979, Meischner *et al.* 1981, Hoyt & Demarest 1981, Fuller & Meisburger 1982, Love *et al.* 1982, Smith 1984, Israel *et al.* 1987). Even conventional rock-drilling equipment has been mounted on an offshore platform to collect sediment samples (Rhodes 1981).

Special corers have also been devised to collect multiple cores. The one used by Barnett *et al.* (1984) consists of an array of plastic tubes mounted on a framework and pushed into the sediment by a hydraulic damper. Many multiple corers are remotely operated (Krogh & Sparck 1936, Willemoes 1964, Buzas 1968, Hamilton *et al.* 1970, Milbrink 1971, Jumars 1975, Nixon 1976, Reise 1979). Diver-operated multiple coring systems for describing small-scale spatial dispersion and patchy habitats have also been used (Jones & Watson-Russell 1984).

One of the most critical steps in collecting sediment cores is the closing of the tube after penetration to avoid sample loss, compaction, and degassing (Aviilov & Trotsyuk 1980). Several mechanisms have been proposed. One simple method to retain the sediments in an unsupported catcher-free plastic barrel is an O-ring (Pedersen *et al.* 1985). Alternatively, more complex mechanisms, such as spring-arm core closers, have been suggested to

Table 1. Characteristics of an Ideal Sediment Sampler

1. permit free water passage during the lowering, to avoid a pressure wave over the bottom
2. minimize frictional resistance and sediment deformation and compaction; that is, it must have a relatively small wall thickness compared to sample area, smooth inside surfaces, sharp edge and small edge angle
3. tightly close *in situ* both the top and the bottom of the sampler after the sampling
4. have at least one transparent side, to record apparent stratification and major features
5. allow for subsampling, to avoid distortion during extrusion
6. easily change weight, to work on different substrata
7. be manually operable, and thus not exceeding 20 kg in weight, for use without special equipment and aids
8. cover a sample area and depth as large as possible, to yield enough material for different kinds of analysis
9. be easy to operate, and not require extensive training of personnel

Source: Håkanson & Jansson (1983).

accommodate various diameter core tubes (Hartwig 1984). The freezing corer avoids the need for this operation because a sediment layer is actually frozen on the outside of the tube, cooled inside by liquid nitrogen and dry ice (Klemens 1984).

To complete the picture, it can be said that a special sampler can be used for collecting the sediment in the benthic boundary layer (Sholkovitz 1970, Shiemer & Schubel 1970, Kajihara *et al.* 1974, Bryant *et al.* 1980) and that an appropriate device should be selected for each type of study to minimize artifacts and facilitate ease of operation and sample recovery.

Sedimentation traps have been used for collecting material representative of the sediments (Sklash *et al.* 1986, Kraus 1987), but obviously this can only be done if the objective of the study requires the measuring of something in the sediment-to-be materials, instead of the already settled particulate matter. As an alternative, instead of collecting the sediments, an *in situ* study can be done using benthic chambers, bell jars, and a box corer (Carlton & Wetzel 1985).

According to Håkanson & Jansson (1983), an ideal sediment sampler should have certain characteristics (Table 1). There are other requirements; the sampler must be inexpensive (there is always the risk of losing it on the bottom) and provided with safety devices, such as, for instance, those that prevent accidental closing over the hand of the operator. In addition, specific purposes for which samples are taken may require the avoidance of some material or the covering of the surface coming in contact with the sediment or the adoption of special techniques to prevent temperature and pressure changes, and so on. Obviously, these requirements are only indicative. For instance, point 7 states that the device must be manually operable, so it must weigh less than 20 kg. This is reasonable if only a few cores have to be taken in a shallow area, but such a device would not be practical in deeper waters.

2.1 Comparisons

Since different coring devices are specially suited for sampling various substrata, at various depths, and yielding different amounts of sediments, many times a researcher may need to use several samplers. For instance, during an exploration campaign of a deep seabed, free fall grabs, dredge baskets, box corers, gravity corers, and free-fall corers can be used (Lawless & Padan 1986). Often, in the same sampling campaign, grab samples are used to describe spatial variability of surficial sediments, while at the same time cores are collected to study vertical variation. Together, these types of sampling can be used to produce three-dimensional sedimentological descriptions (Israel et al. 1987). Thus, it seems obvious that the comparison of the data relating to the different samples can only be done when the different types of sampling tools are taken into consideration. It is advocated that several samplers, each of which is specifically designed for the required type of sampling, be used instead of trying to use one type of sampler for all tasks, which would compromise all samples.

A number of comparisons of the efficiency of different samplers have been based on the estimation of the abundance and composition of the benthic fauna. The efficiency of seven different grabs, four dredges, and three air-lift samplers in sampling benthic invertebrates in the sea floor and rivers has been tested by Word (1976), Elliott & Drake (1981a,b), and Drake & Elliott (1982). Probert (1984) found that the estimates of the soft-bottom macrofaunal composition in samples collected using an anchor-box dredge or a box corer are very similar. A multiple, hand-held corer is more reliable than a suction sampler for estimating the macrofaunal density and composition in a bare sand habitat, according to Stoner et al. (1983). Plocki & Radziejewska (1980) and Jensen (1983) compared different types of corers with the diver-held corer for describing the meiofauna abundance. A flow-through benthic corer that compares favorably with diver coring and is better than a ball-type, check valve corer for recovery of meio- and macro-fauna and particle-bound hydrocarbons has been described by Frithsen et al. (1983). Other comparisons between a Pfleger corer, a Van Veen grab, and scuba divers resulted in a statistically significant difference, at least for meiofauna abundance estimates in coastal silty sediments and on sandy bottoms (Heip et al. 1977, Heip 1984, Vidakovic 1984).

As for the chemical composition of sediments, William & Pashley (1979) compared the piston corer with a diver's performance by measuring the total carbon profiles in two cores, sampled at the same time and place by the lightweight corer and the scuba diver, and found that the two profiles were similar. Also, Evans and Lasenby (1984) provided evidence that a modified Kajak-Brinkhurst corer and divers can collect short, undisturbed cores with the same efficiency in the study of metal profiles in sediment, at least for soft, recently deposited sediment and for tubes having inner diameters of 2.5, 5.1, and 7.6 cm.

It has, however, been pointed out that diver-held corers are not necessarily better than different sampling devices for collecting intact cores (Rutledge & Fleeger 1988). These authors prepared an experimental sediment which contained Sephadex® gel beads, sampled the material by hand, gently tipped the cores to simulate the possible agitation induced by the diver while carrying the tubes to the surface, and then took subsamples from different parts of the cores. The analyses of these subsamples pointed out a significant difference between the inner and outer regions. On the other hand, no statistical difference was seen among replicate samples taken from the larger core while it was in place. The authors concluded that movement during core transport by the diver may concentrate particles in the center of the core and warned against subsampling this type of core.

When a piston corer and a bucket dredge were compared, it was found that the first introduced a considerable sample disturbance (Okusa et al. 1983). It has also been found that the shortening of cores obtained with a gravity corer produce steepened gradients in chemical profiles for pore water (alkalinity, dissolved iron, manganese, and phosphates), when compared with a box corer (Lebel et al. 1982). Comparing by direct observation, measurement, and photographic documentation by a diver the performance of the Axelsson-Håkanson gravity corer, the Kajak gravity corer, the Jenkin bottom sampler, and the Ekman grab, it was concluded that the core shortening is related to the tube size and that in sampling soft sediments there is the risk of collecting a sample unrepresentative of the real stratification (Blomqvist 1985).

The chemical characteristics of sediments collected by a gravity corer and a modified Ekman grab sampler were compared at six locations in Lake Como (Northern Italy) (Baudo et al. 1986). The Ekman grab had a surface area of 500 cm^2 and dug into the sediment to a maximum depth of 10–15 cm. The corer had an inside diameter of 2.6 cm. Three cores were taken and sectioned into 5-cm intervals. The contents of the three corresponding intervals were pooled together. The average chemical characteristics for the top 10 cm were compared to the results of the grab sampler (Table 2). Apart from the difference in subsampling, it must be remarked that the difference in the concentration values reflects both the variation in sampling technique and the local inhomogeneity of sediments: when sampling from the boat at a depth varying from 56 to 415 m, as in this case, the three independently sampled cores and the grab sample can actually be collected a few meters apart. Moreover, the mass of material recovered by the grab is roughly 30

Table 2. Comparison Between Corer and Grab Samples Taken at Six Different Locations in Lake Como[a]

| Element | Concentration mg kg^{-1} | | Difference % |
	Min.	Max.	
As	17	100	15
C	25000	85000	−15
Cd	0.2	3.5	29
Cr	120	300	− 1
Cu	50	250	4
Fe	18000	58000	3
Hg	0.09	0.65	14
Mn	350	10000	−21
N	1800	5000	−10
Ni	80	250	− 2
P	1000	5000	−27
Pb	30	380	−12
Ti	2800	4400	− 1
V	100	180	5
Zn	200	2000	−13

[a]The difference (%) between the two samplers here reported is the average for the six samples; each difference has been calculated as 100 [(concentration core − concentration grab)/concentration grab].

times larger than that of the three pooled cores, thereby contributing much more effectively in averaging the local patchiness.

Nevertheless, as summarized in Table 3, for almost all of the elements analyzed the agreement between the two sampling techniques was satisfactory, since the correlations between grab and core are at least statistically

Table 3. Correlation Coefficients (r) and Their Statistical Significance (Probability P) Between Corer and Grab Samples Taken at Six Different Locations in Lake Como[a]

Element	r	P	Mean S1 mg kg^{-1}	RSD %	Mean S2 mg kg^{-1}	RSD %
As	0.8424	0.035	9.19	31	13	8.3
C	0.8209	0.089	21300	4.9	232300	1.5
Cd	0.9847	0.001	0.36	0.03	1.57	0.08
Cr	0.9768	0.001	320	17	74.6	13
Cu	0.9622	0.002	473	2.9	48.2	24
Fe	0.9943	0.001	20798	2.6	26426	3.8
Hg	0.7863	0.064	0.287	16	1.24	25
Mn	0.9994	0.001	391	7.7	6035	8.8
N	0.6914	0.196	2150	3.2	20600	1.3
Ni	0.9420	0.005	248	16	71.2	12
P	0.9311	0.007	1588	14	5557	3.6
Pb	0.9862	0.001	66	1.1	87.6	1.8
Ti	0.9565	0.003	3815	6.2	2196	3.7
V	0.8443	0.035	112	6.5	78.3	6.3
Zn	0.9931	0.001	365	22	370	3.3

[a]For the same elements, the repeatability (relative standard deviation %) of chemical analyses for two Standard Reference Materials is reported.

Table 4. Comparison of the General Characteristics of the Different Sampler Types

Characteristic		Dredges	Grabs	Corers
Free water passage		Yes	Yes	Yes
Wall thickness/sample area		Low	Low	High
Closing on retrieval		No	Yes	Yes
	At top	No	Lid	Lid
	At bottom	No	Jaw	No
Transparent side(s)		No	Yes	Yes
Subsampling:	From Sampler	No	Top	No
	Extrusion	No	No	Yes
Exchangeable weight		Yes	Yes	Yes
Handling:	Use	Easy	Less easy	Difficult
	Weight	Low	Medium	Low/medium
	Safety	High	Low	Medium
Sample:	Area	Large	Large	Small
	Depth	Small	Medium	High
Cost		Low	Medium	Medium/high
Studies on:	Physics	No	Yes	Yes
	Chemistry	?	Yes	Yes
	Biology	Yes	Yes	Yes/no
	Pore Water	No	Yes/no	Yes

significant ($p < 0.05$). Exceptions are represented by nitrogen and carbon; in these cases, however, very likely the differences between samplers are enhanced by the great degree of heterogeneity of the sediments. Moreover, the mean differences between corer and grab samples compare favorably with the repeatability of the elemental analysis method, from subsampling to the final determination (calculated on ten repeats for each one of two Standard Reference Materials produced by the European Bureau of Standards, and precisely the sediments S1 and S2). That is to say, the variability between samplers was no greater than the analytical error of chemical analyses. At least for Lake Como sediments and the specific chemical analyses reported here, it seems that sampling of surficial sediments grab and corer samplers gives similar results. However, since the literature gives contradictory evidence about the comparisons between different sampling devices, it can be concluded that such a comparison should be done each time two or more of these samplers are to be used during research or if comparison with other samples, collected by somebody else or with other techniques, is foreseen.

As expected, none of the different types of corers described to date meets all of the criteria set by Håkanson & Jansson (1983) for an optimal coring device, but each one has specific advantages and disadvantages (Table 4). For instance, dredges or large area grabs are recommended for sampling bottom flora and fauna, while paleolimnological studies need to be based on long cores, and specific studies of the physical and chemical characteristics can use grabs, corers, or both.

Often it is the subsequent handling of the sample that dictates the choice of sampler. A review of the current literature on sample handling and long-term storage for environmental materials, sediment included, is given by Maienthal & Becker (1976). Generally the sample used for a given determination is only a small part of the bulk sample: if this is a large one, the multiple successive splitting operations (quartering) can be time consuming and, according to Carver (1981), even misleading. On the basis of a statistical comparison, that author claims that small representative subsamples may be obtained simply by pouring the sediment sample on a tray, forming a cone-shaped mound, and taking a part with an ordinary teaspoon inserted at about 45° within the center of the mound.

In some cases, large cores are subsampled after collection (Elmgren 1973, Holopainen & Sarvala 1975). A single subcore was found to be adequate at least to describe the physical and geoacoustic properties of sediment which had been collected with a box corer (Briggs et al. 1985).

Especially with cores, often the subsampling is specifically intended for describing trends in time. Thus, the core must be sectioned either in regularly spaced levels or according to some structure or properties observed in the sediment (Håkanson & Jansson 1983). Corers allow subsampling for stratification studies, but some artifacts may arise. For instance, the profile of uranium in interstitial water is noticeably different if the water is sampled by squeezing core sections or by an *in situ* porewater sampler (Toole et al. 1984). This sampling artifact is very likely due to core decompression.

It is often desirable to retain intact cores for transport to the laboratory, so several methods have been developed to do so. Sometimes cores are frozen in the field with liquid nitrogen (Chandler & Fleeger 1983), or with a mixture of dry ice and acetone (Bell & Sherman 1980), in an attempt to preserve the vertical distribution. In their experiment with artificial sediment, however, Rutledge & Fleeger (1988) found that fast freezing of samples "distorts the sample to such a degree that probably it is useless", and the distortion is strongly related to the temperature. If the cores are frozen in a commercial freezer, however, the perturbation is "probably not distinguishable from that caused by core compacting due to drag of the corer wall upon insertion". As an alternative to freezing, unconsolidated cores can be impregnated with epoxy or polyester resins (Ginsburg et al. 1966, Bouma 1969, Crevello et al. 1981) for preserving sedimentary structures and textures; however, the treatment is likely to modify the chemical characteristics. Splettstoesser & Hoyer (1983) described a portable heat sealer for preserving intact cores within transparent, heat-sealable pouches made of polyester outer backing, with a polyolefin coating.

3.0 SAMPLING STRATEGIES

The investigation of sediment characteristics is generally aimed to produce an indicative value for the ecosystem as a whole, such as a map of
horizontal distribution or a trend in sedimentation through time. The ful-
fillment of these objectives depends very much on the choice of the appropriate sampling strategy, which according to Håkanson & Jansson (1983)
can belong to three different types. These types are:

1. the deterministic system, with a sampling design based on previous information and varying density
2. the stochastic system, when the sampling stations are randomly selected
3. the regular grid system, with the sampling locations randomly or determi-
   nistically selected (Sly 1975, Rapin *et al.* 1978)

Examples of these sampling systems are given in Figure 5.

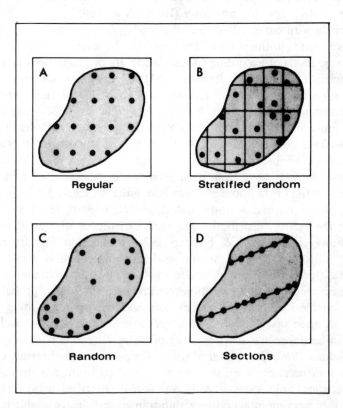

Figure 5.   Examples of different sampling strategies. (Modified from Huijbregts 1975. With
permission.)

3.1 Deterministic Sampling Design

The deterministic sampling design is the more commonly used, but the criteria for choosing locations may differ greatly from author to author. Besides the common use of looking for specific spots on the basis of previous knowledge, some indication could be given by a specific screening of the area. For instance, Håkanson (1986b) developed a sediment penetrometer that can be easily used to establish the prevailing bottom conditions (erosion, transportation, and accumulation; Håkanson 1982) before coring.

As an alternative, one can look for simple chemical markers. The total organic content of sediments has been suggested as an indicator of the spatial variability, since this analysis is inexpensive and quick. The subsequent sampling should then cover all the subbasins which proved to be heterogeneous (Reynoldson & Hamilton 1982). In addition, this helps in tracing the zones of pollutant accumulation, since the greatest contaminant loads are generally associated with the sediment fraction with the smallest particle size and greatest organic carbon content. Volcanic ash layers have also been used as stratigraphic markers to survey the sediment thickness and select representative sampling stations prior to coring (Kimmel 1978). However, such a marker is not common, so the author suggested the use of low-frequency echosounders for determining sediment accumulation patterns.

For specific purposes, sampling has sometimes been coupled with subbottom acoustic scanning in order to select the sampling stations in relation to the roughness characteristics. For instance, the bed texture and morphological properties have been described by using a high-resolution, fully corrected, side-scan sonar mapping system and the results checked against direct *in situ* observations by divers (Wright *et al.* 1987). In another study, the symmetry of the bedforms has been investigated by means of sonographs, and it was found that they cannot be simply explained in terms of depth and current velocity (Goedheer & Misdorp 1985). Alternatively, the substrate variability could be photographically recorded: this method has been used by Nishimura (1984) and Schneider *et al.* (1987) to describe the patchiness of epibenthic megafauna, but it could be employed for chemical investigations as well. A combination of devices such as echosounders, acoustic side-scan and subbottom profilers, and television and stereo camera systems was also used to monitor the sampling stations where sediment samples were taken (Lawless & Padan 1986). However, such studies require quite sophisticated equipment and are more commonly used in seabed prospecting for minerals than for oceanographic or limnological research.

3.2 Stochastic System

In this sampling network a random pattern is created, dividing the sampling area into equal subareas and randomly choosing, on a statistical basis, which ones have to be used for sampling. The stochastic system is highly recommended when a statistical treatment of the data is foreseen and can also be used for estimating the cost/benefit ratio of the project (Bernstein & Zalinski 1981).

3.3 Regular Grid System

The regular grid is obviously the one preferred to cover all possible unknown sources of variation, but at the same time is the one that requires the greatest sampling intensity. In fact, the "regularity" of the network is achieved by using the same distance, in a specific direction, between all pairs of points in all parts of the map (J.C. Davis 1973). Implicitly, this means that to cover equally the large variability in the fast-changing environments and the more homogeneous conditions in other parts of the study area, one has to collect a lot of samples even where just a few would be enough.

Recently, compromises that allow one to plan the sampling network with areas of different density, yet retain statistical power, which is almost equal to the larger complete sampling network, have been proposed (McBratney *et al*. 1981a,b, Kwaadsteniet 1986, van der Gaast 1986). This statistical method, known as "kriging", works on the basis of an estimate of the spatial variability, which is different in the various parts of the investigated area, and permits the determination of the sampling densities required for any prescribed maximum standard error. It has been confirmed that the network design algorithm provided significant gain in information, especially for large sets of data (>50, Hughes & Lettenmaier 1981). In all cases, the true validity of the sampling is given by the number of collected samples, with respect to the overall variability. In fact, the sampling network seems to be more sensitive to sample size than to configuration (Switzer 1979, Lesht 1988).

4.0 NUMBER OF SAMPLES

The estimation of the number of samples necessary to realize a particular statistical power in a study is often an undervalued requisite. In fact, the sediment "natural" heterogeneity is seldom considered, and all differences are regarded as significant. On the contrary, all physical, chemical, and biological characteristics can have an unsuspectedly great variability within a few centimeters of distance. The patchiness of benthic flora and fauna of both freshwater and marine environments is well known. As an example, it has been found that the downcore distribution of algal microfossils shows a variability in stratification in different areas of lakes that appears to be strongly related to their habitat (Dixit & Evans 1986); the animal spatial distribution is usually related to the sediment microtopography, as revealed by using the appropriate statistical treatment of the data (Hogue & Miller 1981). Sedimentation rates in lakes, determined by dating the sediment with ^{210}Pb (Dominik *et al.* 1981, Evans & Rigler 1983) or with ^{137}Cs (Bennett 1987), have been shown to vary from subbasin to subbasin. More specifically, it is often found that the sediment accumulations in different parts of a lake are more or less linearly related to depth (R.B. Davis *et al.* 1969, M.B. Davis 1973, Lehman 1975, Likens & Davis 1975, Sly 1977, Evans & Rigler 1980, Davis & Ford 1982). The heterogeneous sediment distribution is known to be caused by factors such as wave and wind action, seasonal or continuous mixing, slumping and sliding on slopes, and deposition of riverine materials. Reviews on this subject have been prepared by Sly (1978), Håkanson & Jansson (1983), Hilton (1985), Hilton *et al.* (1986), and Bennett (1987).

Apart from the expected variation between shallow and deep parts of a lake (Likens & Davis [1975] coined the term "sediment focusing" to describe the increased sedimentation of finer material in the deeper zones), it is generally assumed that the more uniform zone of a water body is its bathyal platform. Battarbee (1984), on the other hand, describing the diatom variability and using a constrained cluster analysis, concluded that this assumption may be true in the case of small, single-basin lakes, but it does not hold for large and complex water bodies. Downing & Rath (1988), comparing the within-core and among-core variations for eight lakes, pointed out that even in bathymetrically uniform sites quite often ($> 62\%$ of the cases) the sediments are heterogeneous, at least with regard to water content, organic matter, pigment, and phosphorus content. More generally, few published papers on sediment characteristics are based on a sufficiently large sample size, and most relate only to the deepest portions of lakes. Since these areas represent only a small portion of the surface areas of lakes, any extrapolations to the whole-lake characteristics, based on these observations, would be biased. In particular, according to Evans (1988), the establishment of the whole-lake burden from these data is questionable.

These findings can even be extended to marine and fluvial environments: when profiles of trace elements from duplicate cores taken a few centimeters apart at 4000–5000 m in the Northeast Atlantic Ocean were compared (Smith *et al.* 1986), large differences that could be explained by the small-scale spatial heterogeneity due to the bioturbation of the top layers were observed. Also, in a gravel-bed river, the spatial variability of grain

Table 5. Mean Concentrations (mg kg^{-1} dry weight) and Sampling Error % (100 s/$\overline{X}$) for Littoral and Pelagic Sediments from Lake Mezzola

Element	Littoral Sediments N = 51		Pelagic Sediments N = 32	
	Mean	Error	Mean	Error
Ca	11089	15	10788	29
Mg	11896	22	11541	19
Na	18365	10	18160	9.2
K	17035	5.6	16804	5.7
Cr	124	43	130	73
Cu	21.4	15	20.8	17
Zn	82.4	23	82.2	27
Fe	40252	14	40092	15
Mn	621	18	622	18
Ni	44.1	21	45.0	21
% Organic matter	3.02	23	7.44	14

Source: Baudo *et al.* (1981).

size was so pronounced that 228 and 50 samples were required to estimate the mean diameter with a variability of ±10 and 20%, respectively (Mosley & Tindale 1985).

As an alternative to collecting a large number of samples in a single station, the mean variability of replicated samples taken in several different sampling stations can be used to provide an indication of the sediment heterogeneity relative to the whole investigated area. This formula has been given by Henriksen & Wright (1977):

$$e = \sqrt{\Sigma(X_1 - X_2)^2 / n}$$

If X_1 and X_2 are the two measures for samples systematically taken a few centimeters apart, the mean difference between n replicated samples gives a measure of the "sampling error". For Lake Mezzola, 51 duplicate samples taken along the shore by a hand-held corer and 32 duplicate samples collected with a Jenkin corer from a boat in deep waters produced the sampling errors shown in Table 5 (Baudo *et al.* 1981). The rather high value for elements like chromium was expected, since for Lake Mezzola this metal is a pollutant with known, specific point sources and its distribution is largely variable in the different parts of the lake; however, the data clearly indicate that even elements which are not significantly influenced by human activities have a "normal" variability ranging from 5 to 20%.

Moreover, at least for some parameters, a temporal variability must also be taken into consideration. One of the aims of studies of sediments is the description of the variation of sedimentation characteristics as a function of time. Usually in these kinds of studies cores are subdivided into a number of sections that should mark the changes in settled materials. However, the sediment may present not only a year-by-year variation (Dominik *et al.* 1981), but also a marked seasonal variability, which is often the case for concentrations of trace metals (Multer *et al.* 1981), organic materials, iron, and grain size (Thomson-Becker & Luoma 1985). This result is partly due to the variability in composition of the settled material (for instance, in relation to the seasonal succession of the planktonic communities), but also to diagenetic processes controlled by redox, pH, and temperature fluctuations.

Considering all the above mentioned sources of variability for the sediment distribution and composition, it is clear that one single sample cannot be regarded as truly representative of a whole water body. Lacking any theoretical criteria to account for all possible sources of variation, Håkanson & Jansson (1983) proposed an empirical formula that allows the calculation of the needed number of samples (N) on the basis of two standard morphometric parameters, the shore development F, which is an indirect measure of the bottom roughness, and the lake area a:

$$N = 2.5 + 0.5 \sqrt{a \cdot F} \tag{1}$$

F is given by the ratio

$$F = l_0 / 2 \cdot \sqrt{\pi \cdot A} \tag{2}$$

where l_0 is the normalized shoreline length and A the total lake area, including islands.

This empirical equation provides an even area cover and typically will yield an estimate ranging between 4 and 50 for lakes up to 1000 km² in surface area. However, the estimation can be quite different if the calculation is made separately for the topographically distinguished subbasins. For instance, for Lake Hjalmaren as a whole the estimated sample size is 28, but it becomes 44 if the lake is divided into its four subbasins. In addition, as pointed out by Evans & Dillon (1986), the formula has been derived from very large lakes and may not be representative of water bodies of smaller dimensions.

As an alternative, the number of samples could be established on a statistical basis before sampling (Håkanson 1984). Already in 1975, Sly pointed out the need for a statistical evaluation of sediment sampling and used the data obtained by sampling at 843 stations according to a nested grid pattern for assessing the reliability of the estimates as a function of grid spacing. The conclusion was that hydrodynamic characteristics and conservative geochemical characteristics (mean particle size, Na, K, Hg) of the investigated environments were accurately estimated by the use of a grid spacing of about 100 m, but closer spacing of <30 m should be used in shallow waters, and a grid interval of approximately 300 m could be used in deeper waters. There is no need to stress the fact that such a dense sampling network is rather unusual.

Instead of the parametric statistic, which among other things requires that the statistical errors associated with the samples are normally and independently distributed with zero mean and constant variance, Lesht (1988) proposed a nonparametric evaluation of the sampling intervals for limnological studies. In this case, from a given set of data based on 31 stations, it was estimated that a network of 22 stations would provide a mean value accurate to within 10% of the "true" value, that is, the average of all 31 stations, with 80% confidence.

Both examples illustrate what Switzer (1979) pointed out to be "a common statistical conundrum", that the estimation of the required number of sampling points can be determined statistically only after the data have been gathered or some estimate of crucial parameters obtained in preliminary studies.

A simple formula that permits the calculation, on a parametric basis, of the required number of samples (N) needed to calculate the average value for a given area with a given statistical uncertainty has been provided by Kratochvil & Taylor (1981). This procedure requires that the variance of the investigated parameter be known or can be estimated from previously obtained information (Equation 3).

$$N = 10^4 \, (t^2 s^2)/(R^2 \overline{X}^2) \tag{3}$$

where t is the Student's t-table value for the desired confidence level, and $\overline{X}$ and s are the mean and the standard deviation estimated from preliminary measurements or from previous knowledge, respectively. This formulation is valid only when the data are normally distributed. Most of the published statistical treatment of data makes use of the parametric statistic, but to what degree this is justified by the data distribution is unknown. In fact, the few studies based on a number of samples large enough to allow for a detailed statistical analysis indicate that the distribution of most parameters is not described by the normal probability function, but by different theoretical frequency distributions. For instance, several authors have described sediment grain size with truncated normal distributions (Sindowski 1957, Harris 1958, Visher 1969). Other investigators have used a series of overlapping normal distributions to describe the frequency of occurrence of grain sizes in sediments (Tanner 1959, 1964, Fuller 1961, Spencer 1963, Folk 1971, Clark & Clark 1976). In some situations a distinct bimodal distribution of grain sizes has been observed (Folk & Ward 1957), while in others multimodal distributions have been observed for mixed sediments derived from separate sources (Sly *et al.* 1983).

The frequency distributions of the chemical constituents of sediments have also been investigated. Concentrations of 13 elements (Al, Ca, Cd, Cr, Cu, Fe, Mg, Mn, Ni, Pb, Ti, V, and Zn) were measured in the sediments of a bay in Lake Maggiore, Italy (Muntau *et al.* 1986), and the frequency distributions were investigated as suggested by Sokal & Rohlf (1969) and R.G. Davis (1971). The area studied was 3 km long and 1.5 km wide, with a maximum depth of 130 m. The frequency distributions of nine elements were described best by the negative binomial distribution, while one was described by the Poisson distribution. The frequency distributions of only three of the elements could be described by the normal probability function (Muntau *et al.* 1986). It must be said that the negative binomial and Poisson distributions describe discrete variables, while metal concentrations are usually regarded as continuous variables; nevertheless, these two distributions have been used because actually the measures of the concentrations are discontinuous, depending on the method sensitivities. Another statistical probability distribution, the log-normal distribution, is usually suggested as an alternative. In fact, this distribution could probably fit the data as well. However, it has not been used here, for it requires the transformation of the data; thus, the conclusion refers to these transformed data instead of the original data. A detailed description of the statistical properties of the different distributions can be found elsewhere (Bliss & Fisher 1953; Sokal & Rohlf 1969); here it will only be mentioned that a distinctive feature is represented by the relationships between average and variance:

$$\text{Normal distribution: } \overline{X} > s^2$$

$$\text{Poisson distribution: } \overline{X} \simeq s^2$$

$$\text{Negative binomial distribution: } \overline{X} < s^2$$

Formulations for estimating the sample sizes (N) required for a particular statistical power for each of these distributions have been described (Kratochvil & Taylor 1981; Equations 4 and 5).

$$N = 10^4 \, t^2/(R^2\overline{X}) \text{ (Poisson distribution)} \tag{4}$$

$$N = 10^4 \, (t^2/R^2)[(1/\overline{X}) + (1/k)] \text{ (negative binomial distribution)} \tag{5}$$

The symbols are the same as in the normal distribution equation, but in Equation 5 there is in addition a term k, called the index of clumping, computed by Fisher's maximum likelihood method (for more details, see Bliss & Fisher 1953).

Table 6. Number of Samples Necessary to Achieve 95% Confidence Limits (Monvallina Bay)

Element	Normal Distribution	Poisson Distribution	Negative Binomial Distribution
Al	202		
Ca	306		
Cd		278	
Cr			122
Cu			474
Fe	363		
Mg			214
Mn			805
Ni			734
Pb			621
Ti			159
V			90
Zn			340

Source: Muntau et al. (1986).

As summarized in Table 6, from the data of Muntau et al. (1986) it can be inferred that the 115 original samples are enough to describe the spatial heterogeneity of the Monvallina Bay only for V and possibly Cr, but not for most of the elements. As many as 805 samples would be needed to adequately describe the distribution of manganese concentrations.

As another example, the copper content of Lake Orta sediments has been used to illustrate the uncertainty in describing the chemical composition of sediments (Baudo 1989). Following the empirical approach of Håkanson & Jannson (1983; Equation 1), six samples evenly distributed over the lake surface should approximate the average value for the sediment. However, since 57 samples had been collected and the data distribution shown to be significantly skewed and leptokurtic, the negative binomial distribution was used instead of the normal distribution. According to the Kratochvil & Taylor (1981) formula (Equation 5), six samples should be enough to provide an estimate of the average value with a relative standard deviation (RSD) of 100% at the Student's t-test level corresponding to 95% confidence limits. With the actual number of data, the RSD was estimated to be approximately 25%; to decrease this deviation to 10%, as many as 412 samples would be required (Figure 6).

5.0 DATA ELABORATION FOR DESCRIBING SPATIAL VARIABILITY

Many techniques have been suggested to elaborate the original data so that they could be more easily related to physical, chemical, or biological processes which explain the observed distribution. These techniques include

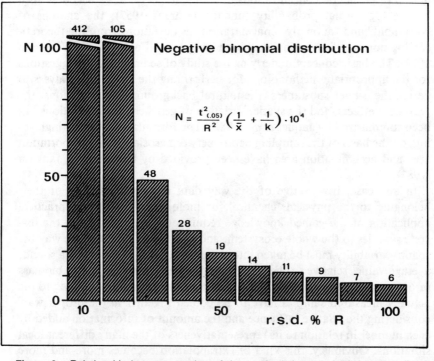

Figure 6. Relationship between number of data and relative standard deviation of the estimate of the average copper content of Lake Orta sediments (Student's *t* level corresponding to 95% confidence limits).

the computation of concentration ratios with specific matrix elements (Foerstner & Wittmann 1979, White & Tittlebaum 1984) or correction for a given size class (Mueller 1979, Robbe 1981) or mineral fraction (Thomas & Jaquet 1976). However, since no general agreement exists in this field, this means that most of the published results are simply not comparable to one another.

Apart from the methodological differences in producing the data, generally the numerical values for the given variables alone are not sufficient for a correct interpretation, so they usually are compared with those of other related variables, looking for significant associations and similarities in trends. Even in this field a lot of different procedures have been suggested, but according to Jones & Bowser (1978), often the best way of using the data is by interpreting their correlations in view of the already known chemical and mineralogical relationships. Symader & Thomas (1978) recommended the use of hierarchical grouping analysis as a procedure for obtaining initial information about a heterogeneous area. Since the distribution of elemental concentrations in sediments seems to be best described by the log-normal probability function (Ahrens 1957), the analysis of "threshold" and "anomaly" concentrations as deviations from the theoretical log-normal distribution has been proposed (Foerstner & Wittmann 1979). The basic concept underlying the study of sediments and suggestions for the appropriate methodology for performing the statistical analyses to define the carrier substances, the natural background concentrations, the grain-size effects, and the spatial variation of geochemical associations has been summarized by Jaquet et al. (1982). The formulations for distinguishing, on the basis of the sampling depth, between erosion plus transportation area and accumulation area have been provided by Håkanson & Jansson (1983).

In any case, irrespective of the way data are produced and of their belonging to the physical, chemical, or biological domain, the practical application of the gained knowledge requires extrapolation from the discrete samples to the whole ecosystem, and this means that in some way the spatial variability must be taken into account. Even for calculating a sound, average value, relative to the whole lake, of the animal or vegetal biomass or of a particular chemical species of an element, the data pertaining to the actually measured samples have to be pooled together in the proper way, considering the relative influence and the amount of information added by each number in relation to its representativeness of the many different local situations. Obviously, this kind of extrapolation becomes more and more difficult when the estimations have to be more detailed and refer to a given subbasin, a single bay, or even just a specific harbor.

There are a number of ways to represent the distributions of sediment parameters cartographically. One method is to use symbol sizes which are proportional to the concentration of particular parameters (Jenks 1975, Viel & Damiani 1987; Figure 7). An alternative method is to connect concentration isolines, interpolating by eye between points. This, however, results in a limited and uneven representation unless a very large number of locations have been sampled. However, statistically valid methods developed by geologists can be used to generate continuous two- or three-dimensional distribution maps (Krumblein & Graybill 1965, J.C. Davis 1973, Davis & McCullagh 1975, Clark 1979).

In principle, mapping can be based either on the trend surface analysis or on a moving average algorithm. The first method uses a multiregression approach to calculate an equation that relates geographical coordinates to the sought-after variable. The second technique is based on the assumption that the estimated point should be more similar to the closer measured points than to more distant observations, and it employs a weighting function for decreasing the influence of the various points according to their distance from the point to be estimated.

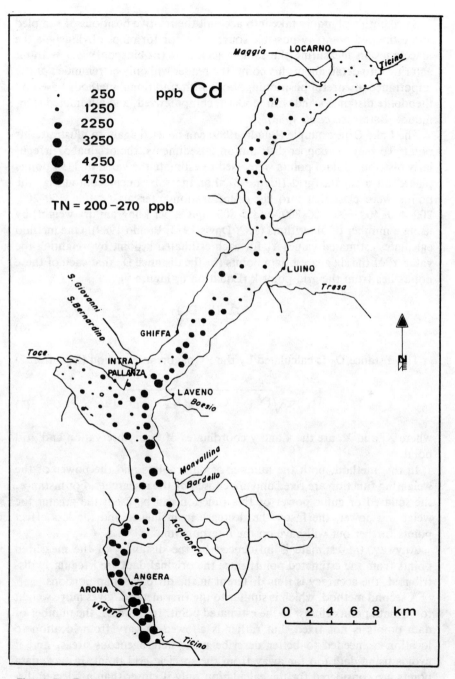

Figure 7. Example of a distribution map which uses proportionally larger symbols for different concentration classes. (Modified from Viel & Damiani 1985. With permission.)

Since both techniques take into account the relative positions of sampled and estimated points, a possible source of error for maps obviously is the inaccuracy in measuring the sample locations (Robinson 1975). Without entering into detail about this point, the reader will only be reminded of the importance of careful positioning, both with traditional methods based on theodolite distancers and with modern, computerized, and automated techniques (Battarbee *et al.* 1983).

The Lake Orta example (Baudo 1989) can be used again to illustrate this point. To plot the copper distribution in sediments, the original 57 irregularly distributed data points were used to estimate the value of 1154 points placed on a regular grid (Figure 8). For the subsequent contouring, the points were classified into six concentration intervals (<100, 100–200, 200–400, 400–600, 600–800, and >800 mg Cu per kilogram dry weight) by using a number of algorithms (J.C. Davis 1973, Baudo 1989). One method calculates estimated values Y_k for each estimated k point by weighting the value Y_i of the six nearest data points for the distance D_{ik} that each of these points lies from the grid point k (Equation 6, Figure 9).

$$Y_k = \frac{\Sigma \ (Y_i/D_{ik})}{\Sigma \ (1/D_{ik}} \tag{6}$$

The distance D_{ik} is calculated by the Pythagorean equation (Equation 7)

$$D_{ik} = \sqrt{(X_{1k} - X_{1i})^2 + (X_{2k} - X_{2i})^2} \tag{7}$$

where X_1 and X_2 are the x and y coordinates of both observation and grid point.

In this method, both the number of data points and the power of the weighting function are fixed and are chosen by the researcher. For instance, the squared or cubic power of this function can be used; the greater the weighting power, the faster the decrease in influence and the less effect points further out will have on the interpolation.

Anyway, the estimate is influenced by the distances of the measured points from the estimated point; since the original data are irregularly distributed, the accuracy is thus different in the different map portions.

A second method, which is similar to the first method, gives more weight to the data points closest to the estimated point. In this case, the number of data points is not fixed, but rather is allowed to vary from location to location as needed to better describe more heterogeneous areas, and it avoids using data too far away from the specific grid point: in fact, data points are considered for the calculation only if closer than a selected distance (Baudo 1989). In addition, the distance D'_{ik} takes into consideration

Figure 8.   Sampling points (squares) and estimated points (dots) for producing distribution
maps of copper content of Lake Orta sediments.

the depth Z_i of the observed data point nearest the estimated value, which
for ease of calculation is assumed to be at a depth of zero (Equation 8). The
distribution of copper in the sediments of Lake Orta estimated in this way is
given in Figure 10.

$$D'_{ik} = \sqrt{D_{ik}^2 + Z_i^2}$$

(8)

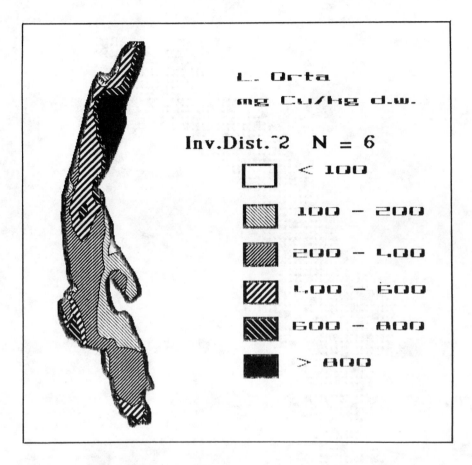

Figure 9. Distribution map calculated with the moving average method and the six nearest sampling points.

The addition of a correction for the bathymetry is especially important when the depth is comparable to the grid spacing.

The third method uses a polynomial equation which relates concentrations to geographic coordinates of observed points to calculate estimated concentrations at other points in the grid (J.C. Davis 1973). This method allows for a test of the goodness of fit by comparing the variance due to the regression to the deviation from the trend. Here the technique is demonstrated by the use of a fifth- and sixth-order polynomial (Figures 11 and 12). These two relationships explained 66 and 81% of the variability in estimated concentrations of copper in Lake Orta, respectively. Both of these maps can be influenced by clustering in some areas or spreading along one axis, since these features affect the regression equations. The results seem generally

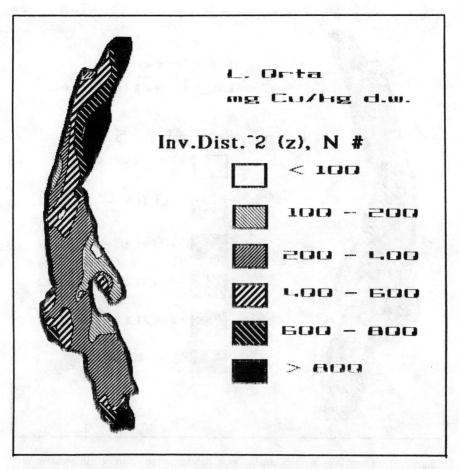

Figure 10. Distribution map calculated with the moving average method and a variable number of sampling points, taking into account the sampling depth.

satisfactory because the geometrically generated shape shows a distinct pattern (waves, concentric rings, and so on), but it is difficult to judge how accurate the representation is.

Another technique has been developed to use a double Fourier transformed series, which represents two interacting sets of two-dimensional sinusoidal waveforms, each containing many harmonics of different amplitude and phase angle, to calculate a trend surface. For the representation of Figure 13, 13 harmonics have been used. This method is especially powerful, for the examination of the power spectrum of the double Fourier series allows one to select those harmonics which are major contributors to the pattern—that is, those harmonics that could be regarded as different

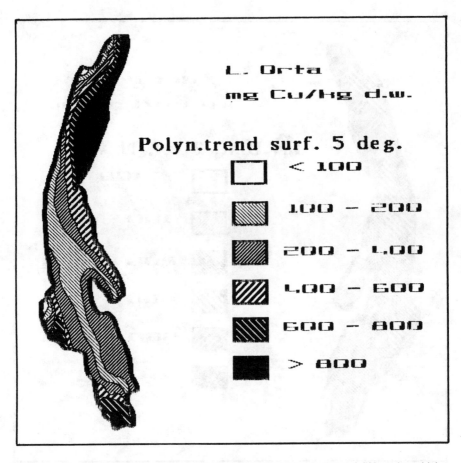

Figure 11. Distribution map calculated as a polynomial equation of fifth order which relates copper content to geographic coordinates.

"sources" of the plotted variables. Also, this algorithm permits one to produce an accurate map with an evident trend, but, again, this is greatly influenced by operator choices.

Finally, the trend surface is calculated using the kriging function (Matheron 1971, Krige 1978, Clark 1979, Gilbert & Simpson 1985). This method is based on the theory of regionalized variables and assumes that spatial variability is expressed quantitatively by the semivariogram, which is a relationship between distance and variability. The predicted distribution of copper in the sediments of Lake Orta estimated by this method is given in Figure 14. This technique differs from other weighted moving average methods since the weighting factor is based on the best linear unbiased

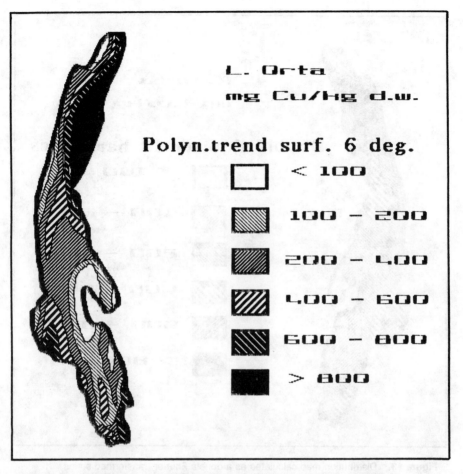

Figure 12.   Distribution map calculated as a polynomial equation of sixth order which relates copper content to geographic coordinates.

estimator and by using local estimation does not require the same mean and covariance function over the entire surface (Gilbert & Simpson 1985).

In addition to the methods discussed here, others have been proposed in the literature. Brodlie (1980) used splines, Foley (1981) used bicubic splines, and Floderus (1988) proposed a filter equation, based on the Gaussian curve, which is independent of a regular sampling grid.

Different treatments of the data will produce different estimated patterns; it is interesting to see what is left if only the common area will be represented—that is, the points for which all six estimations produce the same classification (Figure 15). Unfortunately, there is no simple way to decide beforehand which method will produce the most accurate results,

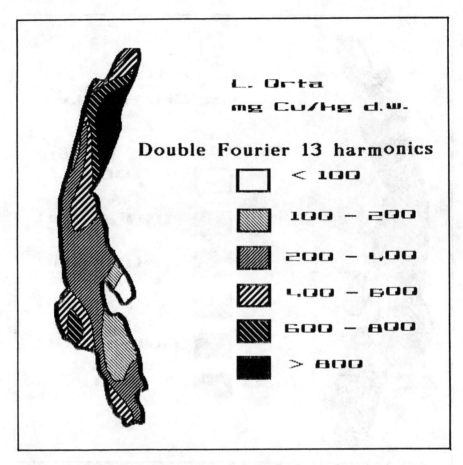

Figure 13.   Distribution map calculated as a double Fourier transformed series.

nor it is possible to establish afterward how close the predictions are to the actual values, unless a new sampling campaign is conducted to verify the accuracy or the original data are so dense that, by comparing the true and estimated values for those points, the mismatch areas can be characterized (Switzer 1975).

It is evident from the Lake Orta example that those mismatch areas can even be very large, with obvious implications for the future uses of those maps. However, as J.C. Davis (1973) pointed out, much of the debate about the relative merits of trend-surface analyses and moving averages can be ascribed to the "misapplication of both techniques and overzealous promotion by a few workers". Without a doubt, the increasing number of studies of this kind will provide more insight into the problem.

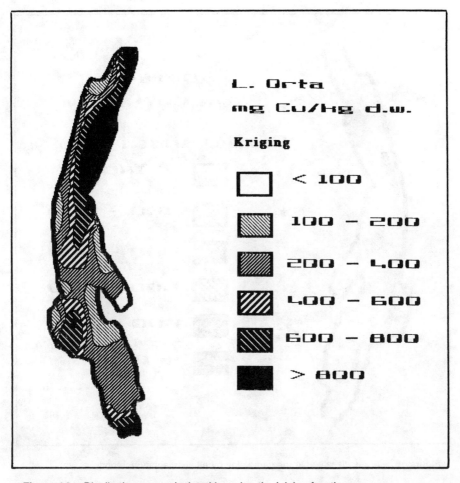

Figure 14.   Distribution map calculated by using the kriging function.

6.0 CONCLUSIONS

From this review of the current literature about the specific subject of sediment sampling and mapping, very few final conclusions can be drawn; they may be considered suggestions, rather than true conclusions, since they are mostly subjective.

First at all, with regard to sampling devices, it is very unlikely that one single sampler could be used for all purposes. The highest priorities should be given to systems which assure good visual interpretation of the sampling spot, such as divers, submersibles, or telecamera scanning. Alternatively, the sampling could be assisted by photographic recording or other methods

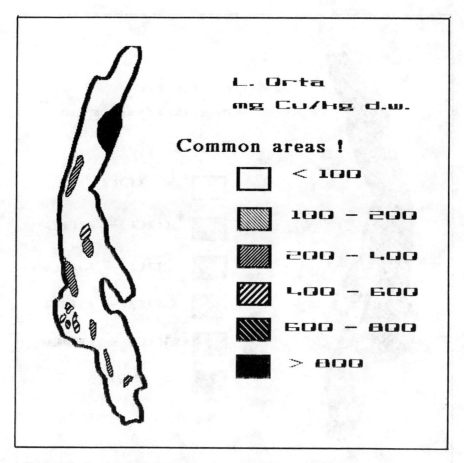

Figure 15. Areas classified in the same way by all six mapping methods.

of prospecting the bottom, either discontinuously, collecting information at given spots, or continuously, for instance with echosounders.

However, after the careful choice of the sampler that best suits the aims, a comparison with other devices that have been used for the same environment or the same objectives is mandatory to allow data comparison. In addition, whenever possible an estimation of the sampling variability should be made, by using either a multiple sampler or repeated sampling at the same station. This evaluation should be made in all markedly different areas with respect to parameters such as slope, depth, sediment texture, and so on.

The sampling strategy, hence both number and place of sampling, must be selected carefully. As a rule of thumb, to minimize the number of samples without decreasing excessively the accuracy of spatial estimation, instead of using a dense regular grid the sampling stations should be selected on the basis of the best possible knowledge obtained from literature, preliminary data, major bottom characteristics evaluated with the Håkanson penetrometer and/or acoustic scanning, etc. The results must then be subjected to an initial statistical evaluation. First the type of distribution must be recognized, then the significant number of samples for the desired confidence level can be calculated and, if proved necessary, additional sampling should be performed with no substantial delay. This is especially important for parameters that are known to undergo seasonal variations.

Since the mapping of the studied variable is likely to be used for practical purposes, such as the estimation of the whole lake burden or sediment treatment by dredging, capping, or other remedial actions, at least two different cartographic techniques should be employed—possibly, one belonging to the trend-surface type, the other to the moving average type.

In the opinion of this author, the two techniques that should be routinely used for mapping bottom sediment characteristics are kriging and a moving average that takes into account depth. The last point is especially important when the horizontal and vertical spacing are of the same magnitude, for in this case the distance between points in the plane can be substantially lower than in the three-dimensional space, with obvious consequences for the weighting factor.

Of course, this does not exclude the possibility that specifically adapted methods can prove to be superior for some applications. For instance, weighting the distribution for other related variables can remove a substantial portion of the observed among-site variability. However, the two methods suggested here are largely applicable without additional information other than geographic coordinates and sampling depth.

In every case, the apparent differences among different zones of the investigated environment should be viewed in relation to the overall "sampling variability", including the natural heterogeneity, and all sample manipulation steps: from sampling, subsampling, storage, preparation for the analysis, and final quantification.

The Modeling Task Force (1988) of the Great Lakes Science Advisory Board pointed out that "in many respects models are like maps; they are of different scales and purposes and should be designed to satisfy the need of the sponsor. The challenge facing the modeler is similar to that facing the cartographer—deciding what should be included and excluded to give a final product that is the optimal tradeoff between excessive complexity and naiveté." As for the models, the correct use of maps depends on the knowledge of their inherent uncertainty, the purpose for which they will be used, and the quality and quantity of input data. Like a fine violin, where the quality of the music depends not only on the instrument but also on the performer, models and predictive mapping systems depend on the experience of the user.

REFERENCES

Ahrens, L.H., 1957. The lognormal distribution of the elements—a fundamental law of geochemistry. *Geochim. Cosmochim. Acta* 30: 1111-1119.

Ali, A., 1984. A simple and efficient sediment corer for shallow lakes. *J. Environ. Qual.* 13: 63-66.

Anderson, F.E. & F.R. Hess, 1969. An inexpensive coring platform for shallow estuarine waters. *J. Sed. Petrol.* 39: 1238-1242.

Anima, R.J., 1981. A diver operated reverse corer to collect samples of unconsolidated coarse sand. *J. Sed. Petrol.* 51: 653-654.

Aviilov, V.I. & V.Ya. Trotsyuk, 1980. Improved means for obtaining hermetically sealed samples of bottom sediment. *Oceanol. Acad. Sci. U.S.S.R.* 20: 365-368.

Barnett, P.R.O., J. Watson & D. Connelly, 1984. A multiple corer for taking virtually undisturbed samples from shelf, bathyal and abyssal sediments. *Oceanol. Acta* 7: 399-408.

Battarbee, R.W., 1984. Spatial variations in the water quality of Lough Erne, Northern Ireland, on the basis of surface sediment diatom analysis. *Freshwater Biol.* 14: 539-545.

Battarbee, R.W., C. Titcombe, K. Donnelly & J. Anderson, 1983. An automated technique for the accurate positioning of sediment core sites and the bathymetric mapping of lake basins. *Hydrobiologia* 103: 71-74.

Baudo, R., 1989. Uncertainty in description of sediment chemical composition. *Hydrobiologia* 176/177: 441-448.

Baudo, R., G. Galanti, P. Guilizzoni & P.G. Varini, 1981. Relationships between heavy metals and aquatic organisms in Lake Mezzola hydrographic system (Northern Italy). 3. Metals in sediments and exchange with overlying water. *Mem. Ist. Ital. Idrobiol.* 39: 177-201.

Baudo, R., G. Galanti, P. Guilizzoni, G. Marengo, H. Muntau, M. van Son & P. Schramel, 1985. Chemical composition of recent sediments from Lake Como (Northern Italy). In *Proc. Int. Conf. Heavy Metals in the Environment*, Vol. 1, Athens, September 1985: 252-254.

Baxter, M.S., J.G. Farmer, I.G. McKinley, D.S. Swan & W. Jack, 1981. Evidence of the unsuitability of gravity coring for collecting sediment in pollution and sedimentation rate studies. *Environ. Sci. Technol.* 15: 843-846.

Bell, S.S. & K.M. Sherman, 1980. A field investigation of meiofaunal dispersal: Tidal resuspension and implications. *Mar. Ecol. Prog. Ser.* 3: 245-249.

Bennett, J.R., 1987. The physics of sediment transport, resuspension, and deposition. *Hydrobiologia* 149: 5-12.

Bernstein, B.B. & J. Zalinski, 1983. An optimum sampling design and power tests for environmental biologists. *J. Environ. Manage.* 16: 35-43.

Blakar, I.A., 1978. A flexible gravity corer based on a plastic funnel closing principle. *Schweiz. Z. Hydrol.* 40: 191-198.

Bliss, C.I. & R.A. Fisher, 1953. Fitting the negative binomial distribution to biological data. *Biometrics* 9: 176–200.

Blomqvist, S., 1985. Reliability of core sampling of soft bottom sediment—An in situ study. *Sedimentology* 32: 605–612.

Bonem, R.M. & J.R. Pershouse, 1981. Inexpensive, portable underwater coring device. *J. Sed. Petrol.* 51: 654–656.

Bothner, M.H. & P.C. Valentine, 1982. A new instrument for sampling flocculent material at the water/sediment interface. *J. Sed. Petrol.* 52: 639–672.

Bouma, A.H., 1969. *Methods for the Study of Sedimentary Structures.* John Wiley & Sons, New York: 458 pp.

Briggs, K.B., M.D. Richardson & D.K. Young, 1985. Variability in geoacoustic and related properties of surface sediments from the Venezuela Basin, Caribbean Sea. *Mar. Geol.* 68: 73–106.

Brinkhurst, R.O., K.E. Chua & E. Batoosingh, 1969. Modification in sampling procedures as applied to studies on the bacteria and tubificid oligochaetes inhabiting aquatic sediments. *J. Fish. Res. Board Can.* 26: 2581–2593.

Brodlie, K.W., 1980. *Mathematical Methods in Computer Graphics and Design.* Academic Press, New York.

Bryant, R., D.J.A. Williams & A.E. James, 1980. A sampler for cohesive sediment in the benthic boundary layer. *Limnol. Oceanogr.* 25: 572–576.

Burke, J.C., 1968. A sediment coring device of 21-cm diameter with sphincter core retainer. *Limnol. Oceanogr.* 13: 714–718.

Burns, R.E., 1966. Free fall behaviour of small light-weight gravity corers. *Mar. Geol.* 4: 1–9.

Buzas, M.A., 1968. On the spatial distribution of Foraminifera. *Contrib. Cushman Found. Foraminifera Res.* 19: 1–11.

Carlton, R.G. & R.G. Wetzel, 1985. A box corer for studying metabolism of epipelic microorganisms in sediment under in situ conditions. *Limnol. Oceanogr.* 30: 422–426.

Carver, R.E., 1981. Reducing sand sample volumes by spooning. *J. Sed. Petrol.* 51: 658.

Chandler, G.T. & J.W. Fleeger, 1983. Meiofaunal colonization of azoic estuarine sediment in Louisiana: Mechanism of dispersal. *J. Exp. Mar. Biol. Ecol.* 69: 175–188.

Clark, I., 1979. *Practical Geostatistics.* Applied Science Publishers, London: 129 pp.

Clark, M.W. & I. Clark, 1976. A sedimentological pattern recognition problem. In D.F. Merriam (Ed.), *Quantitative Techniques for the Analysis of Sediments.* Pergamon Press, Toronto: 121–141.

Craib, J.S., 1965. A sampler for taking short undisturbed marine cores. *J. Cons. Perm. Int. Explor. Mer* 30: 34–39.

Crevello, P.D., J.M. Rine & D.E. Lanesky, 1981. A method for impregnating unconsolidated cores and slabs of calcareous and terrigenous muds. *J. Sed. Petrol.* 51: 658–660.

Dall, P.C., 1981. A new grab for the sampling of zoobenthos in the upper stony littoral zone. *Arch. Hydrobiol.* 92: 396–405.

Davis, J.C., 1973. *Statistics and Data Analysis in Geology*. John Wiley & Sons, New York: 550 pp.

Davis, J.C. & M.J. McCullagh, 1975. *Display and Analysis of Spatial Data. NATO Advanced Study Institute*. John Wiley & Sons, London: 378 pp.

Davis, M.B., 1973. Redeposition of pollen grain in lake sediments. *Limnol. Oceanogr.* 18: 44–52.

Davis, M.B. & M.S. Ford, 1982. Sediment focusing in Mirror Lake, New Hampshire. *Limnol. Oceanogr.* 27: 137–150.

Davis, R.B., L.A. Brewster & J. Sutherland, 1969. Variation in pollen spectra within lakes. *Pollen Spores* 11: 557–572.

Davis, R.G., 1971. *Computer Programming in Quantitative Biology*. Academic Press, London: 492 pp.

de Groot, A.J., K.H. Zschuppe & W. Salomons, 1982. Standardization of methods of analysis for heavy metals in sediments. *Hydrobiologia* 92: 689–695.

Dixit, S.S. & R.D. Evans, 1986. Spatial variability in sedimentary algal microfossils and its bearing on diatom-inferred pH reconstructions. *Can. J. Fish. Aquat. Sci.* 43: 1836–1845.

Dominik, J., A. Mangini & G. Mueller, 1981. Determination of recent deposition rates in Lake Constance with radioisotopic methods. *Sedimentology* 28: 653–677.

Downing, J.A. & L.C. Rath, 1988. Spatial patchiness in the lacustrine sedimentary environment. *Limnol. Oceanogr.* 33: 447–458.

Drake, C.M. & J.M. Elliott, 1982. A comparative study of three air-lift samplers used for sampling benthic macro-invertebrates in rivers. *Freshwater Biol.* 12: 511–533.

Elliott, J.M. & C.M. Drake, 1981a. A comparative study of seven grabs used for sampling benthic macroinvertebrates in rivers. *Freshwater Biol.* 11: 99–120.

Elliott, J.M. & C.M. Drake, 1981b. A comparative study of four dredges used for sampling benthic macroinvertebrates in rivers. *Freshwater Biol.* 11: 245–261.

Elliott, J.M. & P.A. Tullett, 1978. *A Bibliography of Samplers for Benthic Invertebrates*. Occ. Publ. Freshwater Biology Assoc. No. 4, Ambleside, U.K.: 61 pp.

Elliott, J.M. & P.A. Tullett, 1983. *A Supplement to a Bibliography of Samplers for Benthic Invertebrates*. Occ. Publ. Freshwater Biology Assoc. No. 20, Ambleside, U.K.

Elmgren, R., 1973. Methods of sampling sublittoral soft bottom meiofauna. *Oikos* 15 (Suppl.): 112–120.

Emery, K.O. & R.S. Dietz, 1941. Gravity coring instrument and mechanics of sediment coring. *Bull. Geol. Soc. Am.* 52: 1685–1715.

Emery, K.O. & J. Hulsemann, 1964. Shortening of sediment cores collected in open barrel gravity corers. *Sedimentology* 3: 144–154.

Evans, H.E. & D.C. Lasenby, 1984. A comparison of lead and zinc sediment profiles from cores taken by diver and a gravity corer. *Hydrobiologia* 108: 165–169.

Evans, R.D., 1988. A review of the whole-lake burden concept and a comparison of multiple core and single core studies. Paper presented at the Int. Conf. Trace Metals in Lakes, McMaster University, Hamilton, Ontario, Canada, August 14–18, 1988.

Evans, R.D. & P.J. Dillon, 1986. Whole-lake sediment sampling strategies. Paper presented at the EUCHEM Conf. Sampling Strategies and Techniques in Environmental Analysis, RIVM, Bilthoven, The Netherlands, January 20–24, 1986.

Evans, R.D. & F.H. Rigler, 1980. Measurement of whole lake sediment accumulation and phosphorus retention using lead-210 dating. *Can. J. Fish. Aquat. Sci.* 37: 817–822.

Evans, R.D. & F.H. Rigler, 1983. A test of lead-210 dating for the measurement of whole lake soft sediment accumulation. *Can. J. Fish. Aquat. Sci.* 40: 506–515.

Floderus, S., 1987. Statistical evaluation of spatial data — A method for irregular sampling. Paper presented at the 4th Int. Symp. Interaction Between Sediments and Water, Melbourne, Australia, February 16–20, 1987.

Foerstner, U. & G.T.W. Wittmann, 1979. *Metal Pollution in the Aquatic Environment.* Springer-Verlag, Berlin: 486 pp.

Foley, T.A., Jr., 1981. *Off-Site Radiological Exposure Review Project: A Computer-Aided Surface Interpolation and Graphic Display.* Desert Research Institute, University of Nevada System, Las Vegas. DOE/DP/01253–45022.

Folk, R.L., 1971. Longitudinal dunes of the northwestern edge of the Simpson Desert, Northern Territory, Australia. *Sedimentology* 16: 5–54.

Folk, R.L. & W.C. Ward, 1957. Brazos River bar: a study in the significance of grain-size parameters. *J. Sed. Petrol.* 27: 3–26.

Frithsen, J.B., D.T. Rudnick & R. Elmgren, 1983. A new, flow-through corer for the quantitative sampling of surface sediments. *Hydrobiologia* 99: 75–79.

Fuller, A.O., 1961. Size distribution characteristics of shallow marine sands from the Cape of Good Hope, South Africa. *J. Sed. Petrol.* 31: 256–261.

Fuller, J.A. & E.P. Meisburger, 1982. A simple, ship-based vibratory corer. *J. Sed. Petrol.* 52: 642–644.

Gilbert, R.O. & J.C. Simpson, 1985. Kriging for estimating spatial pattern of contaminants: Potential and problems. *Environ. Monit. Assess.* 5: 113–135.

Ginsburg, R.N., H.A. Bernard, R.A. Moody & E.E. Daigle, 1966. The Shell method of impregnating cores of unconsolidated sediments. *J. Sed. Petrol.* 36: 1118–1125.

Goedheer, G.J. & R. Misdorp, 1985. Spatial variability and variations in bedload transport direction in a subtidal channel as indicated by sonographs. *Earth Surf. Process. Landforms* 10: 375–386.

Håkanson, L., 1982. Bottom dynamics in lakes. *Hydrobiologia* 91: 9–22.

Håkanson, L., 1984. Sediment sampling in different aquatic environments: Statistical aspects. *Water Resour. Res.* 20: 41–46.

Håkanson, L., 1986a. Modifications of the Ekman sampler. *Int. Rev. Gesamt. Hydrobiol.* 71: 719–721.

Håkanson, L., 1986b. A sediment penetrometer for *in situ* determination of sediment type and potential bottom dynamic conditions. *Int. Rev. Gesamt. Hydrobiol.* 71: 851–858.

Håkanson, L. & M. Jansson, 1983. *Principles of Lake Sedimentology.* Springer-Verlag, Berlin: 316 pp.

Hamilton, A.L., W. Burton & J.F. Flanagan, 1970. A multiple corer for sampling profundal benthos. *J. Fish. Res. Board Can.* 27: 1867–1869.

Harris, S.A., 1958. Probability curves and the recognition of adjustment to depositional environment. *J. Sed. Petrol.* 28: 151–163.

Hartwig, E.O., 1984. Spring-arm core closer (SACC). *Oceanol. Acta* 7: 359–361.

Heip, C., 1984. Meiofauna of silty sediments in the coastal area of the North Adriatic, with special reference to sampling methods. *Hydrobiologia* 118: 67–72.

Heip, C., K.A. Willems & A. Goossens, 1977. Vertical distribution of meiofauna and the efficiency of the Van Veen grab on sandy bottoms in Lake Grevelingen (The Netherlands). *Hydrobiol. Bull.* 11: 35–45.

Henriksen, A. & R.F. Wright, 1977. Effects of acid precipitation on a small acid lake in Southern Norway. *Nordic Hydrol.* 8: 1–10.

Hilton, J., 1985. A conceptual framework for predicting the occurrence of sediment focusing and sediment redistribution in small lakes. *Limnol. Oceanogr.* 30: 1131–1143.

Hilton, J., J.P. Lishman & P.V. Allen, 1986. The dominant processes of sediment distribution and focusing in a small, eutrophic, monomictic lake. *Limnol. Oceanogr.* 31: 125–133.

Holopainen, I.J. & J. Sarvala, 1975. Efficiencies of two corers in sampling soft-bottom invertebrates. *Ann. Zool. Fenn.* 12: 280–284.

Hogue, E.W. & C.B. Miller, 1981. Effects of sediment microtopography on small-scale spatial distributions of meiobenthic nematodes. *J. Exp. Mar. Biol. Ecol.* 53: 181–191.

Hongve, D. & A. Erlandsen, 1979. Shortening of surface sediment cores during sampling. *Hydrobiologia* 65: 283–287.

Hopkins, T.L., 1964. A survey of marine bottom samplers. In M. Sears (Ed.), *Progress in Oceanography*, Vol. 2. Pergamon-Macmillan, New York: 213–256.

Hoyt, W.H. & J.M. Demarest II, 1981. A versatile twin-hull barge for shallow-water vibracoring. *J. Sed. Petrol.* 51: 656–658.

Hughes, J.P. & D.P. Lettenmaier, 1981. Data requirements for kriging: Estimation and network design. *Water Resour. Res.* 17: 1641–1650.

Huijbregts, C.J., 1975. Regionalized variables and quantitative analysis of spatial data. In J.C. Davis & M.J. McCullagh (Eds.), *Display and Analysis of Spatial Data, NATO Advanced Study Institute*. John Wiley & Sons, London: 38–53.

Inderbitzen, A.L., 1968. A study of the effects of various core samples on mass physical properties in marine sediments. *J. Sed. Petrol.* 38: 473–489.

Irwin, J., R.A. Pickrill & W. de L. Main, 1983. An outrigger for sampling from small boats. *J. Sed. Petrol.*, 53: 675–676.

Israel, A.M., F.G. Ethridge & E.L. Estes, 1987. A sedimentological description of a microtidal, flood-tidal delta, San Luis Pass, Texas. *J. Sed. Petrol.* 57: 288–300.

Jaquet, J.-M., E. Davaud, F. Rapin & J.-P. Vernet, 1982. Basic concepts and associated statistical methodology in the geochemical study of lake sediments. *Hydrobiologia* 91: 139–146.

Jenks, G.F., 1975. The evaluation and prediction of visual clustering in maps symbolized with proportional circles. In J.C. Davis & M.J. McCullagh (Eds.), *Display and Analysis of Spatial Data. NATO Advanced Study Institute*. John Wiley & Sons, London: 311–327.

Jensen, P., 1983. Meiofaunal abundance and vertical zonation in a sublittoral soft bottom, with a test of the Haps corer. *Mar. Biol.* 74: 319–326.

Jones, A.R. & C. Watson-Russell, 1984. A multiple coring system for use with scuba. *Hydrobiologia* 109: 211–214.

Jones, B.F. & C.J. Bowser, 1978. The mineralogy and related chemistry of lake sediments. In A. Lerman (Ed.), *Lakes. Chemistry Geology Physics.* Springer-Verlag, New York: 179–235.

Jumars, P.A., 1975. Methods for measurement of community structure in deep-sea macrobenthos. *Mar. Biol.* 30: 245–252.

Kajihara, M., K. Matsunaga & Y. Maita, 1974. Anomalous distribution of suspended matter and some chemical compositions in seawater near the seabed. Transport processes. *J. Oceanogr. Soc. Jap.* 30: 232–240.

Kemp, A.L.W., C.B.J. Gray & A. Mudrochova, 1971. A simple corer and a method for sampling the mud-water interface. *Limnol.Oceanogr.* 16: 689–694.

Kimmel, B.L., 1978. An evaluation of recent sediment focusing in Castle Lake (California) using a volcanic ash layer as a stratigraphic marker. *Verh. Int. Ver. Limnol.* 20: 393–400.

Klemens, W.E., 1984. How to get quantitative samples out of gravel stream sediments, concerning the benthic macro- and meiofauna. *Jahresber. Biol. Stn. Lunz Oesterr. Akad. Wiss.* 8: 39 pp.

Kratochvil, B. & J.K. Taylor, 1981. Sampling for chemical analysis. *Anal. Chem.* 53: 924A-938A.

Kraus, N.C., 1987. Application of a portable traps for obtaining point measurements of sediment transport rates in the surf zone. *J. Coast. Res.* 3: 139–152.

Krige, D.G., 1978. *Lognormal – De Wijsian Geostatistics for Ore Evaluation.* South African Institute of Mining & Metallurgy, Johannesburg: 50 pp.

Krogh, A. & R. Sparck, 1936. On a new bottom sampler for investigation of the micro fauna of the sea bottom with remarks on the quantity and significance of the benthonic micro fauna. *K. Danske Vidensk. Selsk. Skr.* 13: 3–12.

Krumblein, W.C. & F.A. Graybill, 1965. *An Introduction to Statistical Models in Geology.* McGraw-Hill Book Company, New York: 475 pp.

Kuehl, S.A., C.A. Nittrouer, D.J. DeMaster & T.B. Curtin, 1985. A long, square-barrel gravity corer for sedimentological and geochemical investigation of fine-grained sediments. *Mar. Geol.* 62: 365–370.

Kwaadsteniet, J.W. de, 1986. Strategies for soil sampling. Paper presented at the EUCHEM Conf. Sampling Strategies and Techniques in Environmental Analysis, RIVM, Bilthoven, The Netherlands, January 20–24, 1986.

Lanesky, D.E., B.W. Logan, R.G. Brown & A.C. Hine, 1979. A new approach to portable vibracoring underwater and on land. *J. Sed. Petrol.* 49: 654–657.

Lawless, J.P. & J.W. Padan, 1986. Developing deep seabed minerals. *Sea Technol.* 27: 1–45.

Lebel, J., N. Silverberg & B. Sundby, 1982. Gravity core shortening and pore water chemical gradients. *Deep-Sea Res.* 29 (11A): 1365–1372.

Lehman, J.T., 1975. Reconstructing the rate of accumulation of lake sediment: The effect of sediment focusing. *Quat. Res. (N.Y.)* 5: 541–550.

Lesht, B.M., 1988. Nonparametric evaluation of the size of limnological sampling networks: Application to the design of a survey of Green Bay. *J. Great Lakes Res.* 14: 325–337.

Likens, G.E. & M.B. Davis, 1975. Post-glacial history of Mirror Lake and its watershed in New Hampshire. An initial report. *Int. Ver. Theor. Angew. Limnol. Verh.* 19: 982–993.

Love, F.G., G.M. Simmons, Jr., R.A. Wharton, Jr., & B.C. Parker, 1982. Methods for melting dive holes in thick ice and vibracoring beneath ice. *J. Sed. Petrol.* 52: 644–645.

Mackereth, F.J.H., 1958. A portable core sampler for lake deposits. *Limnol. Oceanogr.* 3: 181–191.

Mackereth, F.J.H., 1969. A short core sampler for subaqueous deposits. *Limnol. Oceanogr.* 14: 145–151.

Maienthal, E.J. & D.A. Becker, 1976. *A Survey of Current Literature on Sampling, Sample Handling, and Long Term Storage for Environmental Materials.* U.S. Department of Commerce, NBS Tech. Note 929: 34 pp.

Martin, E.A. & R.J. Miller, 1982. A simple, diver-operated coring device for collecting undisturbed shallow cores. *J. Sed. Petrol.* 52: 641–642.

Matheron, G., 1971. *The Theory of the Regionalized Variables and its Applications.* Cahiers Centre Morphol. Mathematique de Fontainebleau: 211 pp.

McBratney, A.B., R. Webster & T.M. Burgess, 1981a. The design of optimal sampling schemes for local estimation and mapping of regionalized variables. I. Theory and method. *Comput. Geosci.* 7: 331–334.

McBratney, A.B., R. Webster & T.M. Burgess, 1981b. The design of optimal sampling schemes for local estimation and mapping of regionalized variables. II. Program and examples. *Comput. Geosci.* 7: 335–365.

McCoy, F.W., 1980. Photographic analysis of coring. *Mar. Geol.* 38: 263–282.

McCoy, F.W. & S. Selwyn, 1983/1984. The hydrostatic corer. *Mar. Geol.* 54: M33–M41.

McIntyre, A.D., 1971. Efficiency of benthos sampling gear. In N.A. Holme & A.D. McIntyre (Eds.), *Methods for the Study of Marine Benthos.* Blackwell Scientific, Oxford: 344 pp.

McIntyre, A.D. & R.M. Warwick, 1984. Meiofauna techniques. In N.A. Holme & A.D. McIntyre (Eds.), *Methods for the Study of Marine Benthos.* 2nd ed. Blackwell Scientific: 217–244.

Meischner, D., H. Torunski & G. Kuhn, 1981. High-energy pneumatic vibration corer for subaqueous sediments. *Senckenbergiana Marit.* 13: 179–191.

Milbrink, G., 1968. A microstratification sampler for mud and water. *Oikos* 19: 105–110.

Milbrink, G., 1971. A simplified tube bottom sampler. *Oikos* 22: 260–263.

Modeling Task Force, 1987. Large lake models — Uses, abuses, and future. *J. Great Lakes Res.* 13: 387–396.

Mosley, M.P. & D.S. Tindale, 1985. Sediment variability and bed material sampling in gravel-bed rivers. *Earth Surf. Process. Landforms* 10: 465–482.

Morris, R.J. & R.D. Peters, 1981. *The IOS Multi-Purpose Gravity Corer.* Report of the Institute of Oceanographic Science, Wormley, U.K., no. 124: 19 pp.

Mueller, G., 1979. Schwermetalle in den Sedimenten des Rheins-Veraenderung seit 1971. *Umsch. Wiss. Tech.* 79: 778–783.

Multer, H.G., D.M. Stainken & J.M. McCormick, 1981. Spatial/temporal patterns of macrobenthos, sediments and pollutants in Raritan Bay-lower N.Y. Bay. *Estuaries* 4: 302.

Muntau, H., M. Van Son, R. Baudo, P. Schramel, G. Marengo, A. Lattanzio & L. Amantini, 1986. Heavy metal variability in sediments and related sampling strategies. Paper presented at the EUCHEM Conf. Sampling Strategies and Techniques in Environmental Analysis, RIVM, Bilthoven, The Netherlands, January 20–24, 1986.

Nishimura, A., 1984. Bottom sampling and photographing on the southeastern offshore of the Boso Peninsula. *Cruise Rep. Geol. Surv. Jap.* 19: 54–66.

Nixon, D.E., 1976. Dynamics of spatial pattern for the Gastrotrich *Tetranchyroderma bunti* in the surface sand of high energy beaches. *Int. Rev. Ges. Hydrobiol. Hydrogr.* 61: 211–248.

Okusa, S., T. Nakamura & N. Dohi, 1983. Geotechnical properties of submarine sediments in the Seto Inland Sea. *Mar. Geotechnol.* 5: 131–152.

Patton, K.T. & G.T. Griffin, 1969. An analysis of marine corer dynamics. *Mar. Technol. Soc. J.* 3: 27–40.

Pedersen, T.F., S.J. Malcolm & E.R. Sholkovitz, 1985. A lightweight gravity corer for undisturbed sampling of soft sediments. *Can. J. Earth Sci.* 22: 133–135.

Phleger, F.B., 1951. Ecology of Foraminifera, northwest Gulf of Mexico. 1. Foraminifera distribution. *Geol. Soc. Am. Mem.* 46: 1–88.

Perillo, G.G., M.C. Albero, F. Angiolini & J.O. Codignotto, 1984. An inexpensive, portable coring device for intertidal sediments. *J. Sed. Petrol.* 54: 654–655.

Plocki, W. & T. Radziejewska, 1980. A new meiofauna corer and its efficiency. *Ophelia* Suppl. 1: 231–233.

Probert, P.K., 1984. A comparison of macrofaunal samples taken by box corer and anchor-box dredge. *NZOI Rec.* 4: 149–156.

Rapin, F., E. Davaud & J.-P. Vernet, 1978. *Etude Generale de la Pollution des Sediments du Leman.* Rep. Comm. Int. Prot. Leman, Geneva: 294–309.

Reise, K., 1979. Spatial configurations generated by motile benthic polychaetes. *Helgol. Wiss. Meeresunters.* 32: 263–300.

Reynoldson, T.B., Jr. & H.R. Hamilton, 1982. Spatial heterogeneity in whole lake sediments—Towards a loading estimate. *Hydrobiologia* 91: 235–240.

Rhodes, E.G., 1981. Reef coring techniques, Great Barrier Reef. *J. Sed. Petrol.* 51: 650-652.

Richards, A.F. & G.H. Keller, 1961. A plastic-barrel sediment corer. *Deep-Sea Res.* 8: 306–312.

Robbe, D., 1981. *Pollutions Metalliques du Milieu Naturel. Guide Methodologique de Leur Etude a Partir des Sediments. Rapport Bibliographique.* Ministere de l'Urbanisme et du Logement—Ministere des Transports, Laboratoire Central des Ponts et Chaussees, Paris, Rapport de Recherche LPC N° 104: 83 pp.

Robinson, J.E., 1975. Frequency analysis, sampling, and errors in spatial data. In J.C. Davis & M.J. McCullagh (Eds.), *Display and Analysis of Spatial Data. NATO Advanced Study Institute.* John Wiley & Sons, London: 78–95.

Rofes, G., 1980. *Etude des Sediments. Methodes de Prelevement et d'Analyses Pratiquees au Laboratoire de Sedimentologie.* Ministere de l'Agriculture, CTGREF. Etude N° 47: 50 pp.

Rofes, G. & M. Savary, 1981. Description d'un nouveau modele de carottier pour sediments fins. *Bull. Fr. Piscic.* 283: 102–113.

Ross, D.A. & W.R. Riedel, 1967. Comparison of upper parts of some piston cores with simultaneously collected open-barrel cores. *Deep-Sea Res.* 14: 285–294.

Rutledge, P.A. & J.W. Fleeger, 1988. Laboratory studies on core sampling with application to subtidal meiobenthos collection. *Limnol. Oceanogr.* 33: 274–280.

Satake, K., 1983. A small handy corer for sampling of lake surface sediment. *Jap. J. Limnol.* 44: 142–144.

Schneider, D.C., J.M. Gagnon & K.D. Gilkinson, 1987. Patchiness of epibenthic megafauna on the outer Grand Banks off Newfoundland. *Mar. Ecol. Prog. Ser.* 39: 1–13.

Shiemer, E.W. & J.R. Schubel, 1970. A near bottom suspended sediment sampling system for studies of resuspension. *Limnol. Oceanogr.* 15: 644–646.

Sholkovitz, E.R., 1970. A free vehicle bottom water sampler. *Limnol. Oceanogr.* 15: 641–644.

Sindowski, K.H., 1957. Die synoptische Methode des Kornkurven — Vergleiches zur Ausdentung Fossiler Sedimentations Raume. *Geol. Jahrb.* 73: 235–275.

Sklash, M., S. Mason & C. Pugsley, 1986. Sandusky nearshore-offshore downflux. In *Proc. 29th Conf. Great Lakes Research*, International Association for Great Lakes Research, Ann Arbor, MI: 47.

Sly, P.G., 1969. Bottom sediment sampling. In *Proc. 12th Conf. Great Lakes Research*, International Association for Great Lakes Research, Ann Arbor, MI: 883–898.

Sly, P.G., 1975. Statistical evaluation of recent sediment geochemical sampling. Paper presented at the IX Int. Congr. Sedimentology, Nice, France, 1975.

Sly, P.G., 1977. Sedimentary environments in the Great Lakes. In H.L. Golterman (Ed.), *Interaction Between Sediments and Freshwater.* W. Junk Publishing, The Hague: 76–82.

Sly, P.G., 1978. Sedimentary processes in lakes. In A. Lerman (Ed.), *Lakes. Chemistry Geology Physics.* Springer-Verlag, New York: 65–89.

Sly, P.G., 1981. Equipment and techniques for offshore survey and site investigations. *Can. Geotech. J.* 18: 230–249.

Sly, P.G. & K. Gardener, 1970. A vibro corer and portable tripod-winch assembly for through ice sampling. In *Proc. 13th Conf. Great Lakes Research*, International Association for Great Lakes Research, Ann Arbor, MI: 297–307.

Sly, P.G., R.L. Thomas & B.R. Pellettier, 1983. Interpretation of moment measures derived from water-lain sediments. *Sedimentology* 30: 219–233.

Smith, D.G., 1984. Vibracoring fluvial and deltaic sediments: Tips on improving penetration and recovery. *J. Sed. Petrol.* 54: 660–661.

Smith, J.N., B.P. Boudreau & V. Noshkin, 1986. Plutonium and ^{210}Pb distributions in Northeast Atlantic sediments: Subsurface anomalies caused by non-local mixing. *Earth Planet. Sci. Lett.* 81: 15–28.

Sokal, R.R. & F.J. Rohlf, 1969. *Biometry. The Principles and Practice of Statistics in Biological Research*. W.H. Freeman & Co., San Francisco: 776 pp.

Spencer, D.W., 1963. The interpretation of grain-size distribution curves of clastic sediments. *J. Sed. Petrol*. 33: 180–190.

Splettstoesser, J.F. & M.C. Hoyer, 1983. Heat-sealable pouches for preserving rock cores. *J. Sed. Petrol*. 53: 674–675.

Stoner, A.W., H.S. Greening, J.D. Ryan & R.J. Livingston, 1983. Comparison of macrobenthos collected with cores and suction sampler in vegetated and unvegetated marine habitats. *Estuaries* 6: 76–82.

Switzer, P., 1975. Estimation of the accuracy of qualitative maps. In J.C. Davis & M.J. McCullagh (Eds.), *Display and Analysis of Spatial Data. NATO Advanced Study Institute*. John Wiley & Sons, New York: 1–13.

Switzer, P., 1979. Statistical considerations in network design. *Water Resour. Res*. 15: 1712–1516.

Symader, W. & W. Thomas, 1978. Interpretation of average heavy metal pollution in flowing waters and sediment by means of hierarchical grouping analysis using two different error indices. *Catena* 5: 131–144.

Tanner, W.F., 1959. Sample components obtained by the method of differences. *J. Sed. Petrol*. 29: 408–411.

Tanner, W.F., 1964. Modification of sediment size distributions. *J. Sed. Petrol*. 34: 156–164.

Thomas, R.L. & J.-M. Jaquet, 1976. Mercury in the surficial sediments of Lake Erie. *J. Fish. Res. Board Can*. 33: 404–412.

Thomson-Becker, E.A. & S.N. Luoma, 1985. Temporal fluctuations in grain size, organic materials and iron concentrations in intertidal surface sediment of San Francisco Bay. *Hydrobiologia* 129: 91–107.

Toole, J., J. Thomson, T.R.S. Wilson & M.S. Baxter, 1984. A sampling artefact affecting the uranium content of deep-sea porewaters obtained from cores. *Nature* 308: 263–265.

Twinch, A.J. & P.J. Ashton, 1984. A simple gravity corer and continuous-flow adaptor for use in sediment/water exchange studies. *Water Res*. 18: 1529–1534.

Van der Gaast, N.G., 1986. Sampling strategies for soil pollution: a chemiometric approach to the major problems. Paper presented at the EUCHEM Conf. Sampling Strategies and Techniques in Environmental Analysis, RIVM, Bilthoven, The Netherlands, January 20–24, 1986.

Vidakovic, J., 1984. Meiofauna of silty sediments in the coastal area of the North Adriatic, with special reference to sampling methods. *Hydrobiologia* 118: 67–72.

Viel, M. & V. Damiani, 1985. Sedimentological and geochemical study of recent deposits in Lago Maggiore (North Italy). *Mem. Ist. Ital. Idrobiol*. 43: 181–238.

Visher, G.S., 1969. Grain-size distribution and depositional processes. *J. Sed. Petrol*. 39: 1074–1106.

Weaver, P.P.E. & P.J. Schultheiss, 1983. Detection of repenetration and sediment disturbance in open-barrel gravity cores. *J. Sed. Petrol*. 53: 649–654.

White, K.D. & M.E. Tittlebaum, 1984. Statistical comparison of heavy metal concentration in various Louisiana sediments. *Environ. Monit. Assess*. 4: 163–170.

Willemoes, M., 1964. A ball-stoppered quantitative sampler for the microbenthos. *Ophelia* 1: 235–240.

Williams, J.D.H. & A.E. Pashley, 1979. Lightweight corer designed for sampling very soft sediments. *J. Fish. Res. Board Can.* 36: 241–246.

Word, J.Q., 1976. Biological comparison of grab sampling devices. In *1976 Annu. Rep. Coastal Water Research Project*, El Segundo, CA: 189–194.

Wright, H.E., E.J. Cushing & D.A. Livingstone, 1965. Coring devices for lake sediments. In B. Kummel & D. Raup (Eds.), *Handbook of Paleontological Techniques*. W.H. Freeman & Co., San Francisco: 494–520.

Wright, L.D., D.B. Prior, C.H. Hobbs, R.J. Byrne, J.D. Boon, L.C. Schaffner & M.O. Green, 1987. Spatial variability of bottom types in the lower Chesapeake Bay and adjoining estuaries and inner shelf. *Estuarine Coastal Shelf Sci.* 24: 765–784.

Zhong, W.B., Z.T. Hou & C.S. Zou, 1986. Analysis of the motion pattern of a boomerang sediment corer after being dropped into water. *Collect. Ocean. Works/Haiyang Wenji* 9: 130–139.

Zuellig, H., 1956. Sedimente als Ausdruck des Zustandes eines Gewaessers. *Schweiz. Z. Hydrol.* 18: 7–143.

CHAPTER 3

Inorganic Sediment Chemistry and Elemental Speciation

Ulrich Förstner

1.0 INTRODUCTION

Sediments are both carriers and potential sources of contaminants in aquatic systems, and these materials may also affect groundwater quality and agricultural products when disposed of on land. Such problems had initially been recognized for inorganic chemicals in the early and middle sixties from the studies on artificial radionuclides in the Columbia and Clinch Rivers by Sayre *et al.* (1963) and on heavy metals in the Rhine River system by De Groot (1966). In the early seventies, following the catastrophic events of cadmium and mercury poisoning in Japan, sediment-associated metal contaminants received public attention, for example, with severe effects on aquatic ecosystems in the Wabigoon River, Laurentian Great Lakes, Swedish lakes, and in many past and present mining areas all over the world (Förstner & Wittmann 1979). Now, at the beginning of the nineties, metal pollution in surface waters seems to have peaked in critical examples, e.g., in the Rhine River, partly as a result of recent improvements in wastewater treatment. Problems with metal pollution, however, still exist both on a local and a regional scale: Waters of large areas in southern Scandinavia and eastern North America are severely endangered by the combined effects of airborne pollutants and acidity. Increasing quantities of dredged materials and mine tailings represent long-term reservoirs for the release of metals into sensitive environmental compartments (Salomons & Förstner 1988a).

During the last decade the major objectives of research on metal-polluted waters have changed from the initial surveys of sources and pathways to more detailed investigations of the mechanisms controlling the mobility and bioavailability of different metal species. The general experience that the environmental behavior and toxicity of an element can only be understood in terms of its actual molecular form led to the introduction of the term "speciation", which is used in a vague manner both for the operational procedure for determining typical metal species in environmental samples and for describing the distribution and transformation of such species in various media (Leppard 1983, Bernhard *et al.* 1986, Landner 1987, Patterson & Passino 1987, Batley 1989, Allen *et al.* 1989, Gücer *et al.* 1990). Problems of "speciation" are particularly complex in heterogeneous systems, e.g., in soils, aerosol particles, and sediments; thermodynamic models may give suggestions as to the possible species to expect, but due to the important role of kinetically controlled processes in biogeochemistry the actual speciation is often different from what is expected (Andreae *et al.* 1984). In polluted ("stressed") systems entropy increases and there is a concomitant increase in instability in both the physical and biological contexts which, among others, may cause problems in sample handling and storage and in preparative and analytical techniques (Wood *et al.* 1986).

On the other hand, it is just the "stressed" system where action is immediately needed and where for an assessment or prognosis of possible adverse effects the species and their transformations of pollutants have to be evaluated. The following questions have been raised with respect to the mobility and bioavailability of potentially toxic metals in contaminated systems (Förstner 1987) and will be treated in the present chapter on inorganic elemental speciation in sediments:

- How reactive are the metals introduced via solid materials from anthropogenic activities (atmospheric fallout, hazardous waste, sewage sludge, dredged material) in comparison to the natural compounds?
- Are the interactions of critical elements between solution and solid phases comparable for natural and contaminated systems?
- What are the factors and processes of remobilization to become particularly effective, when either the solid input or the solid/solution interactions lead to weaker bonding of certain elemental species in contaminated compared to natural systems?

In a final section remedial measures for highly contaminated sediments will be described, including techniques for reducing their mobilities and transfer rates into biological systems.

2.0 METAL CONCENTRATIONS IN SEDIMENTS

2.1 Global and Regional Budgets; Toxicity; Pathways

Metals are natural constituents of rocks, soils, sediments, and water. However, over the 200 years following the beginning of industrialization huge changes in the global budget of critical chemicals at the earth's surface

Table 1. Some Hazardous Elements in Industrial Waste Streams

Industry	As	Cd	Cr	Cu	Pb	Hg	Se	Zn
Mining and metallurgy	x	x	x	x	x	x	x	x
Paints and dyes		x	x	x	x	x	x	
Pesticides	x			x	x	x		x
Electrical & electronic				x	x	x	x	
Cleaning & duplicating	x		x	x	x		x	
Electroplating/finishing			x	x				x
Chemical manufacturing			x	x		x		
Explosives	x			x	x	x		
Rubber & plastics						x		
Batteries		x			x	x		
Pharmaceuticals	x					x		
Textiles			x	x				
Petroleum & coal	x				x			
Pulp & paper						x		
Leather			x					

Source: Barnhart (1978).

have occurred, challenging those regulatory systems which took millions of years to evolve (Wood & Wang 1983). For example, the ratio of the annual mining output of a given element to its natural concentration in unpolluted soils, which can be used as an "Index of Relative Pollution Potential" (Förstner & Müller 1973), is particularly high for Pb, Hg, Cu, Cd, and Zn, namely 10 to 30 times higher than for Fe or Mn, respectively.

On a regional scale, for industrialized areas, both wastewater and atmospheric sources occur at about the same order of magnitude. An inventory on Lake Erie sediments by Nriagu *et al.* (1979) revealed that atmospheric inputs account for 20%, 35%, and 50% of the Cu, Zn, and Pb discharges, respectively; the contributions from sewage effluents are 45%, 30%, and 20%.

There are typical connections of metal accumulations in sediments to specific local sources such as discharges from smelters (Cu, Pb, Ni), metal-based industries (e.g., Zn, Cr, and Cd from electroplating), as well as chemical manufacturing plants. Some industries use a wide spectrum of different metals, whereas other production is connected with the release of a few typical metals (Table 1).

As for the mechanisms of toxicity, the most relevant is certainly the chemical inactivation of enzymes. All divalent transition metals readily react with the amino, imino, and sulfhydryl groups of proteins; some of them may compete with essential elements such as zinc and displace them in metalloenzymes. Some metals may also damage cells by acting as antimetabolites, or by forming precipitates or chelates with essential metabolites. Soil biochemical processes considered especially sensitive to heavy metals are mineralization of N and P, cellulose degradation, and possibly N_2 fixation. Ecotoxicological considerations in aquatic systems involve self-

Table 2. Perturbation of the Geochemical Cycles of Selected Metals by Society

Element	Scale of Perturbation			Diagnostic Environments	Mobilizing Mechanisms	Critical Pathway
	Global	Reg.	Local			
Lead	+	+	+	Ice, sediment	Volatilization	Air, food
Aluminum	–	+	–	Water, soil	Solubilization	Water
Chromium	–	–	+	Water, soil	Solubilization	Water
Mercury	(–)	+	+	Fish, sediment	Alkylation	Food (air)
Cadmium	(–)	+	+	Soil, sediment	Solubilization	Food

Source: Examples after Andreae *et al.*, in Nriagu (1984).

purification in surface and groundwater, the effects of heavy metal enrichment on biologic purification treatment, and the influence on crustaceans, fish, and ultimately on man (Moore & Ramamoorthy 1984). Elements such as silver and copper may induce adverse effects on aquatic biota far below the actual limits for drinking water. In addition to the hazard of direct toxicity to organisms, biological uptake of trace metals may lead to modification of food webs and toxicity to man through the consumption of contaminated food. With respect to the aqueous metal species it has been suggested that the "free" or aquo-metal ion form is the most available for organisms compared to the particulate, complexed, or chelated forms. On the other hand, there are physical and chemical processes, such as adsorption, filtration, sedimentation, complexation, precipitation, and redox reactions, which can act as partial or almost complete barriers to the movement of metals along their pathway to man (e.g., Nriagu 1984).

There are many pathways or routes by which aquatic or terrestrial biota (in particular, humans) are exposed to metallic compounds, and these are changing as society uses more or less of the metal or changes the chemical form of the metal in the environment. To assess which elements may be of concern, four criteria have been proposed:

1. Has the geochemical cycle of the element been substantially perturbed by man, and on what scales?
2. Is the element mobile in geochemical processes because of either its volatility or its solubility in natural waters, so that the effect of geochemical perturbations can propagate through the environment?
3. What is the degree of public health concern associated with the element?
4. What are the critical pathways by which the most toxic species of the element can reach the organ in man which is most sensitive to its effect?

Of the elements listed in Table 2, global perturbations are most dramatically seen for lead. Present-day levels of lead in Americans and Europeans are probably two to three orders of magnitude higher than those of pretechnological humans, as evidenced from studies on blood lead concentrations in remote populations. Changes on a regional scale are typically found for aluminum mobilization in soils and waters of low buffer capacity affected by acid precipitation; despite insignificant inputs of aluminum due to human activities, increased solubility will induce toxic effects on both terrestrial and aquatic biota. Chromium usually represents examples of only local significance; here, elemental species exhibit characteristic differences, in that the hexavalent form is more toxic than the trivalent form. Other elements, such as lead and mercury in Table 2, may be mobilized by the biotic or abiotic formation of organometallic compounds. Accumulation of methylmercury in seafood, probably the most critical pathway of a metal to humans, has caused several thousand cases of poisoning incidents in Japan (Takeuchi *et al.* 1959). The first catastrophic event of cadmium pollution, causing the "Itai-Itai" disease in inhabitants in the Jintsu River area of Japan for 20 years after the Second World War, has been caused by effluents from zinc mine wastes, which flooded low-lying rice field areas (Kobayashi 1971). Cadmium pollution has been recorded from sediment studies in different regions of the world and is related to various sources; particularly high concentrations have been measured in the Hudson River Estuary, New York (nickel-cadmium battery factory), in the Hitachi area near Tokyo (braun tube factory), in Palestine Lake, Indiana (plating industry), in Sörfjord, Norway and Derwent Estuary, Tasmania (smelter emissions), and from the Neckar River, FRG (pigment factory).

2.2 Historical Development; Sediments as Indicators

Geochemical investigations of stream sediment have long been standard practice in mineral exploration (Hawkes & Webb 1962); by more extensive sampling and analysis of metal contents in water, soils, and plants, the presumable enrichment zones can be narrowed down and, in favorable cases, localized as exploitable deposits. Generally, the variation in trace metal content of stream sediments can be characterized as a function of the potential controlling factors "influence of lithologic units", "hydrologic effects", "geologic features", "cultural (man-made) influences", "type of vegetational cover", and "effects of mineralized zones" (Dahlberg 1968). Similarly, lake sediment geochemistry has been used as a guide to mineralization, particularly intensive on lakes of the Canadian Shield (Allan 1971). This approach attracted much attention when mineral exploration was followed by large-scale mining and processing activities: "Both the exploration and environmental geochemist can be looking for the same type of areas, those with high metal concentrations, but obviously from a different motivation" (Allan 1974).

Sediments are an important storage compartment for the metals released into surface waters; furthermore, because of their ability to sequester metals, sediments can reflect water quality and record the effects of anthro-

Table 3. The Total Concentrations[a] of Metals in a Variety of Aqueous sediments

Element	River Suspended Solids[b]		River Sediments[c]		Estuarine Sediments[d]		Coastal Sediments[e]		Base-line[f]
	A	B	A	B	A	B	A	B	
Nickel	131	298	51	230	72	174	35	51	35
Chromium	–	–	191	401	111	110	74	48	60
Copper	110	611	70	270	74	251	38	84	56
Lead	141	295	140	960	131	523	94	274	22
Zinc	318	975	263	1420	392	>1000	230	585	92

Source: After Chester (1988).
[a]In mg/kg.
[b]River suspended solids: (A) average of 15 suspended solids from the Yamaska and St. Francois River basins, Canada; (B) a "polluted" sample from the same region. Data from Tessier et al. (1980).
[c]River sediments: (A) average of 29 surficial river sediments from the Detroit River and western Lake Erie (Canada); (B) a "polluted" sample from the same region. Data from Lum & Gammon (1985).
[d]Estuarine sediments: (A) average of 13 surficial estuarine sediments from South Wales (U.K.); (B) a "polluted" sample from the same region. Data from Towner (1984).
[e]Coastal sediments: (A) average of 8 surficial sediments (< 64 μm fraction) from the Severn Estuary (U.K.); (B) a "polluted" sample from the same region. Data from Stoner (1974).
[f]Wedepohl (1969/1978): Average composition shallow water sediments.

pogenic emissions. This is shown in the data in Table 3 (after Chester 1988), in which total metal concentrations are listed for a number of different types of aqueous sediment; the data also serve to demonstrate that the extent to which individual sediments can concentrate metals varies considerably. Thus, it is to be expected that, with respect to metal reactivity, the various types of sediment will have a different impact on the environment via the manner in which they act as metal "sinks" or "sources".

The study of dated sediment cores has proven particularly useful as it provides a historical record of the various influences on the aquatic system by indicating both the natural background levels and the man-induced accumulation of elements over an extended period of time. Marine and – in particular – lacustrine environments have the ideal conditions necessary for the incorporation and permanent fixing of metals and organic pollutants in sediments: reducing (anoxic) and nonturbulent environments, steady deposition, and the presence of suitable, fine-grained mineral particles for pollutant fixation. Various approaches to the dating of sedimentary profiles have been used, but the isotopic techniques, using ^{210}Pb, ^{137}Cs, and $^{239+240}$Pu, have produced the more unambiguous results and therefore have been the most successful (see review on "Historical Monitoring" by Alderton, 1985).

An example is given in Figure 1 of concentrations of heavy metals in a sediment core from 22 m water depth in the Bay of Helgoland, North Sea (Gerlach 1981, after Förstner & Reineck 1974). The natural rate of sedimentation in this area is approximately 50 cm per century. The shadowed verti-

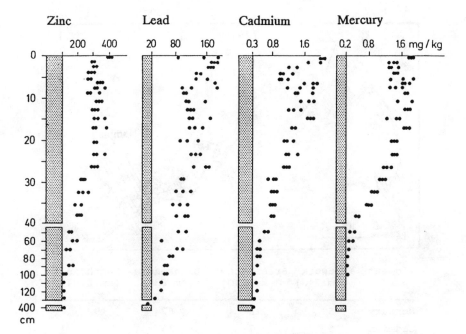

Figure 1. Concentrations of heavy metals in a sediment core taken in the Bay of Helgoland, North Sea. (Reprinted from Förstner & Reineck 1974. With permission.)

cal bars represent as background the heavy metal concentrations before
man and his civilization affected it. At a core depth of 125 cm, corresponding approximately to a period 200–300 years ago, fairly great concentrations of some heavy metals begin to stand out. There are especially great
amounts in the upper 20–40 cm. It has been concluded from the chemical
form of the heavy metals that atmospheric contributions play the most
important role and that a 3.5-fold increase of the lead concentration and a
2.5-fold increase of the cadmium concentration are from this source (Patchineelam & Förstner 1977).

2.3 Concentration Factors and Grain Size Effects

As shown in the previous section, concentration factors of metals in
recent sediments, introduced from civilizational sources, can be evaluated
from comparative values found in deeper parts of core profiles, which are
collected, for example, from alluvial deposits in the catchment areas of
large rivers. An example is given in Figure 2 for a longitudinal section of the
tidal Elbe River upstream and downstream from the city of Hamburg
(Anon. 1989a). In the upper part extremely great factors of enrichment –
compared to metal concentration in the < 63 μm fractions of core sediments

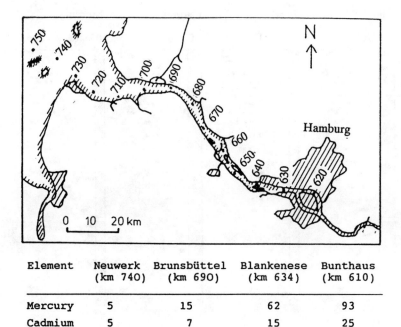

Element	Neuwerk (km 740)	Brunsbüttel (km 690)	Blankenese (km 634)	Bunthaus (km 610)
Mercury	5	15	62	93
Cadmium	5	7	15	25
Zinc	2.5	3.5	13	20
Copper	1	2	8	12
Lead	2.5	3	5	7
Arsenic	1	1	4	6
Chromium	1	1	3	4

Figure 2. Enrichment factors of selected elements in sediments from a longitudinal section of the tidal Elbe River. (After ARGE-Elbe Data; Anon. 1989a.)

from the older Elbe flood plain upstream from Hamburg — were found for mercury, but high concentrations also occurred for cadmium, zinc, and copper. Subsequent to the dilution with cleaner sediments from the outer Elbe River estuary, factors of enrichment were decreasing downstream; however, even in the lower Elbe River section, critical concentrations of mercury and cadmium were measured. These data indicate that the transport of contaminated suspended material from the upper Elbe River is now influencing the coastal zone in the southern German Bight, and this should be understood as an early signal for avoiding any additional transfer process of polluted materials into these sensitive zones, for example, by the practice of "sludge harrowing" in the Hamburg Harbor area.

A large number of sediment analyses which have been performed for the inventory, monitoring, and surveillance of pollution in aquatic systems have

Table 4. Methods for the Reduction of Grain Size Effects in Sediment Samples

	Method	Reference (Example)[a]
I. Separation of grain size fractions (mechanical)	204 μm (sieving)	Thornton et al. (1975)
	175 μm (sieving)	Vernet & Thomas (1972)
	63 μm (sieving)	Allan (1971)
	20 μm (sieving)	Jenne et al. (1980)
	2 μm (settling tube)	Banat et al. (1972)
II. Extrapolation from regression curves	Metal/percent 16 μm	De Groot et al. (1971)
	Metal/percnt 20 μm	Lichtfuβ & Brümmer (1977)
	Metal/percent 63 μm	Smith et al. (1973)
	Metal/specific surface area	Olivet (1973)
III. Correction for inert mineral constituents	Quartz-free sediment	Thomas (1972)
	Carbonate/quartz-free	Salomons & Mook (1977)
IV. Treatment with dilute acids or with complexing agents (determination of "mobile" fraction)	0.1 M hydrochloric acid	Gross et al. (1971)
	0.3 M HCl	Malo (1977)
	0.5 M HCl	Agemian & Chau (1976)
	25% acetic acid	Loring (1977)
	EDTA, DTPA, NTA	Gambrell et al. (1977)
V. Comparison with "conservative" elements	Metal/aluminum	Bruland et al. (1974)
	Sediment enrichment factor (Al_x/Al background)	Kemp et al. (1976)
	Rel. atomic variation	Allan & Brunskill (1977)
	M/Cs, Sc, Eu, Rb, Sm	Ackerman (1980)
	Al_x/standard-Al	Li (1981)
	Sc_x/standard-Sc	Schoer et al. (1982)
		Thomas & Martin (1982)

[a]For references see Salomons & Förstner (1984).

clearly shown that it is imperative, particularly for river sediments, to base these data on a standardized procedure with regard to particle size. Different methods for grain size correction are compiled in Table 4 (after Förstner & Wittmann 1979, De Groot et al. 1982, Förstner & Schoer 1984). These methods will mostly reduce (not eliminate!) the fraction of the sediment that is largely chemically inert, i.e., mostly the coarse-grained, feldspar, and carbonate minerals, and increase the substances active in pollutant enrichment, i.e., hydrates, sulfides, and amorphous and fine-grained organic materials.

Separation of grain size is advantageous because only a few samples from a particular locality are needed. However, it has been inferred that the decrease of pollutant concentrations in the medium grain size range should be even more pronounced if mechanical fractionation would more accurately separate individual particles according to their grain size. One has to consider that "coatings", for example, of iron/manganese oxides, carbonates, and organic substances on relatively inert material with respect to sorption act as substrates of pollutants in coarser grain size fractions

(Förstner & Patchineelam 1980). Nonetheless, the fraction <63 μm has been recommended for the following reasons (Förstner & Salomons 1980):

- Pollutants have been found to be present mainly on clay/silt particles.
- This fraction is nearly equivalent to the material carried in suspension—the most important transport mode by far.
- Sieving generally does not alter pollutant concentrations (for metals, even by wet sieving, when water of the same system is used).
- Numerous pollutant studies, especially with respect to heavy metals, have already been performed on the <63-μm fraction, allowing better comparison of results.

Extrapolation techniques, both for the grain size and specific surface area, require a relative large number of samples $(10-15)$. Further complicating this is the fact that the calculation of the regression line is a tedious and mostly inaccurate procedure. The quartz correction method involves fusion with potassium pyrosulfates, which preferentially removes the layered silicates (clay), organic and inorganic carbon, and sulfides with a residue made up of quartz plus feldspar and resistant heavy minerals such as zircone (Thomas *et al.* 1976).

Generally, five types of elements have been distinguished according to their distribution in sediment cores from Lake Erie (Kemp *et al.* 1976): (i) Diagenetically mobile elements such as Fe, Mn, and S; (ii) carbonate elements, carbonate-C and calcium; (iii) nutrient elements, organic C, N, and P; (iv) enriched elements, such as Cu, Cd, Zn, Pb, and Hg; and (v) conservative elements, e.g., Al, Li, Si, K, Ti, Na, and Mg.

Comparison of group iv elements of environmental concern with "conservative" elements (v) seems to be particularly useful for the reduction of grain size effects, since no separation step is required. After many discussions (e.g., Anon. 1989b) it can be concluded that normalization to aluminum—as a chemical tracer of Al-silicates, particularly the clay minerals—provides several advantages, and it is recommended that this approach will be considered primarily in comparative studies on a regional scale.

2.4 Quantification of Metal Pollution in Sediments

New objectives regarding the improvement of water quality as well as problems with the resuspension and land deposition of dredged materials require a standardized assessment of sediment quality. Numerical approaches are based on (1) accumulation, (2) porewater concentrations, (3) solid/liquid equilibrium partition (sediment/water and organism/water), and (4) elution properties of contaminants (Förstner *et al.* 1989).

On the basis of the background approach (1), a quantitative measure of

Table 5. Comparison of IAWR Water Quality Indices (Based on Biochemical Data) and Index of Geoaccumulation (I_{geo}) of Trace Metals in Sediments of the Rhine River

IAWR Index	IAWR Water Quality (Pollution Intensity)	Sediment Accumulation (I_{geo})	I_{geo} Class	Metal Examples Upper Rhine	Lower Rhine
4	Very strong pollution	>5	6		Cd
3–4	Strong to very strong	>4–5	5		
3	Strongly polluted	>3–4	4		Pb, Zn
2–3	Moderately to strongly	>2–3	3	Cd, Pb	Hg
2	Moderately polluted	>1–2	2	Zn, Hg	Cu
1–2	Unpolluted to moderate pollution	>0–1	1	Cu	Cr, Co
1	Practically unpolluted	<0	0	Cr, Co	

Source: After Müller (1979).

metal pollution in aquatic sediments has been introduced by Müller (1979), which is called the "Index of Geoaccumulation":

$$I_{geo} = \log_2 C_n / 1.5 \times B_n$$

C_n is the measured concentration of the element "n" in the pelitic sediment fraction (<2 μm), and B_n is the geochemical background value in fossil argillaceous sediment (average shale); the factor "1.5" is used because of possible variations of the background data due to lithogenic effects. The Index of Geoaccumulation consists of 7 grades, whereby the highest grade (6) reflects ca. 100-fold enrichment above background values.

In Table 5 an example is given for the River Rhine, and a comparison of these sediment indices with the water quality classification of the International Association of Waterworks in the Rhine Catchment (IAWR) has been made. It should be mentioned that — similar to the sediment standards in Table 6 — no further consideration is given to the ecological relevance of the values.

Table 6. Proposal for Norm Values of Metal Concentrations in Dredged Sediments Based on a 10-Fold Enrichment (Exceptions: Mercury and Cadmium) Compared to Natural Background Concentrations of Elbe River Sediments (Fraction <63 μm)

	Target Values NL Draft	Standard Values NL Draft	Background Elbe River Sediment	Proposal for "Norm Values"
Mercury	0.3	1	0.2–0.4	1 mg/kg
Cadmium	0.8	4	0.3–0.5	2 mg/kg
Arsenic	25	40	3–5	40 mg/kg
Lead	50	125	25–30	250 mg/kg
Copper	25	70	20–30	250 mg/kg
Chromium	100	125	60–80	700 mg/kg
Zinc	180	750	90–110	1000 mg/kg

An example of standard values for sediment quality criteria is given by the Dutch sediment quality draft (Van Veen & Stortelder 1988). Such quality guidelines will be needed after the year 2003, when the present "sludge island" practice for the Rotterdam harbor material is running out, and new solutions have to be based on their environmental compatibility. Nonetheless, the lack of ecological impact considerations in the present draft, which is mostly based on existing differentiations of pollutant concentrations in the Rotterdam area (class 1 to 4), seems to be a major disadvantage and should be modified in the coming years. In the present draft — aimed for disposal of contaminated sediment on land — the pollutant concentrations are normalized to a standard sediment ("underwater soil") consisting of 10% organic matter and 24% clay content (particle size <2 μm). The level for the target value is based on field observations of sediments in surface waters not affected by industrial or other discharges. At the levels for the limit value immediate action will be needed.

An initial attempt is undertaken on the basis of the "background approach" for introducing "norm values" of metal concentrations for estuarine and coastal sediments related to dredging and resuspension activities (Table 6). With respect to the potentially toxic metals (for which a comparison with natural concentrations can be made) a 10-fold enrichment in a defined grain size fraction can be considered as a significant accumulation, at which level — irrespective of dilution effects and temporal implications — a transfer of polluted sediments into less contaminated areas should not take place. Of the concentrations data in the Dutch draft (see above), values of cadmium (4 mg/kg) and arsenic (40 mg/kg) would reflect a 10-fold enrichment compared to natural background values as determined from a sediment core in the Elbe River catchment (Anon. 1989a). The other values of the Dutch draft standard indicate more or less strong deviations from a 10-fold enrichment: A 3-fold enrichment of mercury would be very acceptable, if one considers the high toxicity potential of this element. On the other hand, it is not understandable that the standard values of chromium (125 mg/kg), copper (70 mg/kg), and lead (125 mg/kg) are just double, three-fold, or five-fold the respective background concentrations, although these elements — according to their relative low toxicity potential and/or low solubility — would permit a higher factor of enrichment compared, for example, to cadmium. From these "base values", which have to be normalized to grain size (see above), modifications are needed with respect to mobility and toxicity of individual substances. This is valid in particular for cadmium, which is easily mobilized by the action of saline solutions, e.g., in estuarine zones, or by lowering of pH values (see below). Therefore, the limiting concentration of cadmium in Table 6 is set at a five-fold increase relative to the background concentration.

3.0 ELEMENTAL INTERACTIONS WITH SEDIMENTARY CONSTITUENTS

3.1 Substrate Characterization; Acid-Producing Potential

Particles as substrates of pollutants originate from two major sources. Endogenic fractions of particulate matter include minerals that result from processes occurring within the water column (Jones & Bowser 1978). Enrichment of minerals generated by endogenic processes may be influenced by settling of particulates, filtering organisms, and flocculation. Endogenic processes exhibit a distinct temporal character, often as a result of the variation of the organic productivity. In lakes, the total particulate concentration of trace metals is generally lowest in the hypolimnion due to the decomposition of organic matter. Consequently, net biogenic flux, for example, of metals depends on the lake's capacity to produce organic particulate matter and to decompose it before it is buried definitely in the sediment (Salomons & Baccini 1986). Authigenic (or diagenetic) fractions include minerals that result from processes within deposited sediments. Decomposition of organic matter, which is mediated by microorganisms, generally follows a finite succession in sediments depending upon the nature of the oxidizing agent (see Berner 1981); the successive events are oxygen consumption (respiration), nitrate reduction, sulfate reduction, and methane formation. The composition of interstitial waters in sediments is perhaps the most sensitive indicator of the types and the extent of reactions that take place between pollutant-loaded sediment particles and the aqueous phase that contacts them. The large surface area of fine-grained sediment in relation to the small volume of its trapped interstitial water ensures that minor reactions with the solid phases will be shown by major changes in the composition of the aqueous phase.

Conceptually, a sediment can be considered as a heterogeneous mixture of dissimilar particles. Solid phases interacting with dissolved constituents in natural waters consist of a variety of components including clay minerals, carbonates, quartz, feldspar, and organic solids. The "matrix vehicle" or residual fraction (Jenne 1977) is associated with more labile and thermodynamically unstable components such as carbonates, amorphous aluminosilicates, organic matter, etc. (Martin *et al.* 1987). These fractions are usually coated with Fe- and Mn-oxides and organic material (both living and nonliving; see Figure 3). A prime medium for sorption of inorganic components by sediments is metastable iron and manganese oxides, which have a high degree of isomorphic substitution (Jenne 1977); the natural organic content of particulates is the decisive parameter controlling the "sorption" of organic chemicals. Organic surfaces for metal sorption could form in three possible ways (Hart 1982): (1) from organisms such as bacteria and

Matrix-Vehicle

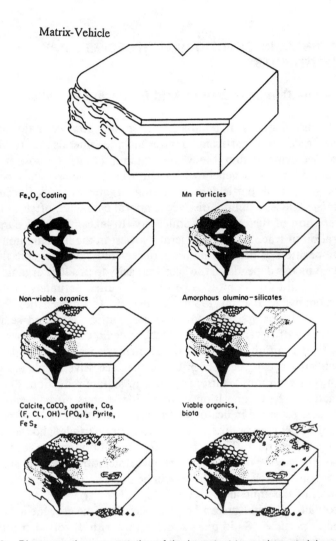

Figure 3.   Diagrammatic representation of the important trace-element sinks on the surface of an idealized kaolinite crystal. (Reprinted from Jenne 1977. With permission.)

algae; (2) by the breakdown of plant and animal material and by the aggregation of lower molecular weight organics; and (3) by organic matter of lower molecular weight sorbed onto clay or metal oxide substrates (Davis & Gloor 1981). Although the difference between these three surface types is not well understood with respect to metal uptake, there is a general agreement that at least one major binding mechanism involves salicylic entities; other strong binding entities (such as peptides) may also be present in some systems. Tipping (1981) suggests that at least part of the organic matter adsorbed onto the particulate matter in natural waters has carboxylic and phenolic functional groups available for binding with trace metals. The trace metal adsorption capacity of organic matter is generally between that for metal oxides and clays.

Regarding the potential release of contaminants from sediments, changing of pH and redox conditions is of prime importance. In practice, therefore, characterization of sediment substrates with respect to their buffer capacity is a first step for the prognosis of middle- and long-term processes of mobilization, in particular of toxic chemicals in a certain milieu.

Evaluation of the pH changes resulting from the oxidation of anoxic sediment constituents can be performed by ventilation of sediment suspensions with air or oxygen. Another approach for evaluating pH effects is titration with acid solutions. For quantifying pH properties and for better comparison of sediment samples it is proposed to use the term $\triangle$ pH, which is characterized by the difference of pH values of 10% sludge suspensions in distilled water (pH_o) and in 0.1 N sulfuric acid after 1 h shaking time (Calmano et al. 1986). Three categories of $\triangle$ pH values can be established ranging from $\triangle$ pH < 2 (strongly buffered) to $\triangle$ pH 2–4 (intermediate) to $\triangle$ pH > 4 (poorly buffered).

For a classification of sludge regarding their acid potential, which can be produced by oxidation of sulfidic components, one can preferentially use the data of calcium and sulfur from the sequential extraction scheme as proposed, for example, by Tessier et al. (1979; see Section 4.1). In anoxic, sulfide-containing sediments the two elements were selectively released during anaerobic experimental procedures (argon or nitrogen atmosphere in glove box) by the Na-acetate step (Ca from carbonates) and the peroxide step (S from oxidizable sulfides, mainly iron sulfide). Reaction of oxygen with 1 mol of iron sulfide will produce three [H+] ions; by reaction with 1 mol of carbonate two [H+] ions are buffered. For an initial estimation, one may compare total calcium and sulfur concentrations in the sediment sample.

Experimental approaches for prognosis of "acid-producing potential" of sulfidic mining residues have been summarized by Ferguson & Erickson (1988). A test described by Sobek et al. (1978) involves analysis of total or pyritic sulfur; neutralization potential is obtained by adding a known amount of HCl, heating the sample, and titrating with standardized NaOH to pH 7. Potential acidity is subtracted from neutralization potential; a negative value below 5 tons $CaCO_3$/1000 tons of rock indicates a potential acid producer. Bruynesteyn & Hackl (1984) calculated acid-producing potential from total sulfur analysis; acid-consuming ability is obtained by titration with standardized sulfuric acid to pH 3.5 (Bruynestein & Duncan 1979). Acid-producing potential is subtracted from acid-consuming ability; a negative value indicates a potential acid producer.

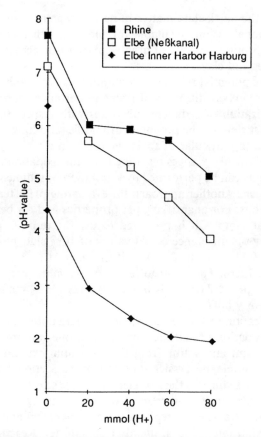

Figure 4. Variations of pH values (titration curves) of suspensions (100 g/L) of sediment samples from the Rhine and Elbe Rivers after addition of 1 *M* nitric acid. (Reprinted from Förstner *et al.* 1989. With permission.)

Results from titration experiments using 1 *M* nitric acid on sediment suspensions of 100 g/L are presented in Figure 4. The titration curve of the Rhine River sediment exhibits a small plateau in the pH range of 5.5 and 6, probably due to a certain fraction of carbonate, which is consumed by addition of 80 mmol of acidity. On the contrary, the titration curves of both Elbe River sediments are continuously decreasing as a result of the low contents of carbonate in these samples. The sediment from the inland harbor basin of Hamburg, originally sulfide-rich material which had been stored for 1 year in a closed bottle, has already reached an initial pH of 4.3; this is probably due to the consumption of the low residual buffer capacity by oxidation of parts of the sulfide fraction. Respective lowering of pH has been found in upland disposal sites of dredged sediments from Hamburg

Table 7. Composition of Anaerobic and Oxidized Porewaters from Sedimentation Polders

	"Anaerobic Porewater"	"Oxidized Porewater"
Nitrate	<3 mg/L	~120 mg/L
Ammonium	125 mg/L	<3 mg/L
Iron	79 mg/L	<3 mg/L
Zinc	<30 μg/L	>5000 μg/L
Cadmium	<0.1 μg/L	~80 μg/L
Arsenic	~150 μg/L	~15 μg/L

Source: Maaβ et al. (1985).

Harbor (Tent 1982). Due to the low carbonate content, which is consumed during several months or years, and subsequent lowering of pH, metals are easily transferred to crops, and permissible limits of cadmium have been surpassed in as much as 50% of wheat crops grown on these materials (Herms & Tent 1982). High concentrations of metals have been measured in porewaters from sedimentation polders in the Hamburg Harbor area in the older, oxidized deposits (Maaβ et al. 1985; Table 7).

It will be shown in Section 4 that similar effects will occur as well in the aquatic system, particularly on tidal areas affected by periodic drying and wetting (Kersten 1989) and at other high-energy sites exhibiting strong resuspension activities. The situation in the Elbe River estuary is particularly critical since low buffer capacities of the sediments coincide with a relative long residence time of suspended particles (Tent 1987).

3.2 Partition Coefficients of Metallic Elements

Mobility of an element in the terrestrial and aquatic environment is reflected by the ratio of dissolved and solid fractions. Evaluation of the current literature indicates at least three major factors affecting the distribution of trace metals between solution and particulates: (1) the chemical form of dissolved metals originating both from natural and civilizational sources (e.g., Förstner & Salomons 1983); (2) the type of interactive processes, i.e., sorption/desorption-or precipitation-controlled mechanisms (Salomons 1985); and (3) concentration and composition of particulate matter, mainly with respect to surface-active phases.

Typical adsorption curves on inorganic substrates, such as iron oxyhydrate, increase from near nil to near 100% as the pH increases through a critical range 1–2 units wide (Benjamin et al. 1982). It is important to note that the location of the pH-adsorption "edge" depends on adsorbent concentration. This effect is due to the existence of a range of "specific" site-binding energies. High-energy adsorption sites, since they are fewer in number than lower energy sites, become limiting first. As lower energy sites are gradually filled, the overall binding constant decreases. In the few cases

Table 8. Factors and Mechanisms Influencing Distribution of Elements between Solid and Dissolved Phases

Factor/Mechanism	Reference
Sample preparation (e.g., drying)	Duursma (1984)[a]
Separation (filtration/centrifugation)	Calmano (1979)[a]
Grain size distribution	Duursma (1984)[a]
Suspended matter concentration	Salomons (1980)
Kinetics of sorption/desorption	Schoer & Förstner (1984)[a]
Nonreversibility of sorption	Lion et al. (1982)

[a]Experiments with artificial radionuclides.

where kinetics of sorption were investigated, surface reactions were not found to be a single-step reaction (e.g., Anderson 1981). Experiments performed by Benjamin & Leckie (1981) showed a rapid and almost complete metal uptake process lasting no more than 1 hr, followed by a second, slower uptake process perhaps lasting days or possibly months; the first effect was thought to be true adsorption and the second to be slow adsorbate diffusion into the solid substrate. For systems rich in organic matter, metal adsorption curves cover a wider pH range than is found for inorganic substrates. Typically, a reduced reversibility of metal sorption has been observed in these organic systems (Lion et al. 1982). Such effects may be important restrictions for using distribution coefficients in the assessment of metal mobility in rapidly changing environments such as rivers, where equilibria between the solution and solid phases often cannot be achieved completely due to the short residence times (for additional information see, e.g., Förstner 1979).

In practice, applicability of distribution coefficients may find further limitations from methodological problems. Sample pretreatment (e.g., dry or wet conditions), solid/liquid separation technique (filtration or centrifugation), and grain size distribution of solid material strongly affects K_D factors of metals (e.g., Calmano 1979, Duursma 1984; Table 8). Such effects also have to take into consideration the interpretations of in situ processes, where the above-mentioned influences of reversibility usually are playing a smaller role than in the case of open-water conditions.

Despite the problematic nature of these relations — as evidenced from various references (Table 8) — it seems that sediment quality criteria of the U.S. Environmental Protection Agency will preferentially be based on the sediment/water equilibrium partitioning approach. This approach is related to a relative broad toxicological basis of water quality data. The distribution coefficient K_D, which is determined from laboratory experiments, is defined as the quotient of equilibrium concentration of a certain compound in sediment (C_s^x; e.g., in mg/kg) and in the aqueous phase (C_w^x; e.g., in mg/ L). For nonpolar organic compounds, which are dominantly correlated to the content of organic carbon in the sediment sample, the partition coefficient K_D can be normalized from this parameter and the octanol/water coefficient (K_{OW}): $K_D = 0.63\ K_{OW}$/content of organic carbon in total dry sediment (0.63 is an empirical value). For these substances, such as PCB, DDT, and PAH, reliable and applicable data can be expected with respect to the development of sediment quality criteria. K_D values of *metals* are not only correlated to organic substances but also with other sorption-active surfaces. Toxicological effects often are inversely correlated with parameters such as iron oxyhydrate. Quantification of competing effects is difficult, and thus the equilibrium partition approach for sediment quality assessment of metals still exhibits strong limitations (see review by Honeyman & Santschi 1988, Shea 1988).

3.3 Effects of Input Sources; Temporal Variations

The ratios between dissolved and solid contaminant fractions are first influenced by the respective *inputs* and subsequently by the *interactions* taking place within the different environmental compartments. Direct emissions, for example of cadmium, into the environment from waste materials are approximately 10-fold higher from solid materials (pigments, phosphate fertilizers, sewage sludge, municipal and mining wastes, smelting residues, etc.) than from dissolved inputs (from lead-zinc mines, sewage treatment plants, effluents from battery factories, electroplating plants, etc.). This is especially valid for river systems, where equilibrium between the solution and the solid phases often cannot be achieved completely due to the short residence times; Bowen (1977) has noted the example of the Thames River flowing 250 km in 14 hr, while chemical equilibrium would require more than 100 hr in some instances. The bulk of detrital trace-element particulates never leaves the solid phase from initial weathering to ultimate deposition. Similarly, metal dust particles (e.g., from smelters) and effluents containing metals associated with inorganic and organic matter undergo little or no change after being discharged into a river. In natural systems, such as in the Amazon River and to a large extent also in the Mississippi River, more than 90% of the metal load is transported by particulate matter (Table 9). A similar sequence of the ratios between particulate and dissolved heavy metals has also been found for polluted systems; typically, however, the dissolved fractions in polluted waters are significantly higher than in less polluted systems, particularly for metals such as Cd, Zn, and Cu.

Metal concentrations in rivers, particularly in solid matter, are strongly influenced by the runoff. From studies on two Cornish estuaries it is suggested by Boyden *et al.* (1978) that higher trace element concentrations in high and low water surface and bottom samples in winter compared to summer probably reflect increased weathering and transport in the catchments during this season. Trace metal concentrations in the Susquehanna

Table 9. Particulate-Bound Metal as Percentage of Total Discharge

	Amazon River	Mississippi River	Polluted Rivers in	
			U.S.A.	F.R.G.
Cadmium	–	88.9	–	30
Zinc	–	90.1	40	45
Copper	93	91.6	63	55
Manganese	83	93.5	–	8–97
Chromium	83	98.5	76	72
Lead	–	99.2	84	79
Iron	99.4	99.9	98	98

Sources: Förstner (1984), after data from Gibbs (1973), Trefry & Presley (1976), Kopp & Kroner (1968), Heinrichs (1975).

River correlated well with the amount of solids discharged (Carpenter *et al.* 1975); when data are calculated for weight concentrations of metals in the solid fraction, it is found that all metals generally peak during December and January and secondary peaks occur for Co, Cr, Ni, Cu, and Mn in July. Troup & Bricker (1975) suggest that this effect is due to decaying organic matter which is abundant in the Susquehanna River during these two periods. Studies performed by Grimshaw *et al.* (1976) on the River Ystwyth in mid-Wales, where strong metal pollution from past mining operations is still obvious, indicate that metal concentrations in solution are highest during low water flow periods, suggesting a dilution effect; for brief periods during initial stages of storm runoff, there is a very significant increase of the metal concentrations in solution, apparently due to a flushing effect.

During the last decade, considerable changes in the pollution status of the Rhine River, particularly with respect to heavy metals, have taken place. Figure 5 presents the example of cadmium and mercury discharges in the Rhine River at the German/Dutch border. Discharges of cadmium decreased from 250 tons in 1971 to approximately 50 tons in 1983. Mercury was even more effectively reduced from 100 tons to approximately 10 tons during this period (Malle 1985). Apart from variations of hydrology, and the effects of the economic crisis, particularly at the end of the 1970s, a significant portion of the reduction should be affected by improvement of wastewater treatment and by the partial replacement of metals in critical applications. It is indicated from these data that a major decrease for cadmium occurred in the dissolved phases, whereas – until 1979 – the reduction of mercury concentrations mainly took place in the solid phases. This is an indication that equilibria between solid and aqueous phases have not been completely established and clearly shows the difficulties involved in the modeling of such processes.

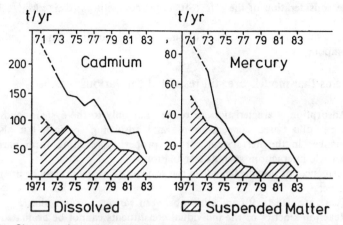

Figure 5. Changes of metal load (particulate and dissolved) in the Rhine River at the Dutch-German border from 1971 to 1983. (After Malle 1985.)

3.4 Metal Transfer from Inorganic to Organic Substrates

The use of equilibrium models has been to some extent successful for anoxic environments. It has been stressed by Salomons (1985) that from an impact point of view it is important to know whether the concentrations of pollutants int he porewaters are determined by adsorption/desorption processes or by precipitation/dissolution processes. If the latter is the case the concentrations of pollutants in the porewaters are independent of the concentrations in the solid phase. There is strong direct (Luther *et al.* 1980, Lee & Kittrick 1984) and indirect (Lu & Chen 1977) evidence that the concentrations of copper, zinc, and cadmium in sulfidic porewaters are determined by precipitation-dissolution processes. The concentration of arsenic and chromium in anoxic porewaters is probably controlled by adsorption-desorption processes and mainly depends on the concentration in the solid phase (Salomons 1985). However, for a full understanding of the behavior of trace metals in porewaters and their potential transfer to biota more experimental speciation studies (Hart & Davies 1977, Batley & Giles 1980, Elderfield 1981) and identification of solid phases are urgently needed.

With respect to the modeling of metal partitioning between dissolved and particulate phases in a natural system, e.g., for estuarine sediments, the following requirements have been listed by Luoma & Davis (1983):

- the determination of binding intensities and capacities for important sediment components
- the determination of relative abundance of these components
- the assessment of the effect of particle coatings and of multicomponent aggregation on binding capacity of each substrate

- the consideration of the effect of major competitors (Ca^{2+}, Mg^{2+}, Na^+, Cl^-)
- the evaluation of kinetics of metal redistribution among sediment components

It seems that models are still restricted for various reasons:

1. Adsorption characteristics are related not only to the system conditions (i.e., solid types, concentrations, and adsorbing species), but also to changes in the net system surface properties resulting from particle/particle interactions such as coagulation.
2. Influences of organic ligands in the aqueous phase can rarely be predicted as yet.
3. There are effects of competition between various sorption sites.
4. Reaction kinetics of the individual constituents cannot be evaluated in a mixture of sedimentary components.

These restrictions have recently been discussed in detail by Honeyman & Santschi (1988), who stated that even for aquatic environments of low particle concentration "the non-deterministic and interactive effects described above generally influence the estimation of an apparent partitioning coefficient by 1 to 3 orders of magnitude in either direction." With respect to environments of moderate to high particle concentration such as in soils and sediments they concluded that these theoretical approaches have failed thus far to provide a sound basis for the prediction of trace element behavior.

At present, experimental studies on the dissolved/solid interactions in such complex systems seem to be more promising. One approach is with a six-chambered device which still permits phase interactions via solute transport of the elements (Calmano *et al.* 1988); in this way, exchange reactions and biological uptake can be studied for individual phases under the influence of pH, redox, ionic strength, and solid and solute concentration.

The laboratory system used in these studies was developed from the experience on sediment/algae interactions with a modified two-chambered device (Ahlf *et al.* 1986). The system is made of a central chamber connected with six external chambers and separated by membranes of 0.45 μm pore diameter (Figure 6A). The volume of the central chamber is several liters and each of the external chambers contains 250 mL. Either solution or suspension can be inserted into the central chamber. Solid components are kept in suspension by magnetic stirring; redox, pH, and other parameters may be controlled and adjusted in each chamber.

In an experimental series on the effect of salinity, i.e., disposal of anoxic dredged mud into sea water, quantities of model components were chosen in analogy to an average sediment composition. In the central chamber, 100 g

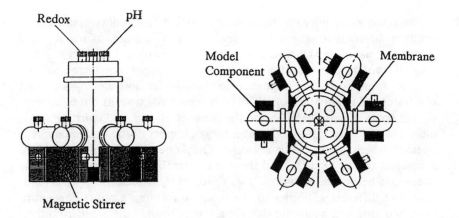

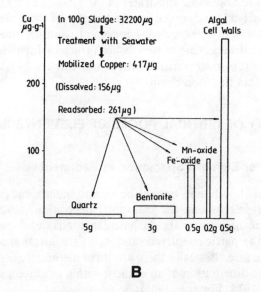

Figure 6. Metal transfer between sedimentary components. 6A: Schematic view of the multichamber device. 6B: Transfer of copper from anoxic harbor mud into different model substrates after treatment with artificial seawater. (Reprinted from Calmano *et al.* 1988. With permission.)

of anoxic mud from Hamburg Harbor was inserted; salts were added corresponding to the composition of seawater. After 3 weeks, solid samples and filtered water samples were collected from each chamber.

The effect of salinity on metal remobilization from contaminated sediments is different for the individual elements. While approximately 16% and 9% of cadmium and zinc, respectively, in the dredged mud from Hamburg Harbor is released, for metals such as copper the factor salinity increase seems to be less important in the transfer both among sediment substrates and to aquatic biota. This is, however, not true, as can be demonstrated from a mass balance for the element copper in Figure 6B: It is indicated that only 1.3% of the inventory of copper of the sludge sample is released when treating with seawater. Only one third stays in solution, equivalent to ca. 40 μg/L, and there is no significant difference from the conditions before salt addition. Two thirds of the released copper is readsorbed at different affinities to the model substrates. Copper concentrations in quartz and bentonite clay are not significantly different from their natural contents. Slight enrichment of copper occurs in the hydrous iron oxide and manganese oxide, whereas the cell walls—a minor component in the model sediment—has accumulated nearly 300 ppm of copper.

The dominant role of organic substrates in the binding of metals such as Cd and Cu is of particular relevance for the transfer of these elements into biological systems. It can be expected that even at relatively small percentages of organic substrates these materials are primarily involved in metabolic processes and thus may constitute the major carriers by which metals are transferred within the food chain.

4.0 ASSESSMENT OF CRITICAL POOLS OF ELEMENTS IN SEDIMENTS

4.1 Techniques for Element Speciation in Sediments

Since adsorption of pollutants onto airborne and waterborne particles is a primary factor in determining the transport, deposition, reactivity, and potential toxicity of these materials, analytical methods should be related to the chemistry of the particle's surface and/or to the metal species highly enriched on the surface. Basically there are three methodological concepts for determining the distribution of an element within or among small particles (Keyser *et al.* 1978, Förstner 1985):

- Analysis of single particles by X-ray fluorescence using either a scanning electron microscope (SEM) or an electron microprobe can identify differences in the matrix composition between individual particles. The total concentration of the element can be determined as a function of particle size. Other physical fractionation and preconcentration methods include density and magnetic separations.
- The surface of the particles can be studied directly by the use of electron microprobe X-ray emission spectrometry (EMP), electron spectroscopy for

chemical analysis (ESCA), Auger electron spectroscopy (AES), and secondary ion-mass spectrometry. Depth-profile analysis determines the variation of chemical composition below the original surface.

- Solvent leaching — apart from the characterization of the reactivity of specific metals — can provide information on the behavior of pollutants under typical environmental conditions. Common single-reagent leachate tests, e.g., U.S. EPA, ASTM, IAEA, and ICES, use either distilled water or acetic acid (Theis & Padgett 1983). A large number of test procedures have been designed particularly for soil studies; these partly used organic chelators such as EDTA and DTPA (Sauerbeck & Styperek 1985).

Laboratory techniques for generating leachate from solid materials are generally grouped into batch and column extraction methods. The *batch extraction method* offers advantages through its greater reproducibility and simplistic design, while the *column method* is more realistic in simulating leaching processes which occur under field conditions (Jackson *et al*. 1984). For batch studies the best results with respect to the estimation of short-term effects can be attained by "cascade" test procedures at variable solid/solution ratios: A procedure of the U.S. EPA (Ham *et al*. 1979) designed for studies on the leachability of waste products consists of a mixture of sodium acetate, acetic acid, glycine, pyrogallol, and iron sulfate. The standard leaching test developed by The Netherlands Energy Research Centre (Van der Sloot *et al*. 1984) for studies on combustion residues combines batch cascade and column procedures; the test column is filled with the material under investigation and percolated by acidified demineralized water (pH = 4; for evaluating the most relevant effects of acid precipitation) to assess short- and medium-term leaching (< 50 years). In the cascade test the same quantity of material is extracted several times with fresh demineralized water (pH = 4) to get an impression of long-term leaching behavior (50–500 years).

One of the potential advantages of leaching methods is seen in obtaining relevant information from a small number of samples, one possibly being sufficient. A number of test protocols in soil science have been designed initially for the assessment of plant-available soil nutrients and speciation of trace metals in sewage sludge-amended soils (Jackson 1958). In the sediment-petrographic field, interest was focussed initially on the differentiation between authigenic and detrital phases in Fe/Mn concretions from deep-sea deposits (e.g., Chester & Hughes 1967). According to Horowitz (1984) two approaches are used in chemical partitioning of sediments: The first is to determine *how* metals are retained on or by sediments — the so-called mechanistic approach; the second determines *where* inorganic constituents are retained on or by sediments (phase or site) — the so-called phase approach. Recent developments in solid-phase differentiations were mainly promoted by environmental studies, both in soil science and in water

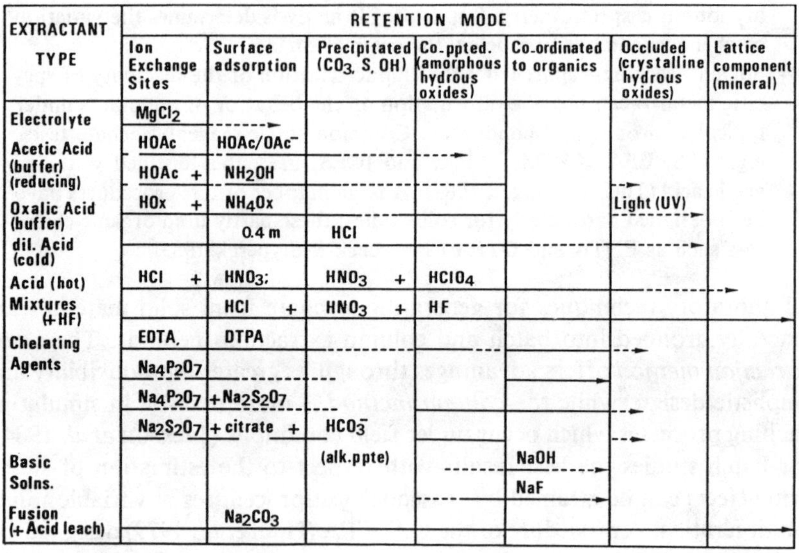

EXTRACTANT TYPE	RETENTION MODE						
	Ion Exchange Sites	Surface adsorption	Precipitated (CO_3, S, OH)	Co-ppted. (amorphous hydrous oxides)	Co-ordinated to organics	Occluded (crystalline hydrous oxides)	Lattice component (mineral)
Electrolyte	$MgCl_2$	------→					
Acetic Acid (buffer) (reducing)	HOAc	HOAc/OAc⁻	-------- ---→				
	HOAc +	NH_2OH					
Oxalic Acid (buffer)	HOx +	NH_4Ox	----			Light (UV)	
dil. Acid (cold)		0.4 m	HCl				
Acid (hot)	HCl +	HNO_3;	HNO_3 +	$HClO_4$		------- --→	
Mixtures (+HF)		HCl +	HNO_3 +	HF			
Chelating Agents	EDTA.	DTPA	-------		---→		
	$Na_4P_2O_7$	-------- ---			--→		
	$Na_4P_2O_7$ +$Na_2S_2O_7$	----			--→		
	$Na_2S_2O_7$ + citrate +	HCO_3^-			--→		
Basic Solns.			(alk.ppte)	--------	NaOH ---→		
				--------	NaF ---→		
Fusion (+ Acid leach)		Na_2CO_3					

Figure 7. Schematic representation of the ability of different extractant solutions to release metal ions retained in different modes or associated with specific soil and sediment fractions. (Reprinted from Pickering 1981. With permission.)

research. For the estimation of relative bonding strength of metals in different phases, extraction procedures have been applied, both as single leaching steps and combined in sequential extraction schemes. A schematic representation of the ability of different extractant solutions to release metal ions from particulate matter is given by Pickering (1981; Figure 7).

In connection with the problems arising from the disposal of solid wastes, particularly of dredged materials, extraction sequences have been applied which are designed to differentiate between the exchangeable, carbonatic, reducible (hydrous Fe/Mn oxides), oxidizable (sulfides and organic phases), and residual fractions (Engler *et al.* 1977). One of the more widely applied extraction sequences of Tessier and co-workers (1979) has been modified by various authors; a version of Kersten & Förstner (1986) differentiates easily and moderately reducible components (Table 10).

Despite the clear advantages of a differentiated analysis over investigations of total sample — sequential chemical extraction is probably the most useful tool for predicting long-term adverse effects from contaminated solid material — it has become obvious that there are many problems associated with these procedures (e.g., Kersten & Förstner 1986, Rapin *et al.* 1986):

Table 10. Sequential Extraction Scheme for Partitioning Sediment Samples

Fraction	Extractant	Extracted Component
Exchangeable	1 M NH$_4$OAc, pH 7	Exchangeable ions
Carbonatic	1 M NaOAc, pH 5 with HOAc	Carbonates
Easily reducible	0.01 M NH$_2$OH HCl with 0.01 M HNO$_3$	Mn-oxides
Moderately reducible	0.1 M oxalate buffer, pH 3	Amorphous Fe-oxides
Sulfidic/organic	30% H$_2$H$_2$ with 0.02 HNO$_3$, pH 2, extracted with 1 M NH$_4$OAc-6% HNO$_3$	Sulfides together with organic matter
Residual	Hot HNO$_3$ conc.	Lithogenic material

Source: Kersten & Förstner (1986, 1987a).

- Reactions are not selective and are influenced by the duration of the experiment and by the ratio of solid matter to volume of extractants. A too high solid content together with an increased buffer capacity may cause the system to overload; such an effect is reflected, for example, by changes of pH values in time-dependent tests.
- Labile phases could be transformed during sample preparation; this can occur especially for samples from reducing sites.

In this respect, earlier warnings have been made by various authors, not to forget changes of the sample matrix during recovery and treatment of the material. The first indicates that even oxic materials are not safe for changes during treatment (Thomson *et al.* 1980). The second relates to the anoxic sediment material, where changes are quite obvious: "The integrity of the samples must be maintained throughout manipulation and extraction" (Engler *et al.* 1977). Although these problems, particularly for anoxic sediments, have been well known for many years, they now become fully evident in the context of process studies.

Partitioning of a sediment sample from Hamburg Harbor which was pretreated in different ways (EPA Standard Elutriate Test, 1:4 sediment/site water for 30 min; freeze-dried sample; oven-drying at 60°C) clearly demonstrates the effect of oxidation in regulating the chemical form of cadmium and other trace metals (Figure 8; Kersten *et al.* 1985): Compared to the original sample (A), which was extracted under an argon atmosphere, there is a typical change from oxidizable phases (mainly Cd-sulfide) to easily reducible forms upon application of the shaking/bubbling test (B); during freeze-drying — which is commonly assumed to present a relatively smooth mode of sample pretreatment — transformation to carbonatic and exchangeable forms takes place (C), and this effect is further enhanced during oven-drying at 60°C (D).

Without going into detailed discussions, it should be mentioned that there is a possibility for standardizing the data from elution experiments with respect to numerical evaluation. In Table 11 examples are given for an "elution index" based on the metal concentrations exchangeable with 1 N

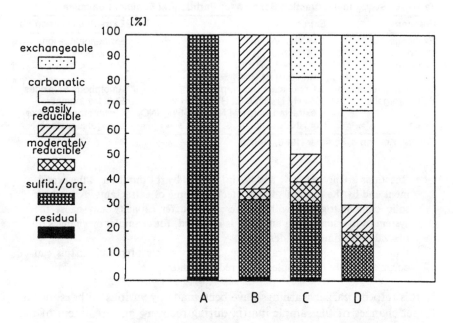

Figure 8. Partitioning of cadmium in anoxic mud from Hamburg Harbor in relation to the pretreatment procedures. A: Control extracted as received under oxygen-free conditions. B: After treatment with elutriate test. C: Freeze-dried. D: Oven-dried (60°C). (Reprinted from Kersten *et al*. 1985. With permission.)

ammonium acetate at pH 7; these metal fractions are considered to be remobilizable from polluted sediments at a relative short term under more saline conditions, for example, in the estuarine mixing zone. Comparison of the release rates from oxic and anoxic sediments clearly indicates that the oxidation of samples gives rise to a very significant increase of the overall mobilization of the element studied here; this effect is particularly important for cadmium. When proceeding further in the extraction sequence, more long-term effects could be estimated (generally with a respective

Table 11. Elution Index for Selected River Sediment, as Determined from Exchangeable Proportions (1 *M* Ammonium Acetate)[a]

	Neckar	Main	Elbe	Weser
Copper	<1	–	1	–
Lead	1	1	1	1
Zinc	7	10	40	10
Cadmium	22	22	25	–
Total oxic	**30**	**33**	**67**	**11**
(Anoxic	0.5	0.3	>4	4)

[a]Calculated relative to background data from Elbe River sediments (Table 6). These values are multiplied by a factor of 100.

reduction of prognostic accuracy). A major disadvantage of the present approach, however, is that the critical element mercury is not yet included in this scheme.

Single-extractant procedures are restricted with regard to prediction of long-term effects in waste deposits, e.g., of highly contaminated dredged materials, since these concepts involve neither mechanistic nor kinetic considerations and therefore do not allow calculations of release periods. This lack can be avoided by controlled significant intensification of the relevant parameters pH value, redox potential, and temperature combined with an extrapolation on the potentially mobilizable "pools", which are estimated from sequential chemical extraction before and after treatment of the solid material. An experimental scheme, which was originally used by Patrick *et al.* (1973) and Herms & Brümmer (1978) for the study of soil suspensions and municipal waste materials, was modified by inclusion of an ion-exchanger system for extracting the metals released within a certain period of time each (Figure 9A; Schoer & Förstner 1987). For an assessment on metal oxide residues, solutions were adjusted to combinations of pH 5/8 and redox 0/400 mV, circulating with 2 L per day through columns which contained 1:4 mixtures of waste material with quartz sand, the latter component to improve permeability; ion-exchanger resins were renewed after 1 week each. In particular for the elements for which the endpoint of release cannot be estimated from the respective cumulative curves of the water concentrations, extrapolations from sequential extraction data on the solid material are needed. Taking the example of zinc in Figure 9B, the more labile "exchangeable" fractions should be released at first ("phase 1"), whereas during "release phase 2" — which is much slower than initial mobilization of acetate-extractable zinc — part of the oxalate-reducible compounds are dissolved. The system can be modified for different intensities of contact between solid materials and solution by using shakers (e.g., erosion of the depot by rivers) or dialysis bags (flow-by conditions).

4.2 Metal Speciation in River Sediments (Examples)

Partition studies on river sediments were first carried out by Gibbs (1973) in the suspended load of the Amazon and Yukon Rivers, which are less affected by civilizational influences, using a four-step leaching sequence. In the case of iron, manganese, and nickel, the most significant bonding occurs, as expected, in hydroxide "coatings", whereas this type of bonding is only secondary for copper and chromium. As the hydroxide bonding decreases, a strong increase in lithogenic ("crystalline") bonding forms can be observed. In the highly polluted Rhine River, sediments have been studied with a five-step extraction sequence (including a step for carbonate-bound elements) by Förstner & Patchineelam (1980). High percentages of

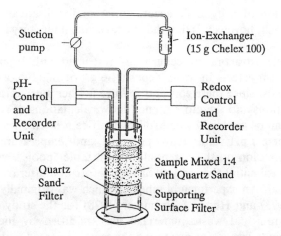

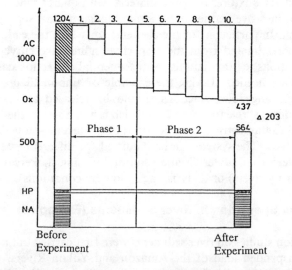

Figure 9.   Prediction of long-term effects of metal release from polluted solids. 9a: Experimental design. 9b: Comparison of zinc "pools" in a sample of heat processing residues before and after treatment with pH 5/400 mV solutions (Ac = ammonium acetate; Ox = oxalate buffer solution; HP = peroxide extraction; NA = residual fraction). (Reprinted from Schoer & Förstner 1987. With permission.)

detrital fractions were found for those elements which are less affected by man's activities, e.g., iron, nickel, and cobalt; on the other hand, lattice-held fractions were very low in metals such as lead, zinc, and cadmium. Lead, copper, and chromium are particularly associated with the hydroxide phases. These findings can be attributed to the specific sorption of lead and copper to Fe-oxide and to the lack of carbonate phases of chromium in natural aquatic systems. The preferential carbonate bonding of zinc and cadmium, on the other hand, can be attributed to the relatively high stability of Zn- and Cd-carbonate under the chemical conditions of the Rhine River.

A scheme consisting of four steps — (1) cation exchange with 1 M ammonium acetete, (2) extraction of reducible phases with 0.1 M hydroxylamine-HCl (pH 2), (3) extraction of oxidizable phases with 30% hydrogen peroxide/ammonium acetate, and (4) HF/HClO$_4$-digestion of detrital minerals — was used to study speciation of trace elements in river sediments from different parts of the world (Salomons & Förstner 1980). The results in Figure 10 show the increase in the relative amount of metals present in the resistant (lithogenous) fraction for less polluted or unpolluted river systems. According to these and other findings, it can be argued that the surplus of metal contaminants introduced into the aquatic systems by human activities usually exists in relatively unstable chemical associations and is, therefore, predominantly available for biological uptake.

Partitioning studies on sediment core profiles are particularly useful, since they provide data on relative changes of elemental phases irrespective of the method applied and thereby an insight into diagenetic processes taking place after deposition of the sedimentary components. Two examples are presented here, both indicating significant changes in the partition of zinc and cadmium during a relatively short period of time.

4.2.1 Metalliferous Floodplain Deposits in the U.K. (Bradley 1988)

A number of river catchments in England and Wales where base-metal deposits were worked in the past have been investigated by Bradley and collaborators (1982, 1984, 1986, 1987). Mining within the catchments occurred from Roman times and then again from the fifteenth century to the early years of this century. In the Hamps and Manifold catchments in Wales the seventeenth and eighteenth centuries were most important. During early mining in the Manifold catchment, contaminated sediments with a range of particle sizes were introduced into the fluvial system. However, later reprocessing of spoils yielded vast quantities of sand-sized sediments and these had a formative influence on sedimentary structures in the catchment.

Floodplains on the River Manifold contain sediments with high metal

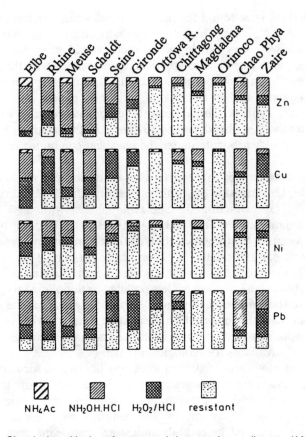

Figure 10.   Chemical partitioning of trace metals in some river sediments. (After Salomons & Förstner 1980.)

concentrations. The sedimentary units described were deposited during concerted mining, and as mining there ceased in 1870, pedogenic processes have influenced these sediments for over 100 years. Figure 11 (right bars) indicates a general decrease of total zinc and cadmium concentrations over time, which can primarily be interpreted as the result of improvements in the processing of ores. Phase differentiations in Figure 11 (left bars) indicate that there is a significant temporal increase of the percentage of exchangeable zinc, and to a lesser extent also of exchangeable cadmium, both mainly at the expense of surface oxide and carbonate bound metals. These changes can be interpreted from pedogenic processes. Since these transformations will result in a lowering of the bonding strength, it cannot be excluded that the reduction of total metal concentrations is partly due to a (diagenetic) remobilization of metals from the floodplain deposit. How-

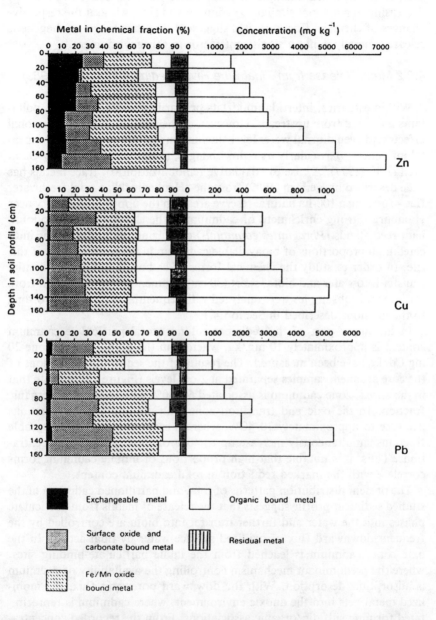

Figure 11. Total concentrations and partitioning of zinc, cadmium, and lead in a flood-plain soil profile in the manifold catchment. (Reprinted from Bradley & Cox 1987. With permission.)

ever, considering the example of lead, where the total concentrations are decreasing in a manner similar to cadmium and zinc, without the respective changes of chemical fractions, it is suggested that the process of diagenetic release generally is not a significant factor in this environment.

4.2.2 Metal Release from Tidal Elbe River Sediments (Kersten 1989)

Within estuaries, intertidal mudflats primarily provide a sink for pollutants imported from upstream. However, at these productive sites, seasonal effects and even diurnal water level fluctuations induce drastic environmental influences, particularly by redox changes, as has been stressed by Gambrell et al. (1977). A case of "oxidative remobilization" of trace metals has been described by Kersten (1989) from the "Heukenlock", a tidal freshwater flat — forming a 100-ha natural reserve area — in the upper Elbe estuary near Hamburg. Strong enrichment of cadmium in the rhizomes of monodominant reed stands (Phragmites communis) colonizing the high flat has indicated high proportions of bioavailable cadmium in the rooting zone of this site. In order to study the chemical forms of cadmium and their potential transfer into water and biota, short (30 cm) sediment cores were taken from this site and subsamples were analyzed with sequential extraction according to the methods described in Section 4.1.

In the upper part of the sediment column, total particulate cadmium content is approximately 10 mg/kg, whereas in the deeper anoxic zone 20 mg Cd/kg have been measured. The results of the sequential extractions of the core sediment samples separated at 2-cm levels (Figure 12) indicate that in the anoxic zone cadmium is associated 60–80% with the sulfidic/organic fraction. In the oxic and transition zone, sulfidic and organic fractions decrease to approximately 30–40%, whereas carbonatic and exchangeable fractions simultaneously increase up to 40% of total cadmium concentrations. Thus, it is notable that high proportions of mobile cadmium forms correlate with the marked reduction in total cadmium content.

The present distribution patterns of total and partitioned cadmium in the studied sediment profile suggests that the release of metals from particulate phases into the water and further transfer into biota are controlled by the frequent downward flux of oxidized surface water by tidal action. In the oxic zone, cadmium is leached from the labile particulate-binding sites, where the predominant mechanism controlling the availability of cadmium is adsorption/desorption. With the downward porewater flux, the mobilized metal gets into the anoxic environment, where cadmium is reprecipitated forming sulfidic/organic associations. From the recorded concentrations, it is expected that long-term removal of up to 50% of the cadmium from the sediment subsurface will take place at the anoxic sedimentary sink located a few centimeters below the sediment-water interface, which gives a

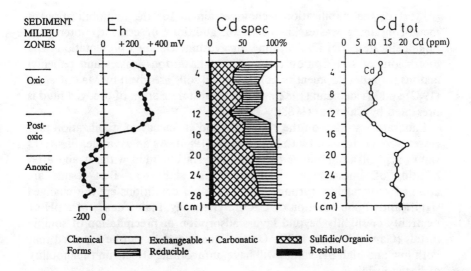

Figure 12. Geochemical characteristics of core sediments from the Heuckenlock intertidal flat, and total contents and chemical forms of particulate cadmium. (After Kersten 1989.)

flux maximum of 0.4 g/m^2 per year in the Heukenlock area. The effect of the process of "oxidative pumping" (Kersten & Förstner 1987) on the release of cadmium and other toxic metals into the overlying water and further biogeochemical cycling within the Elbe River estuary will need further investigation.

5.0 CHEMICAL STABILIZATION OF ELEMENTS IN DREDGED SEDIMENTS

In general, solidification/stabilization technology is considered a last approach to the management of hazardous wastes (Salomons & Förstner 1988b). The aim of these techniques is a stronger fixation of contaminants to reduce the emission rate to the biosphere and to retard exchange processes (Wiles 1987). Two objectives can be distinguished:

- At best, a material is produced which can be used for landscape modification and constructions, e.g., of roads, dikes, and walls for noise retardation; in this case disposal can be avoided.
- In other cases the material could be improved by a stabilization process in order to deposit it safely and/or cheaply.

Most of the stabilization techniques aimed for the immobilization of metal-containing wastes are based on additions of cement, water glass (alkali silicate), coal fly ash, lime, or gypsum. Detailed presentations and discussions of stabilization techniques of hazardous wastes and remedial actions for redevelopment of contaminated soils are given by Malone *et al.* (1982) and Wiedemann (1982). A review on stabilization of dredged mud is presented by Calmano (1988).

Laboratory studies on the evaluation and efficiency of stabilization processes were performed by Calmano *et al.* (1986). As an example, Figure 13 shows acid titration curves for Hamburg Harbor mud without and after addition of limestone and cement/fly ash stabilizers. Best results are attained with calcium carbonate, since the pH conditions are not changed significantly upon addition of $CaCO_3$. Generally, maintenance of a pH of neutrality or slightly beyond favors adsorption or precipitation of soluble metals (Gambrell *et al.* 1983). On the other hand, it can be expected that both low and high pH values will have unfavorable effects on the mobility of heavy metals.

Regarding the various containment strategies it has been argued that upland containment, e.g., on heap-like deposits, could provide a more controlled management than containment in the marine environment. However, contaminants released either gradually from an imperfect imperme-

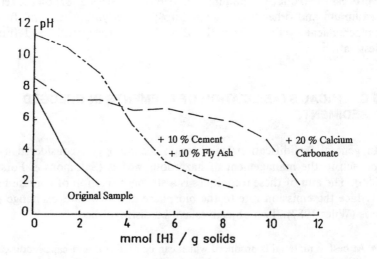

Figure 13.   Effects of calcium carbonate and cement/fly ash additives on chemical stabilization of fine-grained sediment from Hamburg Harbor. (Reprinted from Calmano *et al.* 1986. With permission.)

able barrier (also to groundwater) or catastrophically from failure of the barrier could produce substantial damage (Kester *et al.* 1983). On the other hand, near-shore marine containment, e.g., in capped mound deposits, offers several advantages, particularly with respect to the protection of groundwater resources, since the underlying water is saline and chemical processes are favorable for the immobilization or degradation of priority pollutants. This type of waste deposition under stable anoxic conditions, where large masses of polluted materials are covered with inert sediment, became known as "subsediment deposition"; the first example was planned for highly contaminated sludges from Stamford Harbor in the Central Long Island Sound following intensive discussions in the U.S. Congress (Morton 1980). In some instances it may be worthwhile to excavate a depression for the disposal site of contaminated sediment that can be capped with clean sediment (Kester *et al.* 1983).

REFERENCES

Ackermann, 1980. A procedure for correcting the grain size effect in heavy metal analysis of estuarine and coastal sediments. *Environ. Technol. Lett.* 1: 518–527.

Ahlf, W., W. Calmano & U. Förstner, 1986. The effects of sediment-bound heavy metals on algae and importance of salinity. In P.G. Sly (Ed.), *Sediments and Water Interactions*, Springer-Verlag, New York: 319–324.

Alderton, D.H.M., 1985. Sediments. In *Historical Monitoring*, MARC Technical Report 31. Monitoring and Assessment Research Centre, University of London, London: 1–95.

Allan, R.J., 1971. Lake sediment, a medium for regional geochemical exploration of the Canadian Shield. *Can. Inst. Min. Met. Bull.* 64: 43–59.

Allan, R.J., 1974. Metal contents of lake sediment cores from established mining areas: An interface of exploration and environmental geochemistry. *Geol. Surv. Can.* 74-1/B: 43–49.

Allan, R.J., 1986. *The Role of Particulate Matter in the Fate of Contaminants in Aquatic Ecosystems*. Scientific Series No. 142, National Water Research Institute, Canada Centre for Inland Waters, Burlington, Ontario: 128 pp.

Allen, H.E. *et al.* (Eds.), 1989. *Metal Speciation and Transport in Groundwaters*. Workshop organized by U.S. EPA and American Chemical Society, May 24–26, 1989, Jekyll Island, GA.

Anderson, M.A., 1981. Kinetics and equilibrium control of interfacial reactions involving inorganic ionic solutes and hydrous oxide solids. In *Environmental Speciation and Monitoring Needs for Trace Metal-Containing Substances from Energy-Related Processes*, F.E. Brinckman & R.H. Fish (Eds.). U.S. Department of Commerce, Washington D.C.: 146–162.

Andreae, M.O. *et al.*, 1984 Changing biogeochemical cycles — group report. In J.O. Nriagu (Ed.), *Changing Metal Cycles and Human Health*. Dahlem-Konferenzen, Life Sciences Res. Rep. 28, Springer-Verlag, Berlin: 359–374.

Anonymous, 1985. *Sediment Quality Criteria Development Workshop*, Nov. 28–30, 1985. Battelle Washington Operations, Richland.

Anonymous, 1989a. *Schwermetalldaten der Elbe von Schnackenburg bis zur See, 1984–1988. Arbeitsgemeinschaft für die Reinhaltung der Elbe.* Wassergütestelle Elbe, Hamburg.

Anonymous, 1989b. *Guidelines for Differentiating Anthropogenic from Natural Trace Metal Concentrations in Marine Sediments (Normalization Techniques).* Annex 4 of a Report of the ICES Working Group on Marine Sediments in Relation to Pollution, Savannah, Georgia, U.S.A., 20-23 February 1989. International Council for the Exploration of the Sea, Copenhagen.

Barnhart, B.J., 1978. The disposal of hazardous wastes. *Environ. Sci. Technol.* 12: 1132-1136.

Batley, G.E., 1989. *Trace Element Speciation—Analytical Methods and Problems.* CRC Press, Boca Raton, FL.

Batley, G.E. & M.S. Giles, 1980. A solvent displacement technique for the separation of sediment interstitial waters. In R.A. Baker (Ed.), *Contaminants and Sediments,* Vol. 2., Ann Arbor Science Publishers, Ann Arbor, MI: 101-117.

Benjamin, M.M. & J.O. Leckie, 1981. Multiple-site adsorption of Cd, Cu, Zn, and Pb on amorphous iron oxyhydroxide. *J. Colloid Interface Sci.* 79: 209-211.

Benjamin, M.M., K.L. Hayes & J.O. Leckie, 1982. Removal of toxic metals from power-generated waste streams by adsorption and coprecipitation. *J. Water Pollut. Control Fed.* 54: 1472-1481.

Berner, R.A., 1981. A new geochemical classification of sedimentary environments. *J. Sed. Petrol.* 51: 359-365.

Bernhard, M., F.E. Brinckman & P.S. Sadler (Eds.), 1986. *The Importance of Chemical "Speciation" in Environmental Processes.* Dahlem-Konferenzen, Life Sciences Research Report 33. Springer-Verlag, Berlin: 763 pp.

Boyden, C.R., S.R. Aston & I. Thornton, 1979. Tidal and seasonal variations of trace elements in two Cornish estuaries. *Estuar. Coastal Mar. Sci.* 9: 303-317.

Bradley, S.B., 1982. *Sediment Quality Related to Discharge in a Mineralized Region of Wales.* I.A.H.S. Publ. No. 137. International Association of Hydrological Sciences, Exeter, U.K.: 341-350.

Bradley, S.B., 1984. Flood effects on the transport of heavy metals. *Int. J. Environ. Studies* 22: 225-230.

Bradley, S.B. & J.J. Cox, 1986. Heavy metals in the Hamps and Manifold valleys, North Staffordshire, U.K. — Distribution in floodplain soils. *Sci. Total Environ.* 50: 103-128.

Bradley, S.B. & J.J. Cox, 1987. Heavy metals in the Hamps and Manifold valleys, North Staffordshire, U.K. — Partitioning of metals in floodplain soils. *Sci. Total Environ.* 65: 135-153.

Bruynesteyn, A. & D.W. Duncan, 1979. *Determination of Acid Production Potential of Waste Materials.* Paper A-79-29, Metall. Soc. AIME: 10 pp.

Bruynesteyn, A. & R.P. Hackl, 1984. Evaluation of acid production potential of mining waste materials. *Miner. Environ.* 4: 5-8

Calmano, W., 1979. *Untersuchungen über das Verhalten von Spurenelementen an Rhein-und Mainschwebstoffen mit Hilfe radioanalytischer Methoden.* Doctoral Dissertation, Techniche Hochschule, Darmstadt.

Calmano, W., 1988. Stabilization of dredged mud. In W. Salomons & U. Förstner

(Eds.), *Environmental Management of Solid Waste—Dredged Material and Mine Tailings*. Springer-Verlag, Berlin: 80–98.

Calmano, W., W. Ahlf & U. Förstner, 1988. Study of metal sorption/desorption processes on competing sediment components with a multi-chamber device. *Environ. Geol. Water Sci.* 11: 77–84.

Calmano, W., U. Förstner M. Kersten, & D. Krause, 1986. Behaviour of dredged mud after stabilization with different additives. In J.W. Assink & W.J. Van Den Brink (Eds.), *Contaminated Soil*. Martinus Nijhoff Publishing, Dordrecht, The Netherlands: 737–746.

Carpenter, J.H., W.L. Bradford & V. Grant, 1975. Processes affecting the composition of estuarine water (H_2CO_3, Fe, Mn, Zn, Cu, Ni, Cr, Co, and Cd). In L.E. Cronin (Ed.), *Estuarine Research*, Vol. 1. Academic Press, London: 137–152.

Chen, Y.R., J.N. Butler & W. Stumm, 1973. Kinetic study of phosphate reaction with aluminium oxide and kaolinite. *Environ. Sci. Technol.* 7: 327–332.

Chester, R., 1988. The storage of metals in aquatic sediments. In *Metals and Metalloids in the Hydrosphere: Impact Through Mining and Industry, and Prevention Technology*. Technical Documents in Hydrology. UNESCO, Paris: 81–110.

Chester, R. & M.J. Hughes, 1967. A chemical technique for the separation of ferromanganese minerals, carbonate minerals and adsorbed trace elements from pelagic sediments. *Chem. Geol.* 2: 249–262.

Dahlberg, E.C., 1968. Application of selective simulation and sampling technique to the interpretation of stream sediment copper anomalies near South Mountain, Pa. *Econ. Geol.* 63: 409–417.

Davis, J.A. & R. Gloor, 1981. Adsorption of dissolved organics in lake water environments by aluminium oxide: Effect of molecular weight. *Environ. Sci. Technol.* 15: 1223–1227.

De Groot, A.J., 1966. Mobility of trace metals in deltas. In G.V. Jacks (Ed.), *Meeting Int. Comm. Soil Sciences*, Aberdeen, Trans. Comm. II & IV: 267–297.

De Groot, A.J., K.H. Zschuppe & W. Salomons, 1982. Standardization of methods of analysis for heavy metals in sediments. In P.G. Sly (Ed.), Sediment/Freshwater Interaction, *Hydrobiologia* 92: 689–695.

Duursma, E.K., 1984. Problems of sediment sampling and conservation for radionuclide accumulation studies. In *Sediments and Pollution in Waterways*. IAEA-TecDoc-302, International Atomic Energy Agency, Vienna: 127–135.

Elderfield, H., 1981. Metal-organic associations in interstitial waters of Narragansett Bay sediments. *Am. J. Sci.* 281: 1184–1196.

Engler, R.M. *et al.*, 1977. A practical selective extraction procedure for sediment characterization. In T.F. Yen (Ed.), *Chemistry of Marine Sediments*, Ann Arbor Science Publishers, Ann Arbor: 163–171.

Ferguson, K.D. & P.M. Erickson, 1988. Pre-mine prediction of acid mine drainage. In W. Salomons & U. Förstner (Eds.), *Environmental Management of Solid Waste—Dredged Material and Mine Tailings*. Springer-Verlag, Berlin: 24–43.

Förstner, U., 1985. Chemical forms and reactivities of metals in sediments. In R. Leschber *et al.* (Eds.), *Chemical Methods for Assessing Bio-Available Metals in Sludges and Soils*, Elsevier Applied Science, London: 1–30.

Förstner, U., 1987. Metal speciation in solid wastes—factors affecting mobility. In

L. Landner (Ed.), *Speciation of Metals in Water, Sediment and Soil Systems.* Lecture Notes in Earth Sciences No. 11, Springer-Verlag, Berlin: 13–41.

Förstner, U., 1989. *Contaminated Sediments.* Lecture Notes in Earth Sciences No. 21. Springer-Verlag, Berlin: 157.

Förstner, U. & G. Müller, 1973. Metal accumulation in river sediments: A response to environmental pollution. *Geoforum* 14: 53–61.

Förstner, U. & S.R. Patchineelam, 1980. Chemical associations of heavy metals in polluted sediments from the lower Rhine River. In M.C. Kavaunagh & J.O. Leckie (Eds.), Particulates in Water, *Adv. Chem. Ser. Am. Chem. Soc.* 189: 177–193.

Förstner, U. & H.E. Reineck, 1974. Die Anreicherung von Spurenelementen in den rezenten Sedimenten eines Profilkerns aus der Deutschen Bucht. *Senckenberg. Marit.* 6: 175–184.

Förstner, U. & W. Salomons, 1980. Trace metal analysis on polluted sediments. I. Assessment of sources and intensities. *Environ. Technol. Lett.* 1: 494–505.

Förstner, U. & W. Salomons, 1983. Trace element speciation in surface waters: Interactions with particulate matter. In G.G. Leppard (Ed.), *Trace Element Speciation in Surface Waters and Its Ecological Implications.* Plenum Press, New York: 245–273.

Förstner, U. & G. Wittmann, 1979. *Metal Pollution in the Aquatic Environment.* Springer-Verlag, Berlin: 245–273.

Förstner, U., W. Calmano, W. Ahlf & M. Kersten, 1989. Ansätze zur Beurteilung der Sedimentqualität in Gewässern. *Z. Vom Wasser* 73: 25–42.

Fuller, W.H. & A.W. Warrick, 1985. *Soils in Waste Treatment and Utilization.* CRC Press, Boca Raton, FL.

Gambrell, R.P. *et al.*, 1977. *Trace and Toxic Metal Uptake by Marsh Plants as Affected by Eh, pH, and Salinity.* Tech. Rep. D-77-40. U.S. Army Engineer Waterways Experiment Station, Vicksburg, MS.

Gambrell, R.P., C.N. Reddy & R.A. Khalid, 1983. Characterization of trace and toxic materials in sediments of a lake being restored. *J. Water Pollut. Control Fed.* 55: 1271–1279.

Gerlach, S.A., 1981. *Marine Pollution—Diagnosis and Therapy.* Springer-Verlag, Berlin: 218 pp.

Gibbs, R.J., 1973. Mechanisms of trace metal transport in rivers. *Science* 180: 71–73.

Grimshaw, D.L., J. Lewin & R. Fuge, 1976. Seasonal and short-term variations in the concentration and supply of dissolved zinc to polluted aquatic environments. *Environ. Pollut.* 11: 1–7.

Gücer, S. *et al.* (Eds.), 1990. *Metal Speciation in the Environment.* Proc. NATO Advanced Study Institute, Oct. 9-20, 1989, Izmir/Turkey. Springer-Verlag, Berlin (in preparation.).

Ham, R.K. *et al.*, 1979. *Background Study on the Development of a Standard Leaching Test.* U.S. EPA-600/2-79-109. U.S. Environmental Protection Agency, Cincinnati.

Hart, B.T., 1982. Uptake of trace metals by sediments and suspended particulates.

In P.G. Sly (Ed.), *Sediment/Freshwater Interactions*, Dr. W. Junk Publishing, The Hague: 299–313.

Hart, B.T. & S.H.R. Davies, 1977. A new dialysis-ion exchange technique for determining the forms of trace metals in water. *Aust. J. Mar. Freshwater Res.* 28: 105–112.

Hawkes, H.E. & J.S. Webb, 1962. *Geochemistry in Mineral Exploration*. Harper & Row, New York: 415 pp.

Heinrichs, H., 1975. *Die Untersuchuing von Gesteinen und Gewässern auf Cd, Sb, Hg, Tl, Pb und Bi mit der flammenlosen Atomabsorptions-Spektralphotometrie*. Doctoral Thesis, University of Göttingen: 82 pp.

Herms, U. & G. Brümmer, 1978. Löslichkeit von Schwermetallen in Siedlungsabfällen und Böden in Abhängigkeit von pH-Wert, Redoxbedingungen und Stoffbestand. *Mitt. Dtsch. Bodenkundl. Ges.* 27: 23–34.

Herms, U. & L. Tent, 1982. Schwermetallgehalte im Hafenschlick sowie in landwirtschaftlich genutzten Hafenschlickspülfeldern im Raum Hamburg. *Geol. Jahrb.* F12: 3–11.

Honeyman, B.D. & P.H. Santschi, 1988. Metals in aquatic systems — predicting their scavenging residence times from laboratory data remains a challenge. *Environ. Sci. Technol.* 22: 862–871.

Horowitz, A.J., 1984. *A Primer on Trace Metal-Sediment Chemistry*. Open-File Rep 84–709. U.S. Geological Survey, Doraville, GA: 82 pp.

Jackson, D.R., B.J. Garrett & T.A. Bishop, 1984. Comparison of batch and column methods for assessing leachability of hazardous waste. *Environ. Sci. Technol* 18: 668–673.

Jackson, M.L., 1958. *Soil Chemical Analysis*. Prentice-Hall, Englewood Cliffs, NJ.

Jenne, E.A., 1977. Trace element sorption by sediments and soils — sites and processes. In W. Chappell & K. Petersen (Eds.), *Symposium on Molybdenum*, Vol. 2. Marcel Dekker, New York: 425–553.

Jones, B.F. & C.J. Bowser, 1978. The mineralogy and related chemistry of lake sediments. In A. Lerman (Ed.), *Lakes — Chemistry, Geology, Physics*. Springer-Verlag, New York: 179–235.

Kemp, A.L.W. *et al.*, 1976. Cultural impact on the geochemistry of sediments in Lake Erie. *J. Fish. Res. Board Can.* 33: 440–462.

Kersten, M., 1989. *Mechanismus und Bilanz der Schwermetallfreisetzung aus einem Süßwasserwatt der Elbe*. Dissertation, Technische Universität Hamburg-Harburg, Hamburg, West Germany.

Kersten, M. & U. Förstner, 1986. Chemical fractionation of heavy metals in anoxic estuarine and coastal sediments. *Water Sci. Technol.* 18: 121–130.

Kersten, M. & U. Förstner, 1987a. Cadmium associations in freshwater and marine sediment. In J.O. Nriagu & J.B. Sprague (Eds.), *Cadmium in the Aquatic Environment*, John Wiley & Sons, New York: 51–88.

Kersten, M. & U. Förstner, 1987b. Effect of sample pretreatment on the reliability of solid speciation data of heavy metals — implications for the study of early diagenetic processes. *Mar. Chem.* 22: 299–312.

Kersten, M., U. Förstner, W. Calmano & W. Ahlf, 1985. Freisetzung von Metallen bei der Oxidation von Schlämmen. *Vom Wasser* 65: 21-35.

Kester, D.R. *et al.* (Eds.), 1983. *Wastes in the Ocean,* Vol. 2: *Dredged-Material Disposal in the Ocean.* John Wiley & Sons, New York: 299 pp.

Keyser, T.R. *et al.*, 1978. Characterizing the surface of environmental particles. *Environ. Sci. Technol.* 12: 768-773.

Kobayashi, J., 1971. Relation between the "Itai-Itai" disease and the pollution of river water by cadmium from a mine. In *Proc. 5th Int. Conf. on Advanced Water Pollution Research*, Vol. I-25. Pergamon Press, Oxford: 1-7.

Kopp, J.F. & R.C. Kroner, 1968. *Trace Metals in Waters of the United States.* Federal Water Pollution Control Administration, Division of Pollution Surveillance, Cincinnati, OH.

Landner, L. (Ed.), 1987. *Speciation of Metals in Water, Sediment and Soil Systems.* Lecture Notes in Earth Sciences No. 11. Springer-Verlag, Berlin: 190 pp.

Lee, F.Y. & J.A. Kittrick, 1984. Elements associated with the cadmium phase in a harbor sediment as determined with the electron beam microprobe. *J. Environ. Qual.* 13: 337-340.

Leppard, G.G. (Ed.), 1983. *Trace Element Speciation in Surface Waters and its Ecological Implications.* Proc. NATO Advanced Research Workshop, Nervi, Italy, Nov. 2-4, 1981. Plenum Press, New York: 320 pp.

Lion, L.W., R.S. Altman & J.O. Leckie 1982. Trace-metal adsorption characteristics of estuarine particulate matter: Evaluation of contribution of Fe/Mn oxide and organic surface coatings. *Environ. Sci. Technol.* 16: 660-666.

Lu, C.S.J. & K.Y. Chen, 1977. Migration of trace metals in interfaces of seawater and polluted surficial sediments. *Environ. Sci. Technol.* 11: 174-182.

Lum, K.R. & K.L. Gammon, 1985. Geochemical availability of some trace and major elements in surficial sediments of the Detroit River and western Lake Erie. *J. Great Lakes Res.* 11: 328-338.

Luoma, S.N. & J.A. Davis, 1983. Requirements for modeling trace metal partioning in oxidized estuarine sediments. *Mar. Chem.* 12: 159-181.

Luther, G.W., A.L. Meyerson, J.J. Krajewski & R. Hires, 1980. Metal sulfides in estuarine sediments. *J. Sed. Petrol.* 50: 1117-1120.

Maaβ, B., G. Miehlich & A. Gröngröft, 1985. Untersuchungen zur Grundwassergefährdung durch Hafenschlick-Spülfelder. II. Inhaltsstoffe in Spülfeldsedimenten und Porenwässern. *Mitt. Dtsch. Bodenkundl. Ges.* 43/I: 253-258.

Malle, K.-G., 1985. Metallgehalt und Schwebstoffgehalt im Rhein. *Z. Wasser Abwasser Forsch.* 18: 207-209.

Malone, P.G., L.W. Jones & R.J. Larson, 1982. *Guide to the Disposal of Chemically Stabilized and Solidified Waste.* Office of Water and Waste Management, SW-872. U.S. Environmental Protection Agency, Washington, D.C.

Martin, J.M., P. Nirel & A.J. Thomas, 1987. Sequential extraction techniques: Promises and problems. *Mar. Chem.* 22: 313-342.

Moore, J.W. & S. Ramamoorthy, 1984. *Heavy Metals in Natural Waters — Applied Monitoring and Impact Assessment.* Springer-Verlag, New York: 268 pp.

Morton, R.W., 1980. "Capping" procedures as an alternative technique to isolate contaminated dredged material in the marine environment. In *Dredge Spoil Dis-*

posal and PCB Contamination: Hearings before the Committee on Merchant Marine and Fisheries. U.S. House of Representatives, 96th Congress, 2nd Session, on Exploring the Various Aspects of Dumping of Dredged Spoil Material in the Ocean and the PCB Contamination Issue, March 14, May 21, 1980. USGPO Ser. No. 96-43, Washington D.C.: 623-652.

Müller, G., 1979. Schwermetalle in den Sedimenten des Rheins — Veränderungen seit 1971. Umschau Wiss. Technik 79: 778-783.

Nriagu, J.O. (Ed.), 1984. Changing Metal Cycles and Human Health. Dahlem-Konferenzen, Life Sci. Res. Rep. 28. Springer-Verlag, Berlin: 367 pp.

Nriagu, J.O., A.L.W. Kemp, H.K.T. Wong & N. Harper, 1979. Sedimentary record of heavy metal pollution in Lake Erie. Geochim. Cosmochim. Acta 43: 247-258.

Patchineelam, S.R. & U. Förstner, 1977. Bindungsformen von Schwermetallen in marinen Sedimenten. Untersuchungen an einem Sedimentkern aus der Deutschen Bucht. Senckenberg. Marit. 9: 75-104.

Patrick, W.H., B.G. Williams & J.T. Moraghan, 1973. A simple system for controlling redox potential and pH in soil suspensions. Soil Sci. Soc. Am. Proc. 37: 331-332.

Patterson, J.W. & R. Passino (Eds.), 1987. Metals Speciation, Separation, and Recovery. Lewis Publishers, Chelsea, MI: 779 pp.

Pickering, W.F., 1981. Selective chemical extraction of soil components and bound metal species. CRC Crit. Rev. Anal. Chem. Nov.: 233-266.

Rapin, F. et al., 1986. Potential artifacts in the determination of metal partitioning in sediments by a sequential extraction procedure. Environ. Sci. Technol. 20: 836-840.

Salomons, W., 1980. Adsorption processes and hydrodynamic conditions in estuaries. Environ. Technol. Lett. 1: 356-365.

Salomons, W., 1985. Sediments and water quality. Environ. Technol. Lett. 6: 315-368.

Salomons, W. & P. Baccini, 1986. Chemical species and metal transport in lakes. In M. Bernhard et al. (Eds.), The Importance of Chemical "Speciation" in Environmental Processes., Springer-Verlag, Berlin: 193-216.

Salomons, W. & U. Förstner, 1980. Trace metal analysis on polluted sediments. II. Evaluation of environmental impact. Environ. Technol. Lett. 1: 506-517.

Salomons, W. & U. Förstner, 1984. Metals in the Hydrocycle. Springer-Verlag, Berlin: 349 pp.

Salomons, W. & U. Förstner (Eds.), 1988a. Chemistry and Biology of Solid Waste — Dredged Material and Mine Tailings. Springer-Verlag, Berlin: 305 pp.

Salomons, W. & U. Förstner (Eds.), 1988b. Environmental Management of Solid Waste — Dredged Material and Mine Tailings. Springer-Verlag, Berlin: 396 pp.

Salomons, W., N.M. De Rooij, H. Kerdijk & J. Bril, 1987. Sediments as a source for contaminants? In R.L. Thomas et al. (Eds.), Ecological Effects of In-Situ Sediment Contaminants. Hydrobiologia 149: 13-30.

Sauerbeck, D. & P. Styperek, 1985. Evaluation of chemical methods for assessing the Cd and Zn availability from different soils and sources. In R. Leschber et al. (Eds.), Chemical Methods for Assessing Bio-Available Metals in Sludges and Soils. Elsevier Applied Science, London: 49-66.

Sayre, W.W., H.P. Guy & A.R. Chamberlain, 1963. Uptake and transport of radio-nuclides by stream sediments. *U.S. Geol. Surv. Prof. Paper* 433-A: 23 pp.

Schoer, J. & U. Förstner, 1985. Chemical forms of artificial radionuclides in fluvia-tile, estuarine and marine sediments compared with their stable counterparts. In E.K. Duursma (Ed.), *Proc. Seminar Behaviour of Radionuclides in Estuaries*, Commission of the European Communities, Luxembourg: 27–54.

Schoer, J. & U. Förstner, 1987. Abschätzung der Langzeitbelastung von Grundwas-ser durch die Ablagerung metallhaltiger Feststoffe. *Vom Wasser* 69: 23–32.

Shea, D., 1988. Developing national sediment quality criteria—equilibrium parti-tioning of contaminants as a means of evaluating sediment quality criteria. *Envi-ron. Sci. Technol.* 22: 1256–1261.

Sobek, A.A., W.A. Schuller, J.R. Freeman & R.M. Smith, 1978. *Field and Labora-tory Methods Applicable to Overburden and Mine Spoils.* Report EPA-600/2-78-054, U.S. Environmental Protection Agency, Washington, D.C.

Stoner, J.H., 1974. Ph.D. Thesis, University of Liverpool, Liverpool, U.K. Cited in Chester (1988).

Stumm, W., 1986. Water, an endangered ecosystem. *Ambio* 15: 201–207.

Takeuchi, T. *et al.*, 1959. Pathological observations of the Minamata disease. *Acta Pathol. Jpn. Suppl.* 9: 769–783.

Tent, L., 1982. Auswirkungen der Schwermetallbelastung von Tidegewässern am Beispiel der Elbe. *Wasserwirtschaft* 72: 60-62.

Tent, L., 1987. Contaminated sediments in the Elbe estuary: Ecological and eco-nomic problems for the Port of Hamburg. In R.L. Thomas *et al.* (Eds.), Ecologi-cal Effects of In-Situ Sediment Contaminants. *Hydrobiologia* 149: 189–199.

Tessier, A., P.G.C. Campbell & M. Bisson, 1979. Sequential extraction procedure for the speciation of particulate trace metals. *Anal. Chem.* 51: 844–851.

Tessier, A., P.G.C. Campbell & M. Bisson, 1980. Trace metal speciation in the Yamaska and St. Francois River (Quebec). *Can. J. Earth Sci.* 17: 90-105.

Theis, T.L. & L.E. Padgett, 1983. Factors affecting the release of trace metals from municipal sludge ashes. *J. Water Pollut. Control Fed.* 55: 1271–1279.

Thomas, R.L. *et al.*, 1976. Surficial sediments of Lake Erie. *J. Fish. Res. Board Can.* 33: 385–403.

Thomson, E.A. *et al.*, 1980. The effect of sample storage on the extraction of Cu, Zn, Fe, Mn, and organic material from oxidized estuarine sediments. *Water Air Soil Pollut.* 74: 215–233.

Tipping, E., 1981. The adsorption of aquatic humic substances by iron hydroxides. *Geochim. Cosmochim. Acta* 45: 191–199.

Towner, J., 1984. Ph.D. Thesis, University of Liverpool, Liverpool, U.K. Cited in Chester (1988).

Trefry, J.H. & B.J. Presley, 1976. Heavy metal transport from the Mississippi River to the Gulf of Mexico. In H.L. Windom & R.A. Duce (Eds.), *Marine Pollution Transfer*. D.C. Heath Publishers, Lexington: 39–76.

Troup, B.N. & O.P. Bricker, 1975. Processes affecting the transport of materials from continents to the ocean. In T.M. Church (Ed.), *Marine Chemistry in the Coastal Environment*. ACS Symp. Ser. 18. American Chemical Society, Wash-ington, D.C.: 133–151.

Van der Sloot, H.A., O. Piepers & A. Kok, 1984. *A Standard Leaching Test for Combustion Residues.* Shell BEOP-31. Studiegroep Ontwikkeling Standaard Uitloogtesten Verbrandingsresiduen, Petten, The Netherlands.

Van Veen, H.J. & P.B.M. Stortelder, 1988. Research on contaminated sediments in the Netherlands. In K. Wolf, W.J. Van Den Brink & F.J. Colon (Eds.), *Contaminated Soil '88.* Kluwer Academic Publishing, Dordrecht, The Netherlands: 1263-1275.

Wedepohl, K.H. (Ed.), 1969-1978. *Handbook of Geochemistry.* Springer-Verlag, Berlin.

Wiedemann, H.U., 1982. *Verfahren zur Verfestigung von Sonderabfällen und Stabilisierung von verunreinigten Böden. Berichte Umweltbundesamt 1/82.* Erich Schmidt Verlag, Berlin.

Wiles, C.C., 1987. A review of solidification/stabilization technology. *J. Hazardous Mat.* 14: 5-21.

Wood, J.M. & H.-K. Wang, 1983. Microbial resistance to heavy metals. *Environ. Sci. Technol.* 17: 582A-590A.

Wood, J.M. *et al.*, 1986. Speciation in systems under stress – group report. In M. Bernhard, F.E. Brinckman & P.S. Sadler (Eds.), *The Importance of Chemical "Speciation" in Environmental Processes.* Dahlem-Konferenzen, Life Sciences Research Report 33, Springer-Verlag, Berlin: 425-441.

CHAPTER 4

Spatial and Temporal Variations in Microbial Processes in Aquatic Sediments: Implications for the Nutrient Status of Lakes

Richard G. Carlton and Michael J. Klug

"A direct impact of environmental factors on the form and function of living beings is a law that admits no exceptions."

Winogradsky

1.0 INTRODUCTION

The global expansion of agriculture and the use of fertilizers in concert with increased domestic and industrial discharges have resulted in accelerated eutrophication of many lakes and reservoirs. These enhanced rates of primary production have led to increased loading of particulate organic material to sediments. The flux of matter at the sediment-water interface is, however, not unidirectional. The early studies of Mortimer (1941, 1942), Hutchinson (1957), and Zobell (1942) demonstrated that microbial metabolism of particulate organic matter in sediments regenerates dissolved inorganic nutrients which can be returned to euphotic strata and stimulate primary production. Efforts to reverse eutrophication and restore water quality in lakes have demonstrated that sediments have a long "memory". That is to say that the continued efflux of regenerated nutrients from sediments subsequent to diversion of sewage effluents and other nutrient-rich point sources is sufficient to maintain rates of primary productivity and biomass similar to prediversion conditions for many years (Bengtsson *et al.* 1975, Stauffer 1986). Therefore, limnologists not only must know when and where sediments act as sources or sinks for nutrients, but they must also attempt to understand the controlling physical, chemical, and biological processes and the interactions among them.

Our knowledge of specific metabolic processes in aquatic sediments is substantial. The microbiologically mediated cycling of carbon, nitrogen, sulfur, and phosphorus in aquatic systems has been the subject of several recent reviews (respectively, Krumbein & Swart 1983, Blackburn 1983, Jørgensen 1983, Boström *et al.* 1988) and symposia (e.g., Sly 1986, Golterman 1988, Sly & Hart 1989). Other recent reviews have compared microbial mediation of nutrient dynamics in sediments of freshwater and marine ecosystems (e.g., Howarth *et al.* 1988, Seitzinger 1988, Capone & Kiene 1988, Mann 1988). This paper deals with recent additions to our understanding of microbial mineralization of organic matter in sediments and the resultant fluxes of dissolved nutrients and gases from sediments. Emphasis is placed on interactions among microbial processes at the sediment-water interface and spatial and temporal variations in the factors that control solute fluxes. Central to these advancements are new *in situ* and laboratory incubation methods that allow investigators to observe changes in and results of microbial function under natural conditions either through time or as results of environmental perturbations.

2.0 SEDIMENTATION AND MINERALIZATION OF ORGANIC MATTER

The sedimentation rate of particulate organic carbon (POC) varies widely within and among lakes. POC sedimentation data summarized in Nedwell (1984) range from ca. 30 to 160 g C m^{-2} year^{-1} among several north temperate lakes. In general, greater sedimentation rates occurred under more eutrophic conditions and lower rates occurred in deeper oligotrophic systems. As additional (and perhaps extreme) examples, the sedimentation rate of POC in deep (z_{max} = 410 m), oligotrophic Lake Superior (U.S.A.-Canada) is estimated at 1.3 mg C m^{-2} year^{-1} (Johnson *et al.* 1982), while in shallow (z_{max} = 6.5 m), hypereutrophic Wintergreen Lake (Michigan, U.S.A.), POC deposition varied from ca. 0.6 to 4.1 g C m^{-2} day^{-1} during summer (calculated from data of Molongoski & Klug 1980a). Because settling times for particles are related to lake depth, much of the potential degradation of organic matter can occur in the water column prior to deposition in systems such as Lake Superior. Organic carbon content in surficial sediments has been observed to decrease with increased water depth (e.g., Vigneaux *et al.* 1980). However, a limited study in Lake Superior revealed an opposite trend (Carlton *et al.* 1989). The fact remains that processes involved in regulating the amount and composition of POC reaching the sediment-water interface are complex and highly interactive. Currently our understanding of seston sedimentation is limited by our ability to accurately quantify input rates. This arises principally from several inadequacies in the performance of sediment traps, which can undercollect or overcollect depending on trap design and prevailing hydrodynamic conditions (reviewed by Butman *et al.* 1986).

The sources and relative quantities of organic matter to lake sediments can potentially exert influences on the types and rates of metabolic processes occurring there. For example, terrestrial detritus and phytoplankton detritus that settle to the sediment surface degrade at different rates (Godshalk & Wetzel 1978); terrestrial detritus has a higher lignin content, is more recalcitrant, and thus has a greater chance of becoming buried and processed through anaerobic processes as sediments slowly accumulate. The relative contributions of allochthonous and autochthonous organic matter to sediments in a lake basin vary spatially with distance from shore (Wetzel 1983). On that basis one might predict differences in the vertical distributions and rates of microbial processes in sediments across the lake basin. In recent years analysis of the distributions of relative abundances of the stable isotopes of elements (e.g., ^{12}C and ^{13}C) has become an invaluable methodology for examining sediment biogeochemistry. The majority of studies utilizing these techniques have been in marine systems (e.g., see Rounik & Winterbourn 1986, Shaw *et al.* 1984, Whiticar & Faber 1985, Martens *et al.* 1986). However, this methodology is increasingly being utilized in freshwater systems, where initial studies have revealed important information on the spatial distributions of microbial processes and the fates of organic carbon compounds in sediments (see Appendix for clarification).

Subsequent to the arrival at the sediment surface, particles are subject to resuspension via hydrodynamic forces, burial through bioturbation and sedimentation, and dissolution through chemical or biological processes. In systems with an oxygenated water column, microbial degradation of organic matter at the sediment surface proceeds through aerobic respiratory pathways, which are more rapid and yield more energy than anaerobic metabolism (Jørgensen 1980a). The depth to which aerobic processes dominate is dependent upon the depth distribution and the supply rate of dissolved O_2. In the absence of fluctuating hydraulic conditions, bioturbation, and epipelic algal photosynthesis (which will be discussed later), oxygen in sediments reaches a steady-state distribution which is a function of diffusive influx of O_2 from the overlying water and respiratory consumption within the sediments. Measurements made with polarographic oxygen-sensitive microelectrodes (methodology reviewed by Revsbech & Jørgensen 1986) in unilluminated sediments from several lakes have demonstrated the limited vertical distribution of the oxic zone (Figures 1, 2, 3). The maximum depth of O_2 penetration in undisturbed, unilluminated sediments is usually < 1 cm and almost always < 2 cm. Although factors such as temperature (which influences process rates) and porosity (which affects diffusive transport) vary among sediment systems, a general inverse relation exists with respect to organic carbon content and O_2 penetration depth in sediments.

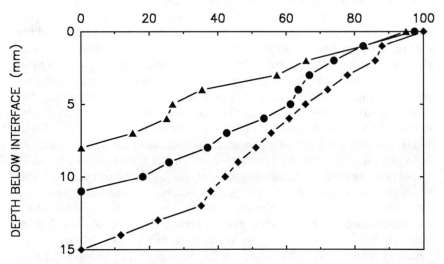

OXYGEN (% saturation)

Figure 1. Vertical microprofiles of oxygen in sediment core samples obtained from depths of 37 m (▲———▲), 82 m (●———●), and 105 m (♦———♦), in Lake Superior with the manned submersible Johnson Sea-Link II. 100% saturation = 384 μmol $O_2 \cdot L^{-1}$. (Modified from Carlton *et al.* 1989.)

Sedimentation rates in lentic systems invariably exceed dissolution and mineralization rates so that particles eventually become buried below the oxic zone, where subsequent degradation occurs only through anaerobic metabolism (Nedwell 1984). Approximately 60–70% of material that is incorporated into anoxic sediments ultimately becomes degraded through fermentation, terminal anaerobic respiration, and methanogenesis (e.g., Kimmel & Goldman 1977, Molongoski & Klug 1980b, Cappenberg *et al.* 1982, Lovely & Klug 1982). Diagenetic models, originally developed by oceanographers (e.g., see Berner 1980), are used to describe the interrelated kinetics of microbial transformations of organic matter in sediments and the subsequent effluxes of regenerated dissolved compounds. Central to such models is the spatial (vertical) distribution of terminal microbial processes which produce either CO_2 or CH_4 and consume specific inorganic electron acceptors (Figure 4). Below the thin oxygenated sediment zone, redox potential is sufficiently low that dissolved organic compounds, largely acetate and other short-chain fatty acid (SCFA) fermentation products and intermediates which are produced throughout the anoxic zone, are metabolized through anaerobic respiratory processes and methanogenesis. The relative thermodynamic yields of the processes, which are carried out

OXYGEN (% saturation)

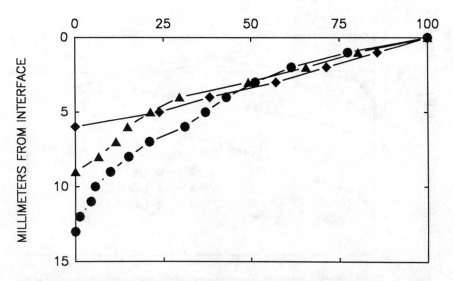

Figure 2. Vertical microprofiles of oxygen in sediment core samples obtained from depths of 115 m (▲——▲), 168 m (♦——♦), and 238 m (●——●) in Lake Michigan with the manned submersible Johnson Sea-Link II. 100% saturation = 384 μmol $O_2 \cdot L^{-1}$.

by different groups of microorganisms, result in sequential (spatial or temporal) consumption of O_2, Mn^{4+}, NO_3^-, Fe^{3+}, SO_4^{2-}, and HCO_3^- (see Nedwell 1984 for further details). Because of the lower thermodynamic yield, NO_3^- and SO_4^{2-} are not consumed appreciably in well-oxygenated zones. This leads to the vertical separations in concentration gradients of O_2, NO_3^-, and SO_4^{2-} that are typically observed in aquatic sediments (e.g., Lake Vechten; Cappenberg 1988).

During May 1984, O_2 penetrated only 1–3 mm in unilluminated sediments of Lawrence Lake, Michigan, while nitrate and sulfate penetration depths exceeded 5 cm (Figure 5). However, in contrast to the more commonly observed vertical order of gradients of oxygen, followed by nitrate and sulfate, the higher concentration of NO_3^- relative to SO_4^{2-} in Lawrence Lake water and the occurrence of nitrification in surficial sediments resulted in deeper penetration of NO_3^-. In sediments of deep, oligotrophic Lake Superior, where organic carbon content was generally <2% by mass, >90% of the microbial demand for terminal electron acceptors was met by oxygen; NO_3^- respiration was low, and SO_4^{2-} respiration rates were below the limit of detection (Carlton *et al.* 1989). These low rates of microbial activity resulted in deep penetration of both oxygen (10–25 mm, the deepest yet observed in

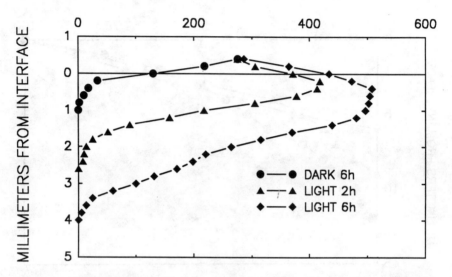

Figure 3. Effect of photosynthetic O_2 production by epipelic microalgae demonstrated by oxygen microprofiles in a box-corer sediment sample obtained from a depth of 8 m in Lawrence Lake. PPFD = 30 $\mu E \cdot m^{-2} \cdot sec^{-1}$; atmospheric equilibrium saturation concentration = 302 $\mu mol\ O_2 \cdot L^{-1}$. (Reprinted from Carlton & Wetzel 1988. With permission.)

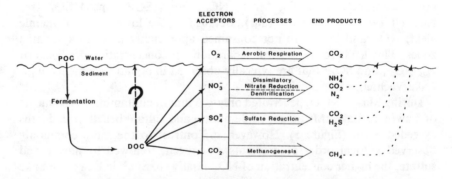

Figure 4. Schematic representation of anaerobic metabolic pathways (solid arrows) of particulate organic carbon (POC) and subsequent flow of dissolved organic carbon (DOC) compounds through the vertically separated terminal pathways in aquatic sediments. Diffusive efflux of dissolved organic and inorganic compounds from the sediments to the overlying water are indicated by dashed arrows.

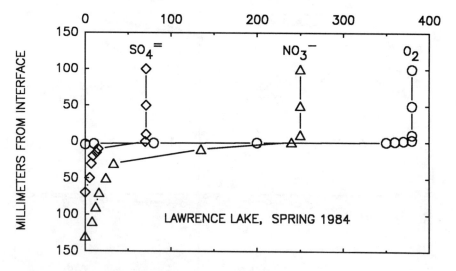

Figure 5. Distributions of SO_4^{2-}, NO_3^-, and O_2 in interstitial water of Lawrence Lake sediments at a depth of 12 m.

undisturbed lake sediments) and NO_3^- (Figure 6), with essentially no down-core depletion of SO_4^{2-} (data not shown). The site with the greatest organic carbon content also had the most rapid nitrate depletion and the shallowest O_2 penetration (core A18; Figure 6).

Anaerobic fermentation of large polymers in sediments yields a variety of intermediary compounds, mostly SCFA, among which acetate typically dominates both in concentration and as substrate for sulfate and nitrate respiration and methanogenesis (Molongoski & Klug 1980b, Lovely & Klug 1982, Capone & Kiene 1988). The fate of acetate in anaerobic sediments is largely a result of the competition between methanogens and sulfate reducers. A model of carbon and electron flow in the sulfate-reducing and methanogenic zones of anoxic lake sediments (Lovely & Klug 1986) predicts the segregation of the two zones based on the ability of sulfate reducers to deplete acetate (and presumably H_2) below levels required by methanogens (ca. 21 μM) as long as the SO_4^{2-} concentration exceeds ca. 30 μM. Based on the shallow depth of sulfate penetration in Lawrence Lake sediments (Figure 5) one would predict that methanogenesis dominated terminal carbon flow throughout most of the sediment profile. Indeed, largely because of the low SO_4^{2-} concentrations in most freshwater environments, methanogenesis is typically reported to dominate terminal carbon metabolism in anaerobic lake sediments (Cappenberg *et al.* 1982, Phelps & Zeikus 1984,

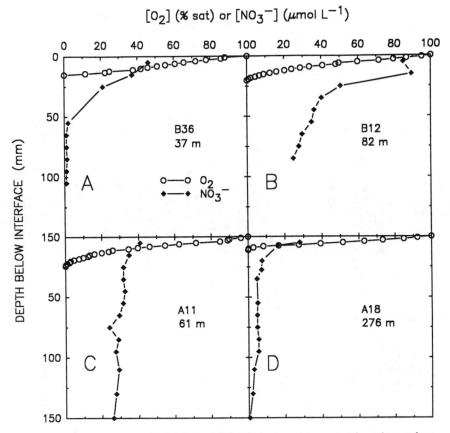

$[O_2]$ (% sat) or $[NO_3^-]$ (μmol L^{-1})

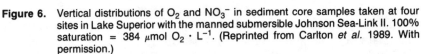

Figure 6.   Vertical distributions of O_2 and NO_3^- in sediment core samples taken at four sites in Lake Superior with the manned submersible Johnson Sea-Link II. 100% saturation = 384 μmol $O_2 \cdot L^{-1}$. (Reprinted from Carlton *et al.* 1989. With permission.)

Lovely & Klug 1986). Conversely, oxidation of organic carbon through sulfate reduction is a much more important process in marine sediments (Capone & Kiene 1988).

3.0 SEDIMENT FUNCTION WITHIN THE EUPHOTIC ZONE OF LAKES

Microbial regeneration of inorganic nutrients (e.g., NH_4^+ and PO_4^{3-}) within sediments results in concentration gradients that force diffusion of the solutes from interstitial water in the sediments to the overlying water (Berner 1980). The efflux of regenerated nutrients from sediments located within the euphotic zone of a lake provides a favorable environment for the growth of epipelic algae. Possibilities for distinct relationships between epipelic algal productivity and the concentrations of nutrients in sediment interstitial water were originally hypothesized by Lund (1942) and discussed later by Round (1957). The epipelic periphyton community exists as a biofilm consisting of a cohesive mucoid matrix secreted by, and in which are suspended, algae, bacteria, and other microorganisms, together with particulate detritus and mineral precipitates. Primarily because of light limitation, the areal photosynthetic rates of benthic periphyton are far exceeded by the phytoplankton (Wetzel 1983). However, because diffusion of oxygen is slow within periphyton and sediments, even low photosynthesis rates can produce oxygen supersaturation within the upper few millimeters of sediment. Carlton and Wetzel (1987) found that epipelic diatoms, with biomass of ca. 30 mg chlorophyll $a \cdot$ m^{-2}, on profundal sediments in Lawrence Lake required a photosynthetic photon flux density (PPFD) of only 10 μE $\cdot$ m^{-2} $\cdot$ sec^{-1} to produce [O$_2$] exceeding that in the overlying water. Sundbäck (1986) found 20 μE $\cdot$ m^{-2} $\cdot$ sec^{-1} to be the lower limit for net photosynthesis in estuarine sediments from Laholm Bay, Denmark. The direct relationship between water depth and PPFD at the sediment-water interface imparts a basin-wide spatial component to any processes affected by benthic photosynthesis.

Results of recent work at two profundal sites in Lawrence Lake (Site A at 12 m and Site B at 8 m; Figure 7) provide several examples of spatial and temporal variation in important aerobic and anaerobic processes. Sites A and B are separated vertically by only 4 m and horizontally by only 33 m. The vertical distribution of O$_2$ in sediments varies on a diel basis at each site as a result of photosynthetic O$_2$ production by epipelic algae. This is demonstrated in Figure 8, which is a composite presentation of data obtained over several years. During darkness there is always a negative gradient of [O$_2$] in the thin (<1 mm) diffusive boundary layer above the sediment-water interface. In spring and early summer greater light availability and algal biomass at Site B result in higher [O$_2$] at the interface and deeper O$_2$ penetration. By midsummer O$_2$ in the deep hypolimnion (below 10 m depth in the lake) is becoming depleted and light availability at both sites is decreasing, resulting in less photosynthesis, with this effect being more severe at the deeper site. By late summer diatoms have been replaced by photosynthetic sulfur bacteria at Site A, and the overlying water is nearly anoxic. Concomitantly at Site B the overlying water is still well oxygenated, and sufficient light penetrates through the 8 m of overlying water to support benthic algal photosynthesis. These data demonstrate that oxygen dynamics must be examined on appropriate spatial (millimeters vertically, meters horizontally) and temporal (from minutes to months) scales in order to fully understand sediment biogeochemistry in lakes.

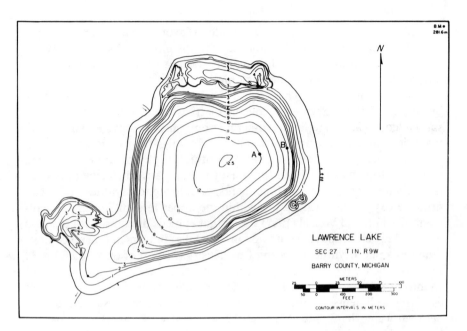

Figure 7. Bathymetric map of Lawrence Lake showing locations of profundal sediment sampling stations A (12 m) and B (8 m). (Courtesy of R. G. Wetzel.)

Metabolic processes occurring deeper in the sediment also differ distinctly between the two sites. This is illustrated by the value of the respiration index (RI) along vertical profiles in the sediment (Figure 9). RI is determined by inoculating horizontal sections sliced from sediment core samples with ^{14}C-labelled acetate and quantifying the relative production rates of $^{14}CO_2$ and $^{14}CH_4$, thus indicating the proportion of carbon terminally metabolized through CO_2 production versus methanogenesis; a high value (maximum of 1.0) indicates a greater proportion of methanogenesis (see equation, Figure 9 inset). The RI profiles at Site B in August and September indicated that while organic carbon was respired to CO_2 in surficial sediments it was metabolized largely through methanogenesis below 5 cm. In August less relative CO_2 production occurred throughout the profile at Site A compared to Site B. In September the seasonal reduction in photosynthetic O_2 production by epipelic algae and the depletion of O_2 in overlying hypolimnetic water led to an increase in the relative rate of CH_4 production (i.e., increased RI) in the top 3 cm at Site A. Determination of RI distribution in sediments of Lake Vechten at one site and time revealed that the transition from CO_2 production to methanogenesis occurred at a depth of ca. 7 cm (Cappenberg 1988). This methodology has also been used to index microbial activity in examinations of the effects of environmental

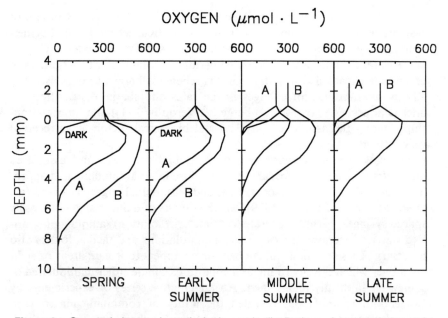

Figure 8. Seasonal changes in vertical microscale distributions of oxygen during night and day in profundal sediments of Lawrence Lake at Sites A (12 m) and B (8 m).

perturbations on microbial function in sediments (e.g., shifts in pH; Phelps & Zeikus 1984).

As seasonal anoxia develops in lake hypolimnia, gradients of reduced solutes (acetate, other SCFA, NH_4^+, HS^-, CH_4) can move from the sediments into the overlying water. Under these conditions the array of processes illustrated in Figure 4 becomes translocated to the oxic-anoxic inter-

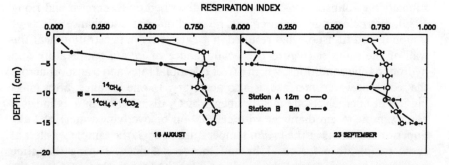

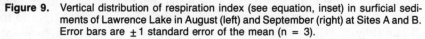

Figure 9. Vertical distribution of respiration index (see equation, inset) in surficial sediments of Lawrence Lake in August (left) and September (right) at Sites A and B. Error bars are ±1 standard error of the mean (n = 3).

face within the water column, where microbial biomass is 1–3 orders of magnitude less per volume than in sediments (Kuznetsov 1970). Furthermore, the change from molecular diffusion in sediments to eddy diffusion in the water column can increase mixing rates of solutes by ca. 2–3 orders of magnitude (Crank 1975). Although this scheme is known to occur seasonally in lakes, studies of changes in the rates and spatial distributions of microbial processes (e.g., Steenbergen & Verdouw 1984) and interrelationships among the associated physical and biological factors have received very little attention.

The close spatial arrangement of microbial processes and associated pools of substrates and products require close interval sampling and appropriate consideration of temporal (diel and seasonal) changes. The remotely operated device of Blakar (1979), which consists of a linear array of closely spaced syringes, permits accurate (centimeter-resolution) sampling of stratified waters. Increased use of *in situ* manipulations and incubations is also necessary. These methods utilize benthic chambers to isolate areas of sediment over which the environmental conditions can be either manipulated or somewhat faithfully maintained. An imaginative series of experiments by Lindeboom *et al.* (1985) revealed the utility of combining macro- and microelectrode technology with the use of benthic chambers. Their methodology permitted determination of O_2 distributions in darkened and illuminated sediments, photosynthetic O_2 production by benthic algae, the fate (up or down) of photosynthetically produced oxygen, the determination of apparent diffusion coefficients, and the respiratory demand of the sediment for oxygen. *In situ* techniques and direct sampling (often requiring SCUBA) are superior to surface-supported sampling and laboratory incubations. Samples can be taken with the utmost care, but "edge effects" (which affect all enclosed samples) can cause significant changes. For example, in a 25 cm x 25 cm (horizontal) x ca. 40-cm-deep box corer sample (25 cm of sediment + 15 cm of overlying water) retrieved in Lake Superior using the manned submersible Johnson Sea-Link II (vessel described by Eisenreich and Long 1989), microelectrode measurements revealed that the disturbance effects of sampling and transport had resulted in greater oxygen penetration near the wall of the sampler (Figure 10). Maintenance of *in situ* temperature and hydrodynamic conditions, which affect microbial rates and transport across the sediment-water interface, is also necessary. In one instance, after making initial microelectrode measurements of O_2 distribution in a cylindrical corer sample (6 cm diameter with ca. 300 ml of overlying water) of Lake Superior sediments at the *in situ* temperature of 6°C, the sample was left at room temperature for ca. 1 hr and then reexamined. During that time period the temperature of the overlying water had warmed to 18°C and the thickness of the oxygenated layer in the sediment had decreased from 15 to 5 mm (Figure 11). In another case, oxygen microprofiles were determined

OXYGEN (% saturation)

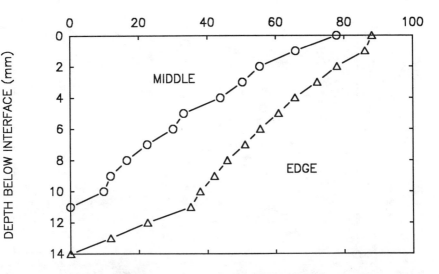

Figure 10. Oxygen microprofiles measured near the middle and within 2 cm of the edge of a 25 cm × 25 cm (horizontal) box-corer sample retrieved from a depth of 276 m in Lake Superior with the Johnson Sea-Link II. Symbols and bars represent means and ranges, respectively, of duplicate measurements made within 1 cm of each other.

before and after incubation of a sealed, unstirred core sample to determine the depletion of O_2 from the enclosed 300 ml of overlying water. During the 13-hr incubation period at 6°C, all O_2 in the sediment and much of the O_2 in the overlying water was consumed (Figure 12), with the result that aerobic metabolism in the sediment had ceased at some point during the incubation period. The excessive incubation period and cessation of natural water movements above the sediment-water interface led to perturbations of the natural oxygen distributions, prohibiting accurate estimation of the O_2 consumption rate of the sediments. The rate of O_2 consumption by sediments has been reported to be unaffected by [O_2] at concentrations exceeding ca. 100 μM (Hall et al. 1989). However, Hall et al. (1989) also point out that reproduction of water movements inside benthic chambers is necessary for accurate determination of O_2 transport across the sediment-water interface. Maintenance of *in situ* conditions in benthic chamber experiments is crucial because of the effect of oxygen on biological and chemical processes in surficial sediments. Whether used in *in situ* enclosures or in the laboratory, oxygen microelectrodes have great utility for monitoring the millimeter-scale distributions and rapid changes of [O_2] that can occur in sediments and in the diffusive boundary layer (Jørgensen & Revsbech 1985). It should

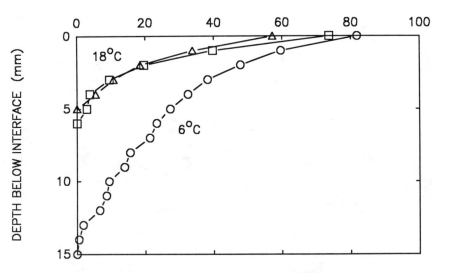

Figure 11. Oxygen distributions in a sediment corer sample retrieved from a depth of 105 m in Lake Superior with the Johnson Sea-Link II. Initial microprofile measured shortly after retrieval of the sample at *in situ* temperature of 6°C (100% saturation = 389 μmol $O_2 \cdot L^{-1}$). Subsequent measurements were made < 1 hr later after the sample had warmed to 18°C in the laboratory (100% saturation = 296 μmol $O_2 \cdot L^{-1}$).

be noted that O_2 microelectrodes are appropriate for measuring O_2 concentrations across velocity gradients in laminar flow regions only if the microelectrode is not sensitive to water velocity (Gust *et al.* 1987, Revsbech 1988).

The rapid rates of photosynthetic O_2 production and respiratory O_2 consumption at the sediment-water interface in the euphotic zone of lakes make O_2 a very dynamic compound in that microenvironment. Changes in light intensity, temperature, or water movements near the sediment-water interface result in rapid changes in the concentration and distribution of oxygen. At a depth of 9 m in Lawrence Lake, during late summer when maximum midday irradiance at that depth was 10 to 20 μE $\cdot$ m^{-2} $\cdot$ sec^{-1}, the sediment surface was well colonized by a thin film of pennate diatoms, mostly *Navicula*. In the lake, water currents close to the sediment-water interface were negligible. Therefore, a box-corer sample (methodology of Carlton & Wetzel 1985) set up as a microcosm in the laboratory for the purpose of studying the effects of photosynthesis on sediment O_2 dynamics was maintained without circulation of the overlying water. After 12 hr of darkness, $[O_2]$ was < 10 μM at the sediment-water interface and O_2 was absent below a depth of

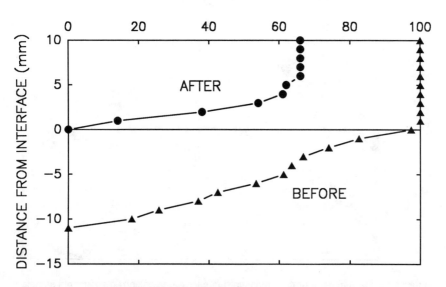

Figure 12. Unreplicated oxygen microprofiles measured before and after a 13-hr incubation (at 6.5°C) of a sediment corer sample retrieved from a depth of 105 m in Lake Superior with the Johnson Sea-Link II. Overlying water (300 ml) was sealed against atmospheric exchange and was not stirred during incubation. 100% saturation = 384 μmol $O_2 \cdot L^{-1}$.

1 mm (Figure 13, left). After 1 hr of illumination with 10 $\mu E \cdot m^{-2} \cdot sec^{-1}$, $[O_2]$ at the interface had increased to 203 μM but had not penetrated farther into the sediment. A total of 8 hr of illumination resulted in a marked increase in $[O_2]$, which was maximal at the sediment-water interface (467 μM), and in penetration of oxygen to a depth of 5.5 mm. The boundary layer was apparently thinner than after the previous dark period; it is possible that convection currents induced during illumination reduced the thickness of the boundary layer. The direct effect of water currents on boundary layer thickness and oxygen dynamics was then assessed with a horizontal velocity of ca. 0.5 cm sec^{-1} above the sediments. After 12 hr of darkness, $[O_2]$ at the interface was 136 μM and the boundary layer was ca. 0.5 mm thick (Figure 13, right). On 3 successive days the microcosm system was exposed for 12 hr to PPFD of 10, 30, and 45 $\mu E \cdot m^{-2} \cdot sec^{-1}$, with a 12-hr dark period each night. After 8 hr at 10 $\mu E \cdot m^{-2} \cdot sec^{-1}$, $[O_2]$ at the interface was only slightly supersaturated at 330 μM and O_2 penetrated to only 1.8 mm, which were, respectively, 137 μM and 3.7 mm less than under stagnant water and equal irradiance. At 30 and 45 $\mu E \cdot m^{-2} \cdot sec^{-1}$, the maximum $[O_2]$ occurred at a depth of 0.2 mm in the sediment, rather than at the interface, and O_2 penetrated to 2.4 and 3.4 mm, respectively, which were considerably

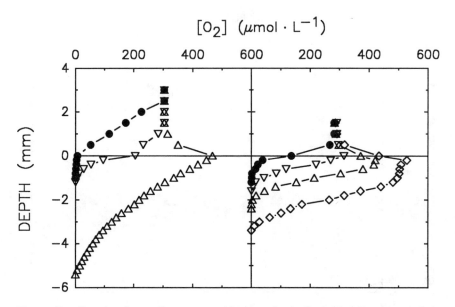

Figure 13. Oxygen microprofiles measured in microalgae-colonized sediments sampled from a depth of 9 m in Lawrence Lake. Left: During darkness ($\bullet$————$\bullet$) and after 1 hr ($\triangledown$————$\triangledown$) and 8 hr ($\triangle$————$\triangle$) of exposure to PPFD of 10 $\mu E \cdot m^{-2} \cdot sec^{-1}$; overlying water not stirred. Right: During darkness ($\bullet$————$\bullet$) and after 8 hr of exposure to PPFD of 10 ($\triangledown$————$\triangledown$), 30 ($\triangle$————$\triangle$), and 45 $\mu E \cdot m^{-2} \cdot sec^{-1}$ ($\diamond$————$\diamond$); overlying water stirred. Air saturated equilibrium = 308 μmol $O_2 \cdot L^{-1}$. (Modified from Carlton & Wetzel 1987.)

less than under unstirred conditions. Conclusions from these observations are that increased water movements produce thinner diffusive boundary layers and enhance transport of solutes across the sediment-water interface. Compared to stagnant conditions, higher water velocity resulted in lower O_2 concentrations and decreased penetration depths during daylight hours and in greater penetration during darkness (Figure 13). At some point along the continuum from zero to high water velocities, oxygen dynamics in sediments become dominated by advective transport of oxygenated overlying water into the sediments (e.g., see King *et al.* 1990), but such conditions probably do not occur in the profundal of most lakes.

The presence or absence of oxygen determines the distribution of essentially all of the microbial processes responsible for the cycling of elements in sediments (Jones 1982). However, very little is known about the effects of photosynthetic O_2 production by epipelic algae on the cycling and fluxes of elements. One of the first reports of the influence of benthic algae on nutrient flux was provided by Henriksen *et al.* (1980), who observed that sediments with benthic algae released less NH_4^+ during daytime, presumably as a result of uptake supported by photosynthesis. Although our knowledge is far from complete, the effects of benthic algal photosynthesis on the flux of phosphorus from sediments have received more attention. On a time scale of days, Sundbäck (1986) demonstrated that illuminated microcosms with algal-colonized sediments released less P and consumed NH_4^+ from the overlying water. Similarly, Hansson (1989) and Kelderman *et al.* (1988) have demonstrated that benthic algae consume phosphorus from lake sediment and overlying water, producing an overall result of a decreased flux of P from the sediments. On a finer temporal scale, Carlton and Wetzel (1988) demonstrated hourly changes in the extent of sediment oxygenation (using microelectrodes) and in phosphorus efflux rates (^{32}P release) from sediments as a direct result of epipelic algal photosynthesis at low ($< 30 \ \mu E \cdot m^{-2} \cdot sec^{-1}$) light intensities. It was suggested by Carlton and Wetzel (1988) that the mechanism consisted of uptake of P by microorganisms during oxygenated conditions, followed by release during the anoxic dark period. Recent studies by Gächter *et al.* (1988) and Hansson (1989) have provided evidence that phosphorus release can indeed be mediated in various ways by sediment microbiota and that chemical sorption/desorption reactions responding to shifts in redox potential and oxygen concentrations may be less important than is commonly realized.

4.0 SUMMARY

Several lines of evidence strongly indicate that microbial activity in sediments is a highly integrated component in the overall function of lake ecosystems: sediments affect properties of the overlying water, and they, in turn, affect the sediments. Sediments cannot be simply viewed as zones of deposition and decomposition operating at slow steady-state rates. Sediment function, i.e., the combined result of sedimentation, microbial metabolism, hydrodynamics, diffusion, and benthic photosynthesis (plus bioturbation and sediment resuspension, which have not been considered here), is spatially variable across a lake basin at any moment in time. Therefore, a "snapshot" view of sediment function and its relationship to a lake ecosystem can only be gained through a well-designed sampling effort. A mathematical treatment of this aspect of limnological research is given by Baudo in Chapter 2 of this volume.

In contrast to the theme of variation across space at a single time, an increasing body of data indicates that sediment function at a given site can vary considerably through time. Although seasonal variations in sediment function were identified over 50 years ago by Mortimer (1941, 1942), the capacity for sediment function to vary significantly on a scale of hours to days has only recently become appreciated (e.g., Carlton & Wetzel 1987, 1988, Kelderman *et al.* 1988). Based on these and other recent studies (e.g.,

King *et al.* 1990, Nielsen *et al.* 1990, Revsbech *et al.* 1990) it is now apparent that the photosynthetic activity of attached algal communities must be considered in essentially any biogeochemical study that takes place within the euphotic zone of an aquatic ecosystem. Benthic algae have also been shown to have remarkable tolerance levels to some metals (Takamura *et al.* 1989); thus, it appears that the function of epipelic periphyton communities must be considered in studies of metal contamination in euphotic sediments. The potential for diel and seasonal shifts in the distributions of oxygen and associated aerobic microbial processes to dominate sediment function necessitates the use of high-resolution techniques and increased *in situ* methodology. The importance of hydrodynamics to diffusive flux across the sediment-water interface is becoming widely recognized. However, quantitative treatments of this complex relationship between physics and biology are rare and, at this time, somewhat limited by the availability of appropriate techniques. Improvements in instrumentation for quantification of flows and in designs for *in situ* and laboratory sediment enclosures will greatly enhance our abilities to manipulate environmental variables in meaningful ways. In sum, there is an inescapable connection between spatial and temporal changes in sediment function on all scales and the associated interactions with overlying waters, all of which must be taken into account in essentially any research program concerning sediment biogeochemistry.

ACKNOWLEDGMENTS

We wish to acknowledge Greg Walker and Jeff Gelwicks for their analytical work in several of the cited studies. Our most sincere thanks go to the editors of this volume and to CNR—Istituto Italiano di Idrobiologia for financial and logistical support of our involvement in this workshop.

This work was supported by subventions from the National Science Foundation (BSR-8407078, BSR-8407189, BSR-8705342) and the National Undersea Research Center, University of Connecticut. Contribution no. 666 of the W. K. Kellogg Biological Station of Michigan State University.

APPENDIX

Of the two stable isotopes of carbon, ^{13}C and ^{12}C, the lighter comprises 98.89% of the total. Because the isotopes each have 6 protons their chemical properties are essentially the same. However, the larger total nuclear mass of ^{13}C (6 protons + 7 neutrons) exerts a slight influence on its chemical and physical behavior relative to ^{12}C. The main processes causing isotope fractionation are described in detail by Hoefs (1980). Briefly, isotope fractionation during transformations of isotopic substances results in small, but distinct, variations in the $^{13}C/^{12}C$ ratio between chemical reservoirs. These differences are expressed relative to a known standard, which is most commonly belemnite carbonate from the Cretaceous PeeDee formation in South Carolina, U.S.A. (Craig 1953). The mathematical expression is:

$$\delta^{13}C = \frac{^{13}C/^{12}C_{sample} - ^{13}C/^{12}C_{standard}}{^{13}C/^{12}C_{standard}} \times 1000$$

with $\delta^{13}C$ in units of 0/00 (per mil). A sample with a negative value contains a greater proportion of ^{12}C relative to the standard (PDB) and is referred to as being isotopically light. The result of isotope fractionation by biochemical and geochemical reactions in nature is the enrichment of ^{13}C in reservoirs of reactant and depletion of ^{13}C in pools of product. This has led to some characteristic isotope ratios (signatures) certain carbon pools in the biosphere (Figure 14). Knowledge of the natural stable isotopic signatures of the pools of reactants and products in a system affords valuable deductive power concerning processes therein. This knowledge is not suitable for estimating reaction rates, but is appropriate for determining sources and

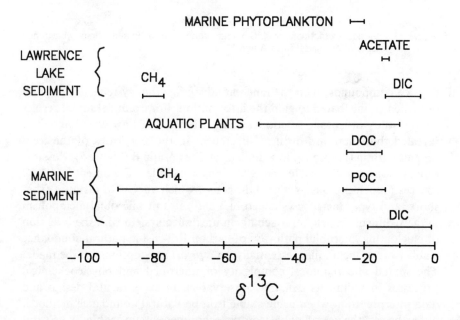

Figure 14. Representative ranges of $\delta^{13}C$ values for organic carbon species from several marine and freshwater environments (After Craig 1953, Whiticar & Faber 1985; Gelwicks & Klug, unpublished data.)

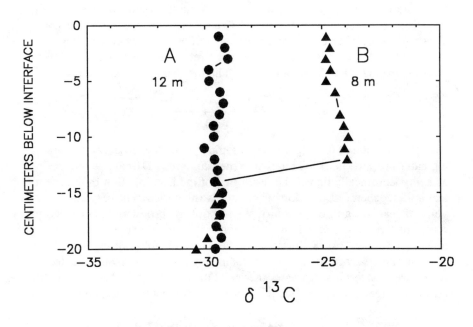

Figure 15. Vertical distributions of $\delta^{13}C$ of total organic carbon in sediments of Lawrence Lake at profundal Sites A and B.

fates of compounds, with the inherent advantage that systems need not be disturbed or incubated to gain the information. For example, initial results from a study of isotope fractionation in sediments of Lawrence Lake have revealed that there are distinct differences in the signatures of the total organic carbon (Figure 15) in sediments at Sites A and B (Figure 7). Possible explanations for the different $\delta^{13}C$ signals include ^{13}C content of POC sources (allochthonous vs autochthonous), which vary with distance from shore, and types (aerobic vs anaerobic) and rates of microbial metabolism in the sediment. When approaching an unstudied system with the intention of elucidating microbial pathways of carbon flow using natural abundance stable isotope techniques, essentially no *a priori* assumptions can be made. The spatial and temporal complexity of microbial and physicochemical processes in sediments demand conservatism in experimental design and data interpretation when using stable isotope natural abundance methods. Nevertheless, the use of stable isotope techniques will undoubtedly be valuable to freshwater biogeochemists in the near future.

REFERENCES

Bengtsson, L., S. Fleischer, G. Lindmark & W. Ripl, 1975. Lake Trummen restoration project. I. Water and sediment chemistry. *Verh. Int. Verein. Limnol.* 19: 1080–1087.

Berner, R. A., 1980. *Early Diagenesis. A Theoretical Approach.* Princeton University Press, Princeton, NJ.

Blackburn, T. H., 1983. The microbial nitrogen cycle. In W. E. Krumbein (Ed.), *Microbial Geochemistry*, Blackwell Scientific, Oxford: 63–90.

Blakar, I. A., 1979. A close-interval water sampler with minimal disturbance properties. *Limnol. Oceanogr.* 24: 983–988.

Boström, B. & K. Pettersson, 1982. Different patterns of phosphorus release from lake sediments in laboratory experiments. *Hydrobiologia* 92: 415–429.

Boström, B., J. M. Andersen, S. Fleischer & M. Jansson, 1988. Exchange of phosphorus across the sediment-water interface. *Hydrobiologia* 170: 229–244.

Butman, C. A., W. D. Grant & K. D. Stolzenbach, 1986. Predictions of sediment trap biases in turbulent flows: A theoretical analysis based on observations from the literature. *J. Mar. Res.* 44: 601–644.

Capone, D. G. & R. P. Kiene, 1988. Comparison of microbial dynamics in marine and freshwater sediments: Contrasts in anaerobic carbon metabolism. *Limnol. Oceanogr.* 33: 725–749.

Cappenberg, T. E., K. A. Hordijk, G. J. Jonkheer & J. P. M. Lauwen, 1982. Carbon flow across the sediment-water interface in Lake Vechten, The Netherlands. *Hydrobiologia* 91: 161–168.

Cappenberg, T. E., 1988. Quantification of aerobic and anaerobic carbon mineralization at the sediment-water interface. *Arch. Hydrobiol. Beih. Ergebn. Limnol.* 31: 307–317.

Carlton, R. G. & R. G. Wetzel, 1985. A box-corer for studying the metabolism of epipelic microorganisms in sediment under in situ conditions. *Limnol. Oceanogr.* 30: 422–426.

Carlton, R. G. & R. G. Wetzel, 1987. Distributions and fates of oxygen in periphyton communities. *Can. J. Bot.* 65: 1031–1037.

Carlton, R. G. & R. G. Wetzel, 1988. Phosphorus flux from lake sediments: effect of epipelic algal photosynthesis. *Limnol. Oceanogr.* 33: 562–570.

Carlton, R. G., G. R. Walker, M. J. Klug & R. G. Wetzel, 1989. Relative values of oxygen, nitrate, and sulfate to terminal microbial processes in the sediments of Lake Superior. *J. Great Lakes Res.* 15: 133–140.

Craig, H., 1953. The geochemistry of the stable carbon isotopes. *Geochim. Cosmochim. Acta* 3: 53–92.

Crank, J., 1975. *The Mathematics of Diffusion*, 2nd ed. Oxford University Press, London.

Eisenreich, S. J. & D. Long (Eds.), 1989. Benthic process research in Lake Superior. *J. Great Lakes Res.*, Vol. 15.

Gächter, R., J. S. Meyer & A. Mares, 1988. Contribution of bacteria to release and fixation of phosphorus in lake sediments. *Limnol. Oceanogr.* 33: 1542–1558.

Godshalk, G. L., & R. G. Wetzel, 1978. Decomposition of aquatic angiosperms. II. Particulate components. *Aquat. Bot.* 5: 301–327.

Golterman, H. L. (Ed.), 1988. *Sediment Water Interaction*. Dr. W. Junk, The Netherlands.

Gust, G., K. Booij, W. Helder & B. Sundby, 1987. On the velocity sensitivity (stirring effect) of polarographic oxygen microelectrodes. *Neth. J. Sea Res.* 21: 255–263.

Hall, P. O. J., L. G. Anderson, M. M. Rutgers van der Loeff, B. Sundby & S. F. G. Westerlund, 1989. Oxygen uptake kinetics in the benthic boundary layer. *Limnol. Oceanogr.* 34: 734–736.

Hansson, L.-A., 1989. The influence of a periphytic biolayer on phosphorus exchange between substrate and water. *Arch. Hydrobiol.* 115: 21–26.

Henriksen, K., J. I. Hansen & T. H. Blackburn, 1980. The influence of benthic infauna on exchange rates of inorganic nitrogen between sediment and water column. *Ophelia* Suppl. 1: 249–256.

Hickey, C. W., 1985. Quantitative addition of dissolved oxygen to in situ benthic chamber systems by use of catalase and hydrogen peroxide. *Appl. Environ. Microbiol.* 49: 462–464.

Hoefs, J., 1980. *Stable Isotope Geochemistry*, 2nd ed. Springer-Verlag, New York.

Howarth, R. W., R. Marino & J. J. Cole, 1988. Nitrogen fixation in freshwater, estuarine, and marine ecosystems. 2. Biogeochemical controls. *Limnol. Oceanogr.* 33: 688–701.

Hutchinson, G. E., 1957. *A Treatise on Limnology. I. Geography, Physics, and Chemistry.* John Wiley & Sons, Inc., New York.

Johnson, T. C., J. E. Evans & S. J. Eisenreich, 1982. Total organic carbon in Lake Superior sediments: Comparisons with hemipelagic and pelagic marine environments. *Limnol. Oceanogr.* 27: 481–491.

Jones, J. G., 1982. Activities of aerobic and anaerobic bacteria in lake sediments and their effect on the water column. In D. B. Nedwell & C. M. Brown (Eds.), *Sediment Microbiology*. Academic Press, London: 107–146.

Jørgensen, B. B., 1980. Mineralization and the bacterial cycling of carbon, nitrogen and sulphur in marine sediments. In D.C. Ellwood, J.N. Hedger, M.J. Latham, J.M. Lynch & J.H. Slater (Eds.), *Contemporary Microbial Ecology*. Academic Press, London: 239–252.

Jørgensen, B. B., 1983. The microbial sulfur cycle. In W. E. Krumbein (Ed.), *Microbial Geochemistry*. Blackwell Scientific, Oxford: 91–124.

Jørgensen, B. B. & N. P. Revsbech, 1985. Diffusive boundary layers and the oxygen uptake of sediments and detritus. *Limnol. Oceanogr.* 30: 111–132.

Kelderman, P., H. J. Lindeboom & J. Klein, 1988. Light dependent sediment-water exchange of dissolved reactive phosphorus and silicon in a producing microflora mat. *Hydrobiologia* 159: 137–147.

Kimmel, B. L. & C. R. Goldman, 1977. Production, sedimentation and accumulation of particulate carbon and nitrogen in a sheltered subalpine lake. In H. L. Golterman (Ed.), *Interactions Between Sediments and Freshwater*. Dr. W. Junk B.V. Publishers, The Hague: 148–155.

King, G. M., R. G. Carlton & T. W. Sawyer, 1990. Anaerobic metabolism and oxygen distribution in the carbonate sediments of Salt River Canyon. *Mar. Ecol. Prog. Ser.* 58: 275–285.

Krumbein, W. E. & P. K. Swart, 1983. The microbial carbon cycle. In W. E. Krumbein (Ed.), *Microbial Geochemistry.* Blackwell Scientific, Oxford: 2–62.

Kuznetsov, S. I., 1970. In C. Oppenheimer (Ed.), *The Microflora of Lakes and its Geochemical Activity.* University of Texas Press, Austin.

Lemmin, U. & D. M. Imboden, 1987. Dynamics of bottom currents in a small lake. *Limnol. Oceanogr.* 32: 62–75.

Lindeboom, H. J., A. J. J. Sandee & H. A. J. de Klerk-v.d. Driessche, 1985. A new bell jar/microelectrode method to measure changing oxygen fluxes in illuminated sediments with a microalgal cover. *Limnol. Oceanogr.* 30: 693–698.

Lovely, D. R. & M. J. Klug, 1982. Intermediary metabolism of organic matter in the sediments of a eutrophic lake. *Appl. Environ. Microbiol.* 43: 552–560.

Lovely, D. R. & M. J. Klug, 1986. Model for the distribution of sulfate reduction and methanogenesis in freshwater sediments. *Geochim. Cosmochim. Acta* 50: 11–18.

Lund, J. W. G., 1942. The marginal algae of certain ponds, with special reference to the bottom deposits. *J. Ecol.* 30: 245–283.

Mann, K. H., 1988. Production and use of detritus in various freshwater, estuarine, and coastal marine ecosystems. *Limnol. Oceanogr.* 33: 910–930.

Martens, C. S., N. E. Blair, C. D. Green & D. Des Marais, 1986. Seasonal variations in the stable carbon isotopic signature of biogenic methane in a coastal sediment. *Science* 233: 1300–1303.

Molongoski, J. J. & M. J. Klug, 1980a. Quantification and characterization of sedimenting particulate organic matter in a shallow hypereutrophic lake. *Freshwater Biol.* 10: 496–506.

Molongoski, J. J. & M. J. Klug, 1980b. Anaerobic metabolism of particulate organic matter in the sediments of a hypereutrophic lake. *Freshwater Biol.* 10: 507–518.

Nedwell, D. B., 1984. The input and mineralization of organic carbon in anaerobic aquatic sediments. *Adv. Microb. Ecol.* 7: 93–131.

Nielsen, L. P., P. B. Christensen, N. P. Revsbech & J. Sorensen, 1990. Denitrification and oxygen respiration in biofilms studied with a microsensor for nitrous oxide and oxygen. *Microb. Ecol.* (in press).

Phelps, T. J. & J. G. Zeikus, 1984. Influence of pH on terminal carbon metabolism in anoxic sediments from a mildly acidic lake. *Appl. Environ. Microbiol.* 48: 1088–1095.

Revsbech, N. P. & B. B. Jørgensen, 1986. Microelectrodes: Their use in microbial ecology. *Adv. Microb. Ecol.* 9: 293–352.

Revsbech, N. P., 1988. An oxygen microsensor with a guard cathode. *Limnol. Oceanogr.* 34: 474–477.

Revsbech, N. P., P. B. Christensen, L. P. Nielsen & J. Sorensen, 1989. Denitrification in a trickling filter biofilm studied by a microsensor for oxygen and nitrous oxide. *Water Res.* 23: 867–871.

Round, F. E., 1957. Studies on bottom-living algae in some lakes of the English

Lake District. I. Some features of the sediments related to algal productivities. *J. Ecol.* 45: 133-148.

Rounik, J. S. & M. J. Winterbourn, 1986. Stable carbon isotopes and carbon flow in ecosystems. *Bioscience* 36: 171-177.

Seitzinger, S. P., 1988. Denitrification in freshwater and coastal marine ecosystems: ecological and geochemical significance. *Limnol. Oceanogr.* 33: 702-724.

Shaw, D. G., M. J. Alperin, W. S. Reeburgh & D. J. McIntosh, 1984. Biogeochemistry of acetate in anoxic sediments of Skan Bay, Alaska. *Geochim. Cosmochim. Acta* 48: 1819-1825.

Sly, P. G. (Ed.), 1986. *Sediment and Water Interactions.* Springer-Verlag, New York.

Sly, P. G. & B. T. Hart (Eds.), 1989. Sediment and water interactions. *Hydrobiologia* 176/177.

Stauffer, R. E., 1986. Linkage between the phosphorus and silica cycles in Lake Mendota, Wisconsin. *Water Res.* 20: 597-609.

Steenbergen, C. L. M. & H. Verdouw, 1984. Carbon mineralization in microaerobic and anaerobic strata of Lake Vechten (The Netherlands): Diffusion flux calculations and sedimentation measurements. *Arch. Hydrobiol. Beih. Ergebn. Limnol.* 19: 183-190.

Sundbäck, K., 1986. What are the benthic microalgae doing on the bottom of Laholm Bay? *Ophelia* Suppl. 4: 273-286.

Takamura, N., F. Kasai & M. M. Watanabe, 1989. Effects of Cu, Cd and Zn on photosynthesis of freshwater benthic algae. *J. Appl. Phycol.* 1: 39-52.

Vigneaux, M. *et al.*, 1980. Matieres organiques et sedimentation en milieu marin. In R. Daumas (Ed.), *Biogeochemie de la Matiere Organique a l'Interface Eau-Sediment Marin.* CNRS, Paris: 113-128.

Wetzel, R. G., 1983. *Limnology*, 2nd ed. W. B. Saunders, Philadelphia.

Whiticar, M. J. & E. Faber, 1985. Methane oxidation in sediment and water column environments – isotope evidence. *Org. Geochem.* 10: 759-768.

Zobell, C. E., 1942. Changes produced by microorganisms in sediments after deposition. *J. Sed. Petrol.* 12: 127-136.

CHAPTER 5

Mechanisms Controlling Fluxes of Nutrients Across the Sediment/Water Interface in a Eutrophic Lake

René Gächter and Joseph S. Meyer

1.0 INTRODUCTION

In oligotrophic lakes with low nutrient concentrations, primary production is low, hypolimnetic oxygen (O_2) concentration is generally high, and a large fraction of the settled phosphorus (P) is permanently buried in the sediments. In the deep hypolimnion, ammonia (NH_4^+) usually occurs in trace concentrations and nitrate (NO_3^-) neither accumulates nor decreases during summer stagnation. This suggests that the NH_4^+ released during decomposition of organic material is oxidized to NO_3^- and then denitrified to a large extent within the sediment.

With increasing external P loading, primary production and, hence, phytoplankton density increase. Plankton blooms are aesthetically unpleasant. They produce bad odors when large masses of algae decay at the shore, or they may even cause fish kills if they produce toxins. As a consequence of increased primary production, hypolimnetic oxygen consumption increases. When the sediment surface becomes anoxic it releases large quantities of P into the hypolimnion, which at least in shallow lakes may further accelerate eutrophication. In the anoxic hypolimnion, hydrogen sulfide (H_2S) and NH_4^+ accumulate and NO_3^- is consumed by denitrification. Worldwide, many lakes are still eutrophic or are becoming eutrophic because of excessive inputs of P. Thus, P is without doubt a pollutant of aquatic ecosystems, and a considerable amount of money is presently being spent to restore eutrophic lakes.

The P content of a lake depends on: 1) its external P loading, 2) its P export via the effluent, 3) the P sedimentation, and 4) the P release from its sediment. P release from sediments is affected by redox conditions at the sediment surface. Oxic conditions favor formation of inorganic solid phases

Table 1. General Limnological Characteristics of Lake Sempach

Morphometry and hydrology		
Surface area	(A_o)	$14.4 \cdot 10^6$ m^2
Maximum depth	(z_{max})	87 m
Mean depth	(z)	46 m
Flushing rate	(ρ)	0.06 $year^{-1}$
Chemical and physical characteristics		
$[P_{tot}]_{overturn}$ (1989)		130 mg m^{-3}
$[NO_3\text{-}N]_{overturn}$ (1989)		900 mg m^{-3}
$[O_2]_{min}$ in hypolimnion before 1984		0 g m^{-3}
$[O_2]_{min}$ in hypolimnion after 1985		4 g m^{-3}
Transparency (min.) (1988)		1.3 m
Transparency (max.) (1988)		8.3 m
Biological characteristics		
Primary production (1983/84)		380 g C m^{-2} $year^{-1}$

(Mortimer 1971) as well as uptake and storage of P in bacterial biomass (Gächter *et al.* 1988). Thus, the concentration of dissolved P in the interstitial water and, consequently, the transport of P across an oxic sediment/water interface decreases. Often it is assumed that oxic conditions completely prevent release of P from sediments, although Lee *et al.* (1977) and Ryding and Forsberg (1977) noted that P release also occurred in well-aerated water. Boström *et al.* (1982) in their review article discussed mechanisms of P release from sediments under oxic and anoxic conditions.

Eutrophic lakes with anoxic hypolimnia are artificially oxygenated for three reasons:

1. to maintain the whole water body as an aerobic biosphere
2. to permit natural propagation of fish
3. to diminish P release from sediments and, therefore, to decrease the trophic state of lakes

2.0 LAKE SEMPACH

Lake Sempach is a eutrophic lake in central Switzerland (Table 1). It is protected from wind by hills and, thus, often does not mix intensively enough during winter to completely restore O_2 saturation in the deeper water. Since 1984, 370 m^3 hr^{-1} of pressurized air has been introduced between fall and spring in order to enhance vertical mixing. From an ecological perspective, artificial mixing is only practical during winter. Therefore, pure O_2, taken from a tank at the shore filled with liquid O_2, is introduced with 8 diffusors into the hypolimnion during summer at a rate of 3 tons per day (Figure 1). The small bubbles completely dissolve before reaching the thermocline and, therefore, do not destroy the natural density

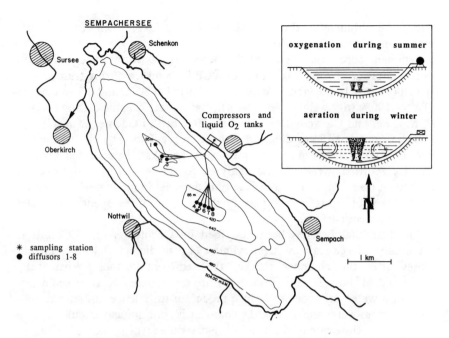

Figure 1.   Lake Sempach with depth contours, sampling position (∗), positions of diffusers (1–8) and of compressors and storage tank for liquid oxygen (■).

stratification of the lake. Due to these restoration measures, the regulatory water quality goal for O_2 ($[O_2] > 4$ mg L^{-1}, anywhere and at any time) can be reached, regardless of the high trophic state of the lake (Stadelmann 1988).

The goal of this study was to investigate the effect of oxygenation of Lake Sempach on its P cycling and to compare results obtained from a whole-lake study based on mass balance calculations, measurements of sedimentation, and investigation of sediment cores with results obtained from flux-chamber measurements and laboratory experiments.

3.0 SAMPLING SITE AND METHODS

Lake Sempach was sampled monthly at depths of 0 m, 2.5 m, 5.0 m, 7.5 m, 10.0 m, 12.5 m, 15.0 m, 20.0 m, 30.0 m, 40.0 m, 50.0 m, 75.0 m, 80.0 m, 82.0 m, and 85.0 m (Figure 1). Concentrations of soluble reactive phosphorus (SRP), total P, O_2, NH_4^+, NO_3^-, Mn, Fe, dissolved organic carbon (DOC), alkalinity, pH, and dissolved inorganic carbon (DIC) were measured as described in Gächter *et al.* (1988). In addition, during summer 1985, O_2 concentration was measured several times in samples taken 10 cm above the sediment in the deepest part of the lake, when flux-chamber measurements were conducted.

Sediment cores were taken at the deepest part of the lake and sliced immediately after sampling. Samples were kept on ice during transport to the laboratory. In the laboratory they were dried at 80°C and dry weight [% DW = 100% (dry weight/wet weight)], particulate organic carbon (POC), P, Mn, Fe, and Ca were determined as described in Gächter et al. (1988).

Sediment traps (9 cm diameter, 70-cm-long cylinders) were exposed at depths of 20 m and 81 m and sampled every 2 weeks. The collected material was freeze-dried, weighed, and analyzed for P, N, Fe, Mn, and organic carbon. A trap designed to collect settling particles in the immediate vicinity of the lake bottom (Figure 2) permitted collection of resuspended particles at the sediment surface.

The flux chamber has been described by Gächter et al. (1988). It is essentially a large stainless steel cylinder (1.5 m diameter) with a lid that closes after the chamber reaches the sediment surface. When fully implanted in the sediment, it encloses approximately 440 L of sediment-overlying water. In order to induce moderate turbulence and ensure uniform mixing, water enclosed in the chamber is continuously circulated with a pump at a flow rate of 5.3 L min^{-1}. No visible resuspension of sediment particles occurs at this low mixing rate. Water samples from inside and outside the flux chamber were collected by pumping them to the lake surface through Tygon® tubing. In order to prevent growth of bacteria in the sample tubings, they were filled with 4% formaldehyde between samplings. Fluxes of various chemical species through the sediment/water interface were calculated from changes in concentration in the sediment-overlying water, the volume of the enclosed water, and the surface area of the sediment enclosed by the flux chamber.

To study release of P and uptake of NO_3^- as a function of O_2 concentration under laboratory conditions, artificial sediment cores were prepared in the following way: 25 sediment cores with a diameter of 112 mm (Ambühl & Bührer 1975) were taken at the deepest location in Lake Sempach. Samples from the surface layer (0–0.5 cm) and from the layer between 0.5 cm and 5.0 cm were collected in separate buckets and homogenized. Subsamples from the lower layer were then transferred into twelve 5.8-cm-diameter Plexiglas® cylinders to a height of about 30 cm and covered with a 0.5-cm layer of surface sediment (Figure 3). The sediments were then overlaid with a 17-cm column of filtered (pore size 0.45 μm) water taken from the hypolimnion in Lake Sempach. This water was continuously renewed at an average rate of about 170 ml/day with filtered hypolimnetic water. Average flux rates through the sediment surface were calculated from concentration differences between inflow and outflow of the tubes, their volumetric flow rates, and from changes of the content in the sediment-overlying water column

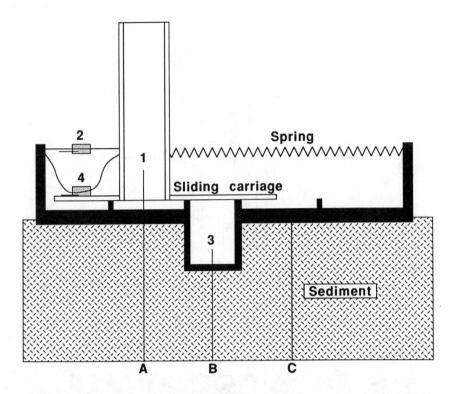

Figure 2. Trap to measure particle flux at the sediment/water interface. The sliding carriage with the sedimentation tube (1) (7 cm diameter, 27.5 cm high) is held in presampling position by wire (2). After this wire is electrically dissolved, the sliding carriage is pulled by the spring and kept in sampling position B by wire (4). In this position the sedimentation tube (1) is aligned exactly over the sediment collection cup (3) (12.5 cm high). At the end of the sampling period, wire (4) is electrically dissolved, the sliding carriage is moved to position C, closing the sampling cup (3) before retrieval of the trap.

between two sampling dates (generally 1 to 2 weeks). All experiments were conducted at 10°C in a dark room. In order to maintain neutral pH and oxic ($[O_2] \approx 4$ mg L^{-1}) or anoxic conditions, six cores were bubbled with a mixture of air, N_2, and CO_2 and the other six were bubbled with a mixture of N_2 and CO_2. Three oxic and three anoxic cores were discontinuously fed in general every 2 weeks with seston collected in sediment traps suspended at a depth of 81 m in Lake Sempach.

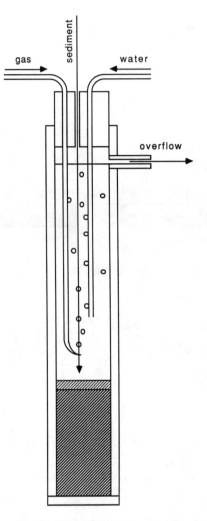

Figure 3. Sediment core prepared to study P release under laboratory conditions.

4.0 RESULTS

4.1 Whole-Lake Analysis

A seasonal pattern of the flux of settling particles was observed. Maximum flux rates occurred during summer and lower values between October and March (Figure 4). Average values for the period 1984 to 1987 are summarized in Table 2.

Flux rates of particles, POC, and PON were similar at 20 m and 81 m.

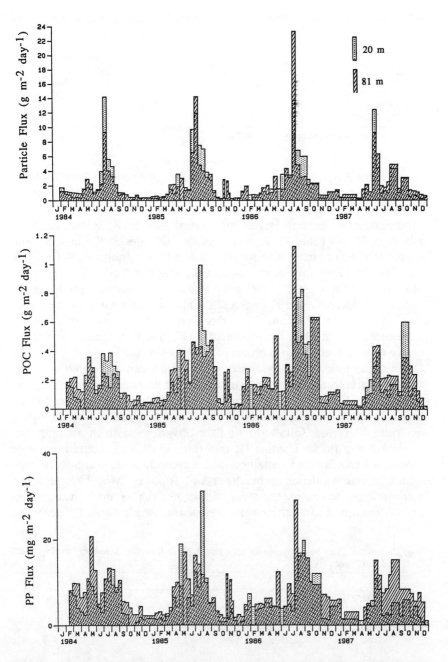

Figure 4. Seasonal variation of particle fluxes and fluxes of particulate phosphorus (PP), particulate nitrogen (PON), and particulate organic carbon (POC) at water depths of 20 m (dotted shading) and 81 m (cross-hatched shading).

Table 2. Average Particulate Flux Rates (g m^{-2} year^{-1}) at Depths of 20 m and 81 m in Lake Sempach

Flux Rate	Depth	
	20 m	81 m
Particle flux	880	880
POC flux	74	78
PP flux	2	2.8
PON flux[a]	13	15

[a]PON fluxes are averages from 1985 to 1987.

However, the PP flux at 81 m exceeded the flux measured at 20 m by 40%, and between 81 m and 87 m it increased by an additional 140% (Tables 2 and 3).

Independent of seasons, larger flux rates of particles were observed at 87 m than at 81 m (Table 3). Average yearly flux rates at the lake bottom exceeded those at 81 m by 60% for particles and Ca, by about 40% for PON and POC, by 110% for Fe, and by 180% for Mn.

In sediment cores, POC, PON, total P, and Mn concentration decreased quickly with increasing sampling depth (Table 4). Compared to these species, concentrations of Fe and Ca seemed to depend less on sampling depth. In the core taken in 1988, water content, POC, and P concentrations of the surface layer exceeded those observed in the 1986 core.

On 13 July 1988, several sediment cores were taken at a depth of 87 m. Oxygen concentration measured in the water immediately overlying the sediment was 5.7 mg L^{-1}. A similar sequence of layers was observed in all cores (Figure 5). The surface was covered with a dense mat of the sulfur bacterium *Beggiatoa*, followed by a light brown layer which was approximately 1.5 mm thick. Then, a 1.5-mm-thick black zone containing large amounts of zooplankton remained, and a second, light, 4-mm-thick zone containing mainly calcium carbonate crystals followed. Most likely, this last layer was deposited during the spring bloom, and the layer rich in zooplankton was deposited after the clear-water phase, which normally occurs in

Table 3. Annual Flux of Particles at 81 m and 87 m in Lake Sempach during 1988 (g m^{-2} year^{-1})

Flux Rate	Depth	
	81 m	87 m
Particle flux	1125	1767
POC flux	117	168
PP flux	3.9	9.3
PON flux	16.6	23.4
PCa flux	190	301
PFe flux	15	32
PMn	14	39

Table 4. Concentration Profiles Observed in Sediment Cores Collected in September 1986 and November 1988

Sediment Core from 1.9.86

Depth cm	POC mg g^{-1}	P mg g^{-1}	Mn mg g^{-1}	Fe mg g^{-1}	Ca mg g^{-1}	DW %
0.0–0.5	92.2	1.9	3.1	12.6	258	8.1
0.5–1.0	74.7	1.1	1.8	10.8	265	9.6
1.0–2.0	63.7	0.8	1.4	14.1	257	10.1
2.0–3.0	64.4	0.8	1.2	16.1	250	10.5
3.0–4.0	52.7	0.7	1.1	13.5	245	13.3

Sediment Core from 30.11.88

Depth cm	POC mg g^{-1}	P mg g^{-1}	Mn mg g^{-1}	Fe mg g^{-1}	PON mg g^{-1}	DW %
0.0–0.5	151.2	5.5	4.2	19.9	12.4	4.9
0.5–1.0	58.7	2.1	3.1	18.2	7.04	10.2
1.0–2.0	60.9	1.3	3.9	15.2	9.98	10.7
2.0–3.0	50.4	0.9	1.8	14.2	5.99	12.7
3.0–4.0	52.5	0.8	1.5	16.5	5.85	12.0
4.0–5.0	43.0	0.8	2.7	18.2	5.03	14.5
5.0–6.0	41.7	0.8	1.9	20.0	5.48	10.9

June. This indicates that during approximately 4 months in spring and summer a sediment layer approximately 7 mm thick was deposited.

Before 1984, when oxygenation and artificial mixing were started, the O_2 concentration at depths greater than 60 m decreased below the water quality goal of 4 mg L^{-1} every summer. Since the lake was sampled only 3 to 4 times per year before 1984, the extent of the anoxic period during that time may be underestimated from Figure 6. Due to artificial mixing and oxygenation, O_2 concentrations no longer decreased below 4 mg L^{-1} at 85 m since 1985. However, in samples taken 10 cm above the sediment surface, O_2 concentration was highly variable and up to 4 mg L^{-1} less than in samples taken from 85 m. The steep concentration gradient suggests that the sediment is an important sink for O_2, and the variability in concentration indicates an unsteady transport of O_2 from the hypolimnion to the sediment surface.

A seasonal pattern of total P concentration in the layer below 20 m was observed. Minima occurred during overturn and maxima at the end of summer. Artificial mixing and oxygenation did not alter this pattern (Figure 6). P accumulation started every year at about the same time and at about the same rate, whether O_2 concentration at 85 m was 0.2 mg L^{-1} (1984) or 10.0 mg L$^-$ (1987). During the period 1979–1983, before the onset of manipulations, accumulation of hypolimnic P averaged 15.8 ± 3.9 tons during summer stratification. From 1984 to 1988, average accumulation was not significantly different (13.8 ± 3.9 tons; $p \leq 0.05$). This result suggests that P release from sediments does not depend only on the O_2 concentration in

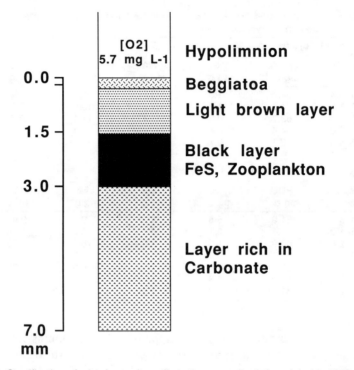

Figure 5. Stratification of microlayers in sediment cores, collected on July 13, 1988.

the overlying water. Obviously, the relationship between O_2 and P cycling cannot be simply described by the statement "more O_2 decreases the release of P from sediments". These findings are in agreement with results reported by McQueen *et al.* (1986), who demonstrated experimentally that hypolimnic aeration did not decrease P release from sediments during summer stratification.

A seasonal pattern of P cycling can also be recognized in the total water mass of the lake. In general, the P content of the lake decreased during early summer and increased between fall and full overturn in February. Before 1986, summer losses were practically compensated during fall and winter. Since 1986 the P content of the lake no longer increased during overturn, indicating that either the lake external P loading decreased or net sedimentation increased during this period.

To sum up the results from direct lake observations, oxygenation had little effect on hypolimnic P release during summer. However, after a delay of 2 years, it may have increased retention of P in sediments from fall to spring.

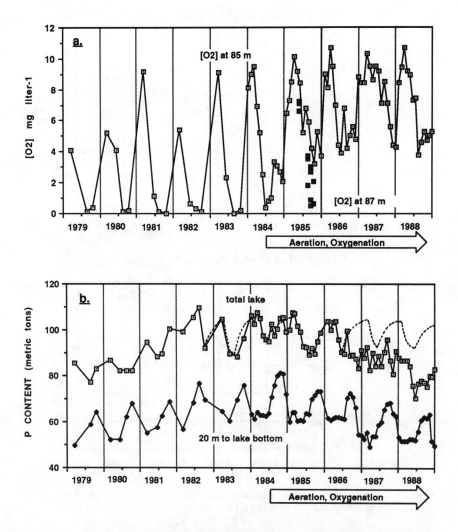

Figure 6. a.) Seasonal variation of O_2 concentration at a depth of 85 m (open boxes), supplemented with some measurements taken 10 cm above the lake bottom (solid boxes). b.) Seasonal variation of the P content of Lake Sempach (upper curve) and seasonal variation of the P content of water layer between 20 m and lake bottom (lower curve).

4.2 Laboratory Experiments with Sediment Cores

Half of the sediment cores prepared as shown in Figure 3 were fed with material collected in sediment traps. Figure 7 shows the cumulative loadings of POC, PON, and PP. The seston load peaked between days 180 and 230.

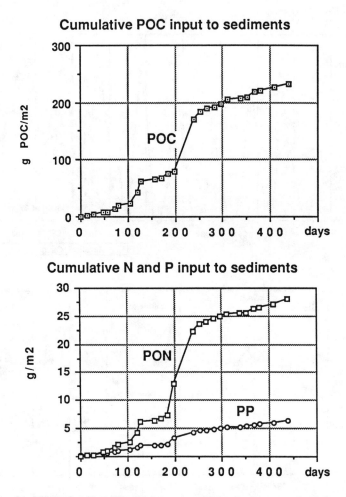

Figure 7. Cumulative loading of sediment cores with particulate organic carbon (POC), particulate organic nitrogen (PON), and particulate phosphorus (PP).

At the end of the experiment, about 230 g POC, 28 g PON, and 6 g of PP were loaded per square meter of sediment area. This load exceeded the natural load in Lake Sempach roughly two- to three-fold.

Different responses were observed depending on whether cores were fed with trapped sediment material and whether the overlying water was oxic or anoxic. Under anoxic conditions, fed cores released 3.5 g P m^{-2} during the first 470 days (Figure 8). This is about 58% of the cumulatively added P. Oxic fed sediments released only about 10% of their load. In fact, during the first 220 days, no P release was observed in oxic cores. However, about 60 days after the onset of the sedimentation peak, from day 250 to day 270,

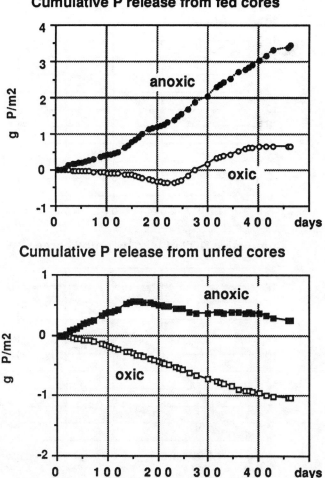

Figure 8. Cumulative release of P from sediment cores.

fed oxic and anoxic cores released approximately equal amounts of P. Oxic sediment cores not fed with seston adsorbed P from the overlying water during the entire experiment. Anoxic unfed cores released P during the first 150 days at nearly the same rate as anoxic fed cores. However, after day 150, even these anoxic sediments took up some P.

From these results we can make several conclusions:

1. A large fraction of the P added to the systems in sestonic form was buried in the sediment under oxic and anoxic conditions.

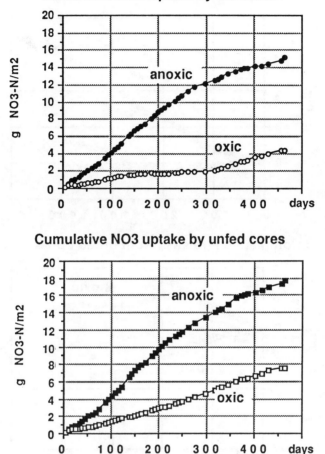

Figure 9.   Cumulative uptake of NO_3^- by sediment cores.

2. P retention by the sediment was greater in fed oxic systems (90%) than in fed anoxic systems (42%).
3. However, under certain conditions, P release rates from fed sediments overlaid with either oxic or anoxic water were about equal.

Cumulative NO_3^- consumption was dependent on whether cores were oxic or anoxic (Figure 9). As expected, NO_3^- consumption rates in anoxic cores exceeded those of oxic cores, and in both cases somewhat more NO_3^- was consumed by the unfed cores. In the oxic fed cores, NO_3^- consumption was minimal between days 180 and 300. That is just after the maximum input of

Table 5. Net Release (+) and Net Uptake (-) of Various Chemical Species by Sediments in Lake Sempach in 1985[a]

Period	O_2	NO_3^-	NH_4^+	SRP	Ptot	DOC	DIC	Fe	Mn	Ca
28.03–31.03	−387	−34	100	10	23	259	325	n.	n.	n.
03.06–07.06	−430	−40	n.	15	40	990	387	81	n.	n.
19.08–23.08	−608	−50	144	18	35	129	375	26	437	n.
16.09–20.09	−470	−64	100	19	24	260	379	n.	n.	250
23.09–26.09	−610	−60	68	10	20	157	485	25	22	320
14.10–17.10	−390	−57	165	28	32	n.	560	32	320	330
21.10–24.10	−700	−91	25	4	5	730	320	29	16	n.
11.11–14.11	−432	−45	115	−6	n.	—	400	70	238	270
18.11–21.11	−600	−83	n.	−2	n.	—	300	n.	n.	160
Average	−514	−58	102	11	26	421	392	44	207	266
SD	116	19	47	11	12	354	83	25	185	68

[a]All fluxes are derived from flux-chamber measurements and expressed as mg m^{-2} day^{-1}.
n. = not measured.

organic nitrogen to the sediments (Figure 7) and during the maximum release of P (Figure 8).

4.3 Flux-Chamber Experiments

In Lake Sempach between March and November 1985, nine flux-chamber experiments of 4 to 6 days duration were conducted. Fluxes of various chemical species through the sediment/water interface were measured (Table 5). In all experiments, the sediment was a net source for NH_4^+, DOC, DIC, Fe, Mn, and Ca and a net sink for O_2 and NO_3^-. The sediment released P from March to October. In late October and in November the sediments released much less P than during spring and summer, or they even took up some P from the water.

If it is assumed that dissolution of $CaCO_3$ is mainly induced by biogenically produced CO_2, then according to Equation 1 and according to the measured Ca flux, about 20% of the released inorganic carbon would be due to dissolution of $CaCO_3$.

$$CaCO_3 + CO_2 + H_2O \rightarrow 2HCO_3^- + Ca^{2+} \qquad (1)$$

5.0 DISCUSSION

The goal of this study was twofold: namely, to investigate the effect of artificially increased O_2 concentrations on P cycling in a eutrophic lake and to compare results of a whole-lake study with results obtained from *in situ* flux-chamber measurements and from laboratory experiments. The results from the lake study will be discussed first and then compared with experimental results.

5.1 Effects of Artificial Mixing and Oxygenation on Phosphorus Cycling Across the Sediment/Water Interface

In lakes, particles are partly of allochthonous origin and partly formed by various autochthonous biogeochemical processes. If their density exceeds the density of water, they settle to the lake bottom and contribute to sediment formation. Sediments may be resuspended, transported by currents or gas ebullition, and then redeposited at another location. Lemmin and Imboden (1987) photographed clouds caused by resuspended particles in the deepest part of Lake Sempach. It is not yet clear if these particles were resuspended in the vicinity of the measuring site or at higher levels where currents are stronger. However, bottom currents in the center of Lake Sempach remain at least strong enough to advect resuspended material along the lake bottom.

While settling through the tropholytic zone, organic particles may be mineralized. As a consequence, inorganic nutrients may be released or microorganisms growing on these particles may even take up nutrients (Gächter & Mares 1985). In steep redox gradients, particles may form biotically or abiotically and adsorb previously dissolved substances (Sigg *et al.* 1987).

The results of the study presented here (Tables 2 and 3) suggest that:

1. In the deepest part of Lake Sempach resuspension and/or advection of sediment is common, but resuspended particles do not increase the gross flux at 81 m (similar fluxes at depths of 20 m and 81 m, but greater rates at 87 m), which indicates that resuspended material is not transported beyond a layer close to the lake bottom.
2. Settling particles are hardly mineralized between depths of 20 m and 81 m (equal POC fluxes in both horizons).
3. Particulate P compounds are formed between depths of 20 m and 81 m either by biological uptake, by abiotic adsorption to surfaces, or by precipitation (Gächter *et al.* 1988).

Resuspension induces a cycling of particles in a layer close to the lake bottom, but it does not alter net sedimentation. Thus, when discussing the transfer of particles from the water to the lake bottom, this recycling can be neglected. Resuspension is, however, indicative of a highly turbulent system. If resuspension of particles occurs, then, the sediment cannot be separated from the turbulent hypolimnion by a stagnant boundary layer. As shall be discussed below, this has implications for the net transfer of dissolved substances through the sediment/water interface.

In order to estimate elimination of P by sedimentation, a constant flux of 2 g P m^{-2} year^{-1} in the layer above 20 m and a flux linearly increasing with depth of up to 2.8 g P m^{-2} year^{-1} at a depth of 81 m have been assumed

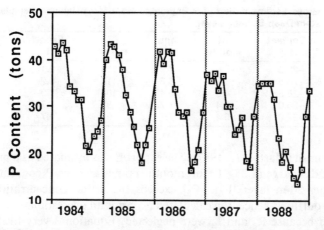

Figure 10. Seasonal changes of P content in the surface layer (0 to 20 m) of Lake Sempach from 1984 to 1988.

(Table 2). Based on these assumptions and on the morphometry of Lake Sempach, a lake-wide P sedimentation of 34 tons year^{-1} was estimated. This estimation assumes horizontally homogeneous sedimentation over the entire lake area. It is, however, likely that larger allochthonous particles settle in the delta region of the inlets and, therefore, will not be collected by sediment traps exposed in the center of the lake (Figure 1). Thus, it is likely that the actual rate of P sedimentation exceeds the estimated value of 34 tons year^{-1}.

Gross sedimentation of P (S) can also be estimated from a mass-balance calculation for the 20-m-deep surface layer during summer stagnation (Equation 2),

$$S = I_D + I_H + R_E - E - \Delta I/\Delta t \qquad (2)$$

where $\Delta I/\Delta t$ is the rate of change of P content in this layer, I_D and I_H are the rates of P input from drainage area and hypolimnion, respectively, R_E is the rate of P release from epilimnic sediments, and E is the rate of P export via the effluent. $\Delta I/\Delta t$ can be estimated (Figure 10). Lacking data concerning P release from epilimnic sediments, it was assumed that $R_E = 0$. Diffusive transport of P from the hypolimnion to the epilimnion (I_H) is negligibly small compared to other terms in Equation 2 and therefore can also be neglected. Thus, Equation 2 can be simplified to Equation 3,

$$S = I_D - E - \Delta I/\Delta t \qquad (3)$$

where Δt denotes the time span between onset and end of P decrease in the upper 20-m layer (Figure 10). To estimate I_D, a monthly input of 1.2 tons

Table 6. Gross Sedimentation of P Estimated from Mass-Balance Calculations of a 20-m-Deep Surface Layer

Year	Δ Content t	Δt months	Input t	$[P_E]$ mg m^{-3}	Export t	Gross Sed. t
1984	−23.0	6.6	7.9	114	2.1	28.8
1985	−25.5	6.9	8.3	114	2.2	31.6
1986	−25.5	4.5	5.4	100	1.3	29.6
1987	−20.0	6.4	7.7	95	1.7	26.0
1988	−22.0	6.4	7.7	95	1.7	28.0

was assumed (Marti *et al.* 1987). The export E was estimated as a product of average discharge ($2.85 \cdot 10^6$ m^3 month^{-1}) times average P concentration in the upper 20-m layer $[P_E]$. $[P_E]$ exceeds the actual concentration of the surface outflow; hence, the calculated export is an overestimate. Because of this and because R_E and I_H were neglected, Equation 3 very likely underestimates gross sedimentation, S.

Calculated gross sedimentation during summer stagnation averaged 28.8 ± 2.1 tons (Table 6). This value is only slightly smaller than the yearly sedimentation rate of 34 tons, estimated from sediment trap measurements. The good agreement of results obtained by two different, completely independent methods indicates that a yearly gross sedimentation of 34 tons of P is a reasonable estimate.

The P load of Lake Sempach in the period May 1984 to April 1986 has been estimated by Marti *et al.* (1987) . According to their measurements, the P load (14.3 tons year^{-1}) was practically identical to the load estimated for the period 1976/77 by Gächter and Stumm (1979) and the P export via the effluent was 4.6 tons year^{-1}. Between spring overturn 1984 and spring overturn 1986 P content of Lake Sempach decreased by about 2 tons. From this information a net P deposition of 10.7 tons year^{-1} can be estimated. Comparison of the estimated gross sedimentation S (34 tons year^{-1}) with the net deposition N (10.7 tons year^{-1}) yields an average yearly release rate of 23.3 tons year^{-1}. Thus, regardless of aeration and oxygenation of the hypolimnion, about 70% of the settled PP was still released from the sediments in 1984 and 1985, and internal P loading of the lake (23.3 tons year^{-1}) exceeded the external P load (14.3 tons year^{-1}) by about 70% regardless of artificial oxygenation of the hypolimnion.

According to these calculations, sediments released about 47 tons of P between 1984 and 1986 (2 years · 23.3 tons year^{-1}). Is this reasonable? Can it be confirmed by other, independent measurements? During the summers of 1984 and 1985 it is estimated that 19 tons and 13.7 tons of P, respectively, accumulated in the hypolimnion (Figure 6). This yields a total hypolimnetic release of 34.2 tons during the two summers. Between the onset of partial overturn (marked in Figure 6 by the onsetting decrease in hypolimnic P content) and the end of complete overturn (marked by the onsetting

Table 7. P Release from Sediments between Spring Overturns 1984 and 1986

Hypolimnic P accumulation in summer 1984	19.0 tons
Hypolimnic P accumulation in summer 1985	13.7 tons
P release in winter 1984/85	5.5 tons
P release in winter 1985/86	7.4 tons
Total release spring 1984 to spring 1986	45.6 tons

decrease of P content in the surface layer; Figure 10) in both years, the P content of the lake increased by about 3 tons within about 150 days. During this period, according to Marti *et al.* (1987), input (I_D) and export (E) were about 6 tons and 2 tons, respectively. Thus, a mass-balance calculation yields for both periods a net sedimentation of roughly 1 ton. During this period, average gross sedimentation rates were 3 mg P m^{-2} day^{-1} (1984/85) and 3.9 mg P m^{-2} day^{-1} (1985/86) at 81 m (Figure 4). Extrapolated to the whole lake surface area, this yields a gross sedimentation of 6.5 tons in 1984/85 and of 8.4 tons in 1985/86. The difference between gross and net sedimentation predicts that the sediments must have released 5.5 tons of P in winter 1984/85 and 7.4 tons in winter 1985/86. Hence, the sum of P released from sediments between spring 1984 and spring 1986 equals 45.6 tons, which is in excellent agreement with the 47 tons estimated independently above (Table 7).

For historical comparison, the P content of Lake Sempach increased by about 2.5 tons year^{-1} between spring 1975 and spring 1977. Average I_D and E were estimated as 14.7 tons year^{-1} and 2.2 tons year^{-1}, respectively (Gächter & Stumm 1979). A mass-balance calculation yields a net retention of 10 tons year^{-1} which is very close to the 10.7 tons year^{-1} retention estimated for the period 1984/86. From this we conclude that aeration and oxygenation did not increase net retention of P in the sediments for the years 1984 to 1986.

In the period 1975–1977 the P content of the lake was only about half the content of 1984–1986. In addition, in 1984–1986 the loading with allochthonous particulate P exceeded the value observed in the 1970s by about 3 tons year^{-1}. Thus, it is very likely that in the period 1975–1977 gross sedimentation was smaller than in 1984–1986. However, similar net retentions of P were estimated for both periods. As a consequence, P release must have been smaller in the anoxic hypolimnion in 1975–1977 than in the oxygenated hypolimnion in 1984–1986, indicating that P release from sediments does not necessarily decrease when oxygen concentration increases in the sediment-overlying water. In other words, release of P at the sediment/water interface obviously does not depend only on the oxygen concentration of the overlying water, but also on other variables, such as the P content of the sediment or its P binding capacity.

Net sedimentation can also be estimated from sediment analysis. Since

Table 8. Thickness of Annual Sediment Layers in Lake Sempach, Estimated as Described in the Text

Age of Layer Years	Core from 1.9.86 mm	Core from 30.11.88 mm
1	9	11
2	8	8
3	8	8
4	7	7
5	6	7
6	—	6

the P content of the lake, and hence its trophic state, was nearly constant from 1982 to 1986, it is reasonable to assume that within this period neither gross sedimentation (880 g dry matter m^{-2} day^{-1}, 2.8 g P m^{-2} day^{-1}, Table 2) nor the quality of the settling seston changed systematically.

Since the dry weight of deposits undergoes only small changes during diagenesis (some mineralization of organic compounds and some release of $CaCO_3$), we assumed a yearly dry matter accumulation rate M of 880 g m^{-2} (Table 2).

If in a sediment profile the relative weight of dry matter α (where wet weight = 1) as a function of depth, the specific weights of the dry matter (ρDM) and of the interstitial water (ρW), and the yearly accumulation rate of dry material (M) are known, then the thickness of annual sediment layers (z) can be calculated by the use of Equation 4.

$$z = M \cdot 10^{-4} \cdot [(1/\rho DM) + (1/(\alpha \cdot \rho W)) - (1/\rho W)] \quad (4)$$

In Equation 4, ρDM and ρW are expressed as g cm^{-3}; the number 10^{-4} is a factor to convert cm^2 to m^2. ρW was assumed to be 1 g cm^{-3}; ρDM varies in the range of 2.4 to 2.7 g cm^{-3}. However, from Equation 4 it is evident that at α values of about 0.1, z is relatively insensitive to the exact value of ρDM. In the deepest part of Lake Sempach, the most recent yearly sediment layer extends over about 10 mm, whereas layers 4 to 6 years old are about 6 to 7 mm thick (Table 8). This agrees well with our observation that during the main sedimentation peak in spring and early summer about 7 mm of sediment accumulate at the lake bottom.

If the thickness of subsequent annual layers and the P profile of a sediment core are known, it is possible to estimate the P content of the annual layers as the product M · [P], where M is the mass of dried material and [P] is the P concentration of the layer. The P content of annual layers was largest at the sediment surface and decreased with increasing distance from the surface (i.e., with increasing age) in both sediment cores (Figure 11). In the core taken in late 1988 the uppermost annual layer contained more P than in the core from 1986. By November 1988, the layer deposited between

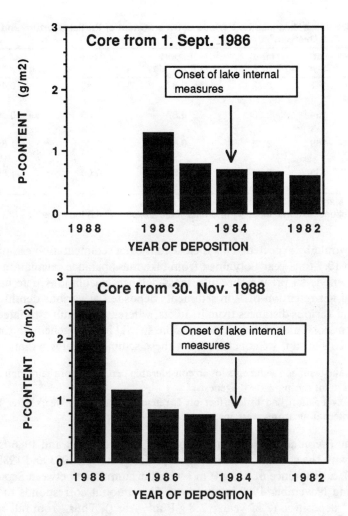

Figure 11. P content of sediment layers collected from the deepest part of Lake Sempach in September 1986 and November 1988.

September 1985 and September 1986 lost about 30% of its original P content. Five years after lake manipulations were initiated, no clear effect of oxygenation on the terminal P fixation had been observed. In both cores, annual layers older than 3 years contained 0.6 to 0.8 g P m^{-2}, regardless of whether they were deposited before or after the onset of lake internal measures to increase oxygen concentration. These values correspond to about 20 to 30% of the P annually deposited at the lake bottom (2.8 g m^{-2} year^{-1}; Table 2). If one extrapolates the terminal rate of P fixation (0.6–0.8 g P m^{-2} year^{-1}) to the whole lake surface area, a net retention of 8.6 to 11.5 tons P

Table 9. P Mass Balances for Periods between Onset of Partial Overturn and End of Full Overturn[a]

Period	Δt days	Input mg P m^{-2} day^{-1}	Export mg P m^{-2} day^{-1}	Δ Content mg P m^{-2} day^{-1}	Net sed. mg P m^{-2} day^{-1}
22.10.84–18.03.85	147	2.72	0.87	1.42	0.43
25.11.85–28.04.86	154	2.72	0.87	1.35	0.50
14.09.86–02.03.87	169	2.72	0.87	0.00	1.85
13.10.87–01.03.88	138	2.72	0.87	0.00	1.85

Source: Marti *et al.* (1987).
[a]Assumptions: external input = 14.3 tons year^{-1}; export = 4.6 tons year^{-1}.

year^{-1} would be expected. This is close to the net sedimentation estimates of 10.0 to 10.7 tons year^{-1} obtained from lake mass-balance calculations. This agreement is surprising, because mass-balance calculations determine an integral net retention of P in sediments deposited at various depths of the lake and various distances from its inlets, whereas the result estimated from sediment core analysis neglected possible spatial inhomogeneity in the sediment. To sum up, we conclude from these sediment analyses that:

1. Lake Sempach sediments lose a considerable percentage of their original P content during early diagenesis.
2. Oxygenation had little effect on P retention the first 2 years after lake internal measures were instituted.

If the P content of sediments deposited between 1983 and 1986 is compared with the content of sediments deposited between 1983 and 1988 (Figure 11), a difference of 3.3 g P m^{-2} which accumulated between September 1986 and November 1988 is observed. This amount corresponds to about 52% of deposition (2.25 years · 2.8 g P m^{-2} year^{-1}). Thus, from fall 1986 to winter 1988 the sediments retained more P than in the previous years in which only 20 to 30% was buried. Such an increase in P retention by the sediments can also be deduced from observations on the P content of the water column (Figure 6). If the P cycling of the lake had not changed—that is, if the cyclic behavior of the years 1982 to 1986 had continued—then in the period 1986 to 1988 the P content would have followed the dotted line (Figure 6). Expected and observed cycling differ mainly during summer and fall. Before 1986 the P content of the lake increased more or less steadily from June/July to the end of spring overturn, compensating for P losses of the previous stratification period. Since 1986, this accumulation during winter has hardly been observed. A mass-balance calculation (Table 9) suggests that net sedimentation during fall and winter has increased three-fold. Of course, this result depends on the external input I_D and would be incorrect if the external loading had been overestimated. However, it seems relatively unlikely that the P load has decreased by a factor of 2, since it was last determined in 1986 by Marti *et al.* (1987). Thus, the postulated increase in net sedimentation of P during winter 1986 and 1987, which also is supported by an increased P content in recent sediments, seems to be real. Thus, it seems that increased hypolimnic O_2 concentration had little effect on P release from sediments during summer, but in the third and fourth year after the onset of lake aeration and oxygenation P retention increased by about 3 tons during winter [(1.8 – 0.5 mg m^{-2} day^{-1}) · 150 days · 14.4 · 10^6 m^2; see Table 9].

In other words, release of P from sediments seemed not to respond on increased hypolimnetic O_2 concentrations in spring, but [O_2] and P release seemed to be related in fall and winter. It follows from Figure 5 that regardless of whether there was sufficient O_2 at the lake bottom ([O_2] = 5.7 mg L^{-1}), black ferrous sulfide formed just 1.5 mm below the sediment surface. Dense mats of the microaerophylic sulfur bacterium *Beggiatoa* indicated that at the sediment surface both H_2S and traces of O_2 were present. According to Jørgensen and Revsbech (1983), *Beggiatoa* reaches maximum density at O_2 concentrations of about 0.2 mg L^{-1}. Thus, the presence of *Beggiatoa* indicated that the sediment surface was nearly anoxic, although 10 cm above, in the overlying water, O_2 concentration reached 5.7 mg L^{-1}, and 2 m above the sediment it even reached 7.3 mg L^{-1}. In other words, the sediment surface may become anoxic even if the water quality objective for O_2 ([O_2] $\geq$ 4 mg L^{-1}) is fulfilled. Jørgensen and Revsbech (1983) recently demonstrated in laboratory experiments using microelectrodes that the O_2 concentration at a sediment surface was distinctly lower than in the surrounding water. Between the well-mixed overlying water and the sediment/ water interface, a steep concentration gradient was observed. The thickness of the so-called diffusive or stagnant boundary layer which separates the well-mixed water from the nonturbulent sediment depends on the hydrodynamics of the "mixed zone" as well as on the roughness of the sediment surface. Within this layer, transport of dissolved substances is only subject to molecular diffusion.

In other words, any water quality objective that does not take into account the O_2 demand of the sediment and the average thickness of the stagnant boundary layer cannot guarantee that the sediment surface will remain oxic. According to Fick's first law, the diffusive flux of a substance is proportional to the diffusion coefficient and the concentration gradient within the considered layer. For obvious reasons, the concentration gradient in the stagnant boundary layer cannot exceed [O_2]$_{Hypo}$/z, where [O_2]$_{Hypo}$ is the O_2 concentration in the overlying water and z is the thickness of the stagnant layer. If the O_2 demand equals or exceeds this maximum possible flux, then the sediment surface becomes anoxic, even if the water quality objective for O_2 in the sediment-overlying water is fulfilled. From this discussion it becomes clear that although an "anoxic system" contains no measurable O_2, it might consume O_2 at a very high rate.

Fish eggs exposed in winter 1987/88 and 1988/1989 at a depth of 8 m on the sediment surface of Lake Sempach failed to develop, although O_2 concentration in the water was close to saturation. If, however, they were kept separated from the sediment, development was not impeded (Ventling 1989). This observation indicates that in eutrophic lakes the sediments rich in organic material are somehow "toxic" to fish eggs. Either the O_2 concentration at the sediment surface is not sufficient, or the burden of toxic substances released (e.g., NH_4^+, NO_2^-, H_2S) is too great for proper development. Thus, oxygenation does not reverse all adverse effects of eutrophication on natural fish populations.

From flux-chamber measurements and from mass-balance calculations, it is known that O_2 consumption of Lake Sempach sediments is approximately 0.5 to 1 g m^{-2} day^{-1}. If this flux is maintained by molecular diffusion, then at a molecular diffusion coefficient of $1.4 \cdot 10^{-5}$ cm^2 sec^{-1} (Broecker and Peng 1974) the concentration gradient must be approximately 4 to $8 \cdot 10^3$ g m^{-4}. Assuming an O_2 concentration of 0.2 mg L^{-1} at the solid/water interface and an O_2 concentration of 5.7 mg L^{-1} at the upper edge of the diffusive layer, an average thickness of 0.7 to 1.3 mm can be calculated for this boundary layer.

Thus, the existence of a very thin boundary layer can explain why the solid sediment surface was nearly anoxic although O_2 concentration in the overlying water was 5.7 mg L^{-1}. At a depth of 85 m (2 m above the sediment surface) O_2 concentration was often as much as 4 mg L^{-1} greater than in the water immediately overlying the sediment (Figure 6). From this it is concluded that even O_2 concentrations of nearly 10 mg L^{-1} at a depth of 85 m cannot safely prevent the sediment surface from becoming anoxic. Since O_2 concentrations this great rarely occur during summer stagnation, it becomes clear why P release rates have not decreased during spring and summer, regardless of increased O_2 concentrations in the hypolimnion.

It is more difficult to explain why, since 1986, oxygenation of the hypolimnion has seemingly decreased the release of P during fall and winter. Since O_2 penetrates the sediment at the maximum to a few millimeters, it is mainly used to mineralize easily decomposable, recently settled organic material and to oxidize reduced substances invading the oxic layer from deeper sediment strata. It seems reasonable to assume that the production of (and thus the supply of the oxic layer with) reduced, mobile substances like H_2S, NH_4^+, or CH_4 is more or less constant throughout the year. However, the supply of the sediment surface with freshly settled organic material follows a strong seasonal pattern (Figure 4).

At the beginning of the stagnation period, after sedimentation has been nearly interrupted for several months, the sediment surface is impoverished with respect to easily oxidizable organic material. As a consequence, the number of heterotrophic bacteria is most likely small during this part of the year. Increased sedimentation and, hence, increased supply of the surface with easily degradable organic substances in spring likely causes bacterial numbers to grow and, therefore, O_2 consumption to increase. As a consequence, O_2 demand eventually exceeds the maximum possible supply. The sediment surface becomes anoxic even in the oxygenated lake. Oxygenation does not seem to influence the release of P. Due to a continually increasing number of bacteria, the rate of decomposition of organic material eventually exceeds its supply. Consequently, substrate concentration, bacterial density and activity, and the demand of the sediment surface for electron acceptors decreases. As O_2 consumption per unit of sediment volume decreases, O_2 will penetrate deeper into the sediment and the probability will increase that an oxic sediment surface can reestablish. If the hypolimnion is artificially enriched with O_2, then in fall and winter O_2 supply to the sediment surface may be sufficient to maintain an oxic surface layer, capable of trapping SRP released from deeper sediment strata. However, under "natural" conditions, i.e., without artificial O_2 supply, the hypolimnion becomes anoxic in May/June and hypolimnion and sediment surface remain permanently anoxic until the turnover reestablishes oxic conditions in the hypolimnion at the end of the year. Under these conditions the sediment releases P permanently, independent of the seasonal variation of O_2 demand by the sediment.

After 5 years of experience with aeration and oxygenation of Lake Sempach, it can be concluded that these lake manipulations had little if any effect on the P release from sediments during spring and early summer. However, they seemed to increase P retention of sediments during fall and winter. Future observation will show if this trend will continue or if it was only a temporary effect. In any case, when the positive effect of aeration/oxygenation (additional elimination of some 3 tons P year⁻¹) is compared with the annual costs of these manipulations (about \$150,000 U.S.), it must be concluded that combined aeration and oxygenation is a very cost-intensive measure to restore this lake.

The laboratory experiments with artificial sediments predicted that oxygenation decreased the release of P drastically. At oxygen concentrations of 4 mg L⁻¹ in the overlying water, about 90% of the settled P was buried in the sediment. If the results of this experiment are extrapolated to the lake, using the external P loading (14.3 tons year⁻¹) and P export (4.6 tons year⁻¹) estimates of Marti *et al.* (1987) and the gross sedimentation estimated earlier (34 tons year⁻¹), aeration/oxygenation would at least in an initial phase drastically decrease the P content of the lake (Figure 12). A comparison of

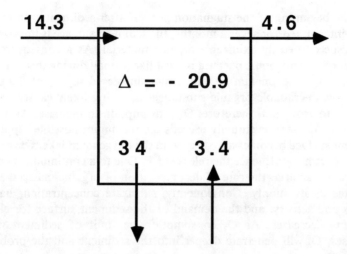

Figure 12.   P mass balance for Lake Sempach, based on P retention measured in the laboratory experiment (values expressed as tons year^{-1}).

this prediction with reality (Figure 6) demonstrates that the laboratory experiment overestimated the positive effect of oxygenation drastically.

Although we do not yet completely understand why the laboratory systems behaved so differently from the lake, this contradiction illustrates the danger of uncritical extrapolation from laboratory results to field conditions. However, in agreement with earlier findings of Tessenow (1972), we conclude from our study that the release of P from sediments is not triggered by a certain minimum O_2 concentration (Figure 6) but rather depends on the O_2 demand of the sediment relative to its supply from the water. If we thus assume that in the laboratory experiment, due to intensive bubbling inducing highly turbulent conditions in the sediment-overlying water, O_2 was supplied at a much faster rate to the sediment surface than at the bottom of Lake Sempach, then we could understand why, in the laboratory, P retention of the sediment exceeded that of Lake Sempach. However, this argument cannot explain the relatively high P retention of anoxic laboratory systems.

5.2 Effects of Mixing on Nutrient Fluxes Across the Sediment/Water Interface

It is generally assumed that *in situ* experiments with flux chambers are much closer to the "real world" than laboratory experiments with "undisturbed" or artificial cores (Gächter *et al.* 1988). The flux chamber permits investigation of flux rates through an undisturbed, natural sediment surface at the ambient temperature and pressure. Thus, this technique overcomes

Table 10. Comparison of Flux Rates [Release (+) and Uptake (−)] Calculated from Mass Balances on Lake Sempach with Average Flux Rates Obtained from Flux-Chamber Measurements[a]

Period	Mn	Fe	DIC	NO₃	NH₄	DOC	Ptot	O₂
05.8.85–28.10.85	−3.3	3.0	133	22	−4	115	6	−790
21.7.86–15.09.86	—	—	392	44	—	—	16	−528
25.5.87–13.10.87	5.4	1.2	460	−5	4	−100	5	−920
20.6.88–10.10.88	−1.8	−0.4	267	−12	−2	+40	7	−1200
Flux chamber	207	44	392	−58	102	421	26	−514

[a]All rates are expressed as mg m⁻² day⁻¹. Averages for the flux chamber are taken from Table 5.

some of the problems with sediment cores brought to the laboratory. It is evident that in artificial cores, concentration profiles and, thus, fluxes are different from those in nature. Even in so-called "undisturbed" sediment cores brought to the laboratory, gas bubble formation induces transport of interstitial water and, thus, changes its concentration profile. The flux-chamber technique overcomes these difficulties to a large extent and, thus, seems to measure fluxes more realistically.

In order to compare flux rates obtained from flux-chamber measurements with flux rates estimated from mass-balance calculations of the undisturbed lake, several assumptions and estimates were made: 1) the "hypolimnion" was arbitrarily defined as the water between the 40-m depth and the bottom; 2) for the period of maximum P accumulation in the hypolimnion (Figure 6), average daily accumulation or loss rates of various chemical species in the hypolimnion were estimated; and 3) the sediment surface was assumed to be the main sink or source for these species. Thus, observed changes in hypolimnetic content were related to the sediment area encompassed in the hypolimnion.

Mn, Fe, NH_4^+, DOC, and P_{tot} exhibited much larger fluxes in the flux chamber than in the open lake (Table 10). Regarding the major electron acceptors, less O_2 but more NO_3^- was consumed by sediments enclosed by the flux chamber.

For several reasons, results obtained from the flux-chamber experiments seem to be unrealistic:

1. If resuspension is neglected, then the average PON flux to the sediment is approximately 40 mg m⁻² day⁻¹ (Table 2). With increasing sediment depth, the N content of the sediment decreases by about a factor of 2 to 3 (Table 4). This means that average N losses from the sediment are on the order of 20 to 30 mg N m⁻² day⁻¹. This is 3 to 4 times less than the average NH_4 flux observed in the flux-chamber experiments.

2. Sediments gained an average of 214 mg POC m⁻² day⁻¹ (Table 2). In the sediment, the POC concentration decreases with increasing sediment age to

about half or one third of its original value (Table 4). Assuming steady state in the sediment profile, about 100 to 140 mg m^{-2} day^{-1} are either mineralized to CO_2, reduced to CH_4, or released as DOC. Flux-chamber measurements, however, resulted in an average release of 420 mg DOC m^{-2} day^{-1} and 390 mg DIC m^{-2} day^{-1}. This again indicates that results obtained from flux-chamber experiments significantly deviate from average natural flux rates.

3. At a depth of 81 m, the downward flux of particulate Mn and, thus, the net gain of sediments is approximately 40 mg m^{-2} day^{-1}. At the lake bottom this flux increased to 107 mg m^{-2} day^{-1} (Table 3), indicating that about 67 mg m^{-2} day^{-1} are continuously recycling either by dissolution and reprecipitation or by resuspension from the sediment to the water and back to the sediment. Flux-chamber measurements suggest an average release rate of 207 mg m^{-2} day^{-1}. If we subtract from this rate 67 mg m^{-2} day^{-1} for immediate reprecipitation, then the lake water should gain 140 mg m^{-2} day^{-1}. However, no significant accumulation of Mn was observed in the hypolimnion.

From this discussion it becomes clear that fluxes of Mn, Fe, NH_4^+ and DOC in the flux chamber reflect transient, rather than undistorted, steady-state fluxes. As mentioned previously, flux chamber and lake might differ in the degree of turbulence in the sediment-overlying water. For example, increased turbulence in the flux chamber would cause the stagnant boundary layer to shrink and, thus, fluxes through this layer to increase. This could explain the observed higher fluxes for Mn, Fe, NO_3^-, NH_4^+, DOC, and P_{tot}, but it could not explain why fluxes of oxygen would be smaller. We therefore suggest that in the flux chamber the thickness of the stagnant boundary exceeded its average thickness under natural conditions. As schematically shown in Figure 13, this results in a smaller O_2 concentration gradient in the stagnant boundary layer and, hence, in a decrease of the O_2 supply to the sediment. The sediment surface becomes anoxic and, as a consequence, NH_4^+ is no longer nitrified. This leads to higher NH_4^+ and lower NO_3^- concentrations at the sediment surface and thus explains the increase in NH_4^+ release and apparent increase in NO_3^- consumption. An anoxic sediment surface favors reduction and release of Fe and Mn and release of P, and/or it hinders precipitation of Fe, Mn, and P, invading the sediment surface from deeper sediment strata. This explains why fluxes of these three elements are accelerated in the flux chamber. In addition, instantaneous anoxia in a previously oxic microlayer kills all obligate aerobic organisms. Their autolysis and the fact that these organisms no longer mineralize DOC released from deeper sediment layers might explain the increased release of DOC. Because anaerobic decomposition typically has a lower yield than oxic decomposition, the postulated increase in thickness of the diffusive boundary layer resulted in an increased release of DIC.

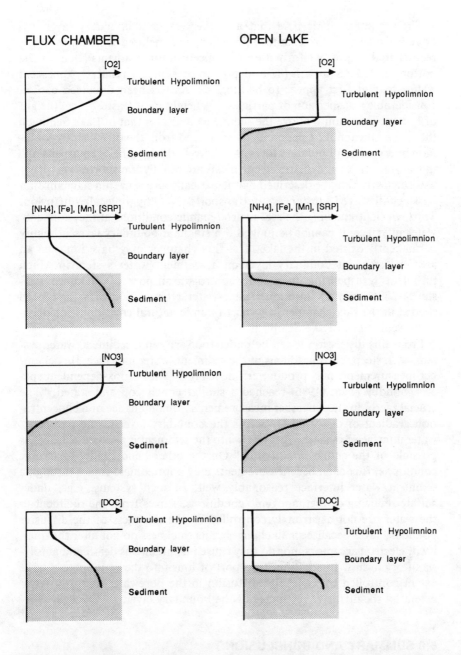

Figure 13.   Effects of increased thickness of the stagnant boundary layer at the sediment/water interface.

The argument made here that a thicker stagnant boundary layer was established in the flux chamber (in which the enclosed water was slowly recirculated to prevent formation of concentration gradients) than at the bottom of Lake Sempach (87-m depth, in presumably relatively quiescent water) may at first appear to be illogical. As has been discussed earlier, considerable resuspension of particles to a height of at least 27.5 cm (height of bottom-trap mouth above the sediment) indicates that in Lake Sempach the sediment cannot be permanently separated from the turbulent hypolimnion by a stagnant boundary layer. A relatively thick "zone of resuspension" in which particles and dissolved chemicals are rapidly transported via turbulent convection can be described mathematically as a very thin stagnant film in which dissolved substances are transported only by molecular diffusion. Thus, without knowing the real hydrodynamic conditions at the sediment/water interface, it cannot be judged if a stagnant boundary layer in reality permanently existed in the lake. The flux chamber may have imposed an artificial stagnant boundary layer on a sediment/water system in which turbulent resuspension frequently controls transport of dissolved substances. Contrary to intuition, faster water recirculation may have been needed in the flux chamber in order to mimic natural conditions for deep-water sediment in Lake Sempach.

From this discussion it can be understood why in a sediment/water system, with steep redox gradients at the sediment/water interface, flux chamber measurements may produce erroneous results for redox-dependent species. Sundby *et al.* (1986) reached a similar conclusion and stated, "The crucial factor in flux chamber measurements is maintenance of a concentration gradient of dissolved O_2 across the boundary layer at the sediment/water interface so that the flux of O_2 into the sediment is as representative as possible of the actual *in situ* flux." On the other hand, under anaerobic conditions, flux chamber measurements may approximate fluxes through a sediment/water interface reasonably well. In such systems, e.g., under sulfate-reducing or methanogenic conditions, fluxes from the sediment to the water are not as strongly controlled by the thickness of the diffusive boundary layer because small changes in its thickness do not alter the quality of electron acceptors, nor do they cause instantaneous death and autolysis of organisms. In addition, transport of ions into the water is to a large extent controlled by molecular diffusion in the porewater and to a lesser extent by the diffusive resistance of the boundary layer.

6.0 SUMMARY AND CONCLUSIONS

Lake Sempach is a highly eutrophic lake in central Switzerland. In its deepest part, it became anoxic during late summer in previous years. During the past 4 years, it has been aerated during winter to increase mixing, and O_2 has been introduced into its hypolimnion during summer at a rate of 3 tons day^{-1}.

The goals of these measures have been to:

- make the whole water body habitable for O_2 dependent organisms
- permit natural propagation of fish
- diminish P release from sediments and, thus, decrease the trophic state of the lake

Although the water quality objective for oxygen ($[O_2] \geq 4$ mg L^{-1}) could be reached, fish eggs exposed on the sediment surface did not develop, and release of P did not decrease during summer. However, aeration and oxygenation seemed to increase P retention of sediments during fall and winter. Future observations will show if this trend will continue or if it was only a temporary effect.

The results presented suggest that much higher O_2 concentrations should be maintained in the hypolimnion in order to decrease P release during summer. Since liquid oxygen is very expensive and O_2 saturation cannot be exceeded by enhanced mixing during overturn, such high concentrations cannot be reached either for economical or for physical reasons.

It is concluded that aeration/oxygenation is not suitable to overcome all symptoms of eutrophication related to low O_2 concentrations. In other words, oxygenation is no quick solution to fight phosphorus pollution.

The O_2 concentration at the sediment surface of lakes does not only depend on the O_2 concentration of the sediment-overlying water, but also on the oxygen demand of the sediment and on the transport of oxygen from the hypolimnion to the sediment. If the goal is to maintain an oxic sediment surface or to permit undisturbed development of fish eggs, then O_2 demand of the sediment, eddy diffusion in the deep hypolimnion, and the thickness of the stagnant boundary layer should as well be taken into account.

Comparison of results obtained from mass-balance calculations with results obtained from flux-chamber measurements suggests that changes in the structure of turbulence in the sediment-overlying water strongly affect transport of redox-dependent species through the sediment/water interface. From this, it is hypothesized that the thickness of the stagnant boundary layer, separating the quiet sediment from the turbulent water, is of great influence for chemical species undergoing redox reactions.

REFERENCES

Ambühl, H. & H. Bührer, 1975. Zur Technik der Entnahme ungestörter Grossproben von Seesediment: Ein verbessertes Bohrlot. *Schweiz. Z. Hydrol.* 37: 175–186.

162   SEDIMENTS: CHEMISTRY AND TOXICITY

Boström, B., M. Jansson & C. Forsberg, 1982. Phosphorus release from lake sediments. *Arch. Hydrobiol. Beih.* 18: 5–59.

Broecker, W. S. & T. H. Peng, 1974. Gas exchange rates between air and sea. *Tellus* 26: 21–35.

Gächter, R. & W. Stumm, 1979. *Gutachten über die Sanierungsmöglichkeiten des Sempachersees.* Auftragsnummer 4564. EAWAG internal report.

Gächter, R. & A. Mares, 1985. Does settling seston release soluble reactive phosphorus in the hypolimnion of lakes? *Limnol. Oceanogr.* 30: 364–371.

Gächter, R. & W. Stumm, 1987. *Kurzbericht über die Resultate der limnologischen Überwachung des Sempachersees im Zeitabschnitt Januar 1986 bis Mai 1987.* Auftrag Nr. 4691. EAWAG internal report.

Gächter, R., J. S. Meyer and A. Mares, 1988. Contribution of bacteria to release and fixation of phosphorus in lake sediments. *Limnol. Oceanogr.* 33: 1542–1558.

Jørgensen, B. B. & N. P. Revsbech, 1983. Colourless sulfurbacteria, *Beggiatoa* spp. and *Thiovulum* spp., in O_2 and H_2S microgradients. *Appl. Environ. Microbiol.* 45: 1261–1270.

Lee, G. F., W. Sonzogni & R. Spear, 1977. Significance of oxic vs. anoxic conditions for Lake Mendota sediment release. In H. Golterman (Ed.), *Interactions between Sediments and Fresh Water.* Dr. W. Junk, The Hague: 249–307.

Lemmin, U. & D. M. Imboden, 1987. Dynamics of bottom currents in a small lake. *Limnol. Oceanogr.* 32: 62–75.

Marti, D., D. Imboden & W. Stumm, 1987. *Sanierung des Sempachersees: Auswertung der Zuflussuntersuchungen Messperiode Mai 1984 bis April 1986.* Auftrag Nr. 4691. EAWAG internal report.

McQueen, D. J., D. R. S. Lean & M. N. Charlton, 1986. The effects of hypolimnetic aeration on iron-phosphorus interactions. *Water Res.* 20: 1129–1135.

Mortimer, C. H., 1971. Chemical exchanges between sediments and water in the Great Lakes—Speculations on probable regulatory mechanisms. *Limnol. Oceanogr.* 16: 387–404.

Ryding, S. O. & C. Forsberg, 1977. Sediments as a nutrient source in shallow polluted lakes. In H. Golterman (Ed.), *Interactions between Sediments and Fresh Water.* Dr. W. Junk, The Hague: 227–235.

Sigg, L., M. Sturm & D. Kistler, 1987. Vertical transport of heavy metals by settling particles in Lake Zürich. *Limnol. Oceanogr.* 32: 112–130.

Stadelmann, P., 1988. Der Zustand des Sempachersees. *Wasser Energ. Luft* 80: 81–96.

Sundby, B., L. G. Anderson, P. O. J. Hall, Å. Iverfeldt, M. M. Rutgers van der Loeff & S. F. G. Westerlund, 1986. The effect of oxygen on release and uptake of cobalt, manganese, iron and phosphate at the sediment-water interface. *Geochim. Cosmochim. Acta* 50: 1281–1288.

Tessenov, U., 1972. Lösungs-, Diffusions- und Sorptionsprozesse in der Oberschicht von Seesedimenten. *Arch. Hydrobiol. Suppl.* 38: 353–398.

Ventling-Schwank, A. & R. Müller, 1989. *Survival of Coregonid (Coregonus sp.) Eggs in Lake Sempach, Switzerland.* Submitted to Verhandlungen Internationale Vereinigung für Limnologie Congr. in Munich, Germany.

CHAPTER 6

The Biomethylation and Cycling of Selected Metals and Metalloids in Aquatic Sediments

Frank M. D'Itri

1.0 INTRODUCTION

Since the 1950s Minamata, Japan has been the bellwether of research on heavy metal contamination of the environment because human disabilities and deaths resulted from the consumption of seafood contaminated with methylmercury (CH_3Hg^+) (D'Itri 1972, D'Itri & D'Itri 1977). The initial hypothesis that the mercury (Hg) might have been methylated in the environment was discarded in 1967 when CH_3Hg^+ was found to have been produced industrially. However, it was soon discovered that Hg could, in fact, be biomethylated and accumulated in the environment as well (Jensen & Jernelov 1969). Since then, methylation and demethylation by both aerobic and anaerobic bacteria have been identified. Less research has been conducted on other metals and metalloids such as arsenic (As) and lead (Pb); however, they also can be methylated, bioaccumulated, and trans-methylated in some circumstances, although they do not appear to biomagnify through the food chain the way Hg does.

1.1 Sedimentation

In an aquatic ecosystem, metals usually enter the sediments by two basic mechanisms. They either bind with or adsorb onto inorganic and organic ligands. Some dissolved organic ligands complex with the metals and precipitate directly into the sediments. Other naturally occurring organic compounds react with the metal ions and remain in solution until they are adsorbed onto suspended matter which then precipitates to the sediments. These processes are affected by the concentration, size, and valence of metal ions on one hand and the redox potential, ionic strength, and pH of the solution on the other. As it occurs, metal-containing colloids may also be destabilized. Depending on redox potential, the complexed or adsorbed organic ligands alter the proportions of oxidized and reduced forms of the metals. This affects their relative availability to aquatic biota and may mitigate or enhance toxicity.

2.0 MERCURY

Mercury can exist in many forms. Elemental ($Hg°$) and mercuric ions (Hg^{2+}) are the predominant natural forms in the atmosphere and water (Kudo et al. 1982), while cinnabar (HgS) is commonly found in mineralized soils and anaerobic sediments. An important attribute of Hg in the ground state is that it exhibits a sufficiently great vapor pressure to be transported as a gas in the atmosphere. Hg is transported to aquatic ecosystems via surface runoff and atmospheric precipitation. Natural sources such as geological weathering contribute about the same order of magnitude of Hg to the environment as do the direct and indirect sources related to human activities (D'Itri et al. 1978). The maximum permissible concentration allowed by the United States Environmental Protection Agency (U.S. EPA) under its drinking water standards is 2 μg Hg/L.

Concentrations of Hg were measured in sediments long before they were identified as an environmental problem more than three decades ago, and these data provide information on background concentrations in many matrices. Historically, depending on their location, background concentrations of Hg in sediments have ranged between 10 and 200 μg Hg/kg. The vast majority of contemporary analyses of sediments indicate that all aquatic systems have received some Hg contamination, and the rate has increased during the past century. Among sites that have been measured, the total concentrations of Hg have usually been from 5 to 10 times greater than background.

Concentrations of Hg in contaminated sediments range from less than 1 mg Hg/kg (dry weight) to 2010 mg Hg/kg (dry weight) in Minamata Bay, Japan, at the confluence of the discharge channel from the Chisso factory which released CH_3Hg^+ (Kitamura et al. 1959, 1960). Evidence of the environmental accumulation of Hg released due to human activities is presented in Figure 1.

The total Hg concentration in a lake is a function of: 1) the Hg content in the soils of the watershed, 2) the efficiency of the transport of the Hg from the soils into the lake, and 3) the humic content of the water in the lake (Håkanson et al. 1988). Hg deposited onto watersheds by precipitation and runoff is usually bound into the humus layer of soils and accompanies this organic material into waterways. This Hg is very stable and, in contrast to other heavy metal ions, seems to increase in stability with acidity over the pH range from 7 to 5 (Jernelov & Johansson 1983). As this organic matter

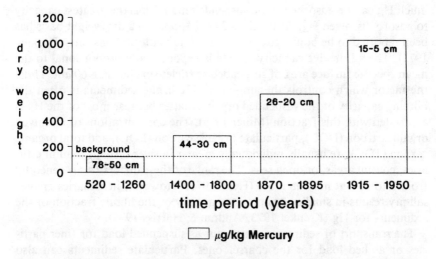

Figure 1. Total mercury concentrations in cored sections of sediments from Lake Winde-mere, England. (After Aston *et al.* 1973.)

settles to the bottom of an aquatic ecosystem, the adsorbed Hg is carried along, adding to what is already in the sediments (Jernelov & Lann 1973). Consequently, the Hg concentration in the water column is a function of the rate of sedimentation. It decreases as the sedimentation rate increases.

Morphological and chemical factors play important roles in determining the rate of adsorption and sedimentation of Hg in aquatic systems. If the amount attached to the organic matter increases, so do the concentrations in the sediments of lakes and streams as long as the biological/chemical parameters do not change substantially. Hg forms complexes with humic materials quite readily. Water-soluble organic complexing agents such as the humates and fulvates can chelate soluble Hg species to form both water-soluble and insoluble complexes (Ogner & Schnitzer 1970). The latter pre-cipitate directly from solution into the sediments. Larger amounts of Hg are adsorbed on humus at low pH. At higher pH values, more Hg is adsorbed by the mineral fraction (Rissanen & Miettinen 1972, Krenkel 1973).

In aerobic environments organic matter oxidizes Hg^o to Hg^{2+} (Colwell & Nelson 1975), while Hg^{2+} can be reduced to Hg^o in anaerobic environments and especially in the presence of humic acid (Alberts *et al.* 1974). Because Hg^o has a low solubility it settles directly to the sediments, whereas Hg^{2+} complexes with dissolved and particulate matter, especially organic matter (Kudo *et al.* 1982). It comes from both autochthonous and allochthonous sources, which may keep Hg in the water column for a longer period of time.

The adsorptive capacity and surface area of particles determine how much Hg can be adsorbed. Fine suspended matter has the greatest capacity to adsorb dissolved Hg. As much as 34 mg Hg/kg on a dry weight basis has been reported to be bound to some particulate matter (Cranston & Buckley 1972). The adsorptive capacity for Hg is exponentially proportional to the mean specific surface area of suspended particles smaller than 60 μm. Thus, the factor which controls the amount of Hg in the sediments is often the binding capacity of the suspended organic matter because most of the Hg is associated with this fraction (Miller 1975). The concentrations of dissolved organic carbon (DOC), particulate organic carbon (POC), and total organic carbon (TOC) are usually individually greater in river systems than in estuaries, coastal areas, or open seas. The range of these differences is generally from 3 to 15 but may be more (Head 1976). However, in estuaries greater salinity results in smaller complexing capacity of the humic fraction of the sediments for Hg (Krenkel 1973, Andren & Harriss 1973).

The transport of sediments may occur as suspended load for finer particles or as bed load for the coarser ones. Particulate sediments can also transport Hg. This transport is greatest in the smallest size fractions of organic matter. These sizes can pass through lake basins to a greater extent. Therefore, lakes with longer turnover times allow the smaller particles with greater concentrations of heavy metals such as Hg more time to settle to the bottom. Thus, the sediments of these lakes are more likely to exhibit greater enrichments of Hg than lakes with shorter turnover times (Cranston & Buckley 1972).

When it reaches the bottom sediments, Hg is strongly held by the following binding mechanisms in natural waters: 1) sorption on hydrated ferric oxide, 2) surface sorption and/or ion exchange with naturally occurring mineral ion exchangers such as montmorillonite, or 3) sorption and/or chemical combination with organic material such as peat and especially sulfur-containing matter. The distribution of Hg is correlated strongly with organic carbon, clay, iron, phosphorus, and sulfur content of particulate sediments in Lake Ontario and with organic carbon, sulfur, and iron in Lake Huron (Thomas 1972, 1973).

Over 90% of the Hg in most lake systems is found in the sediments (Faust & Aly 1981), which leaves only a small fraction available to become associated with the biota. However, the inputs of Hg into a watercourse are not necessarily correlated with the concentrations in the sediments. Hg concentrations in the soils surrounding an oligotrophic lake and a eutrophic lake in a remote, relatively undeveloped region of Northern Michigan were approximately the same; however, the eutrophic lake sediments contained from 0.30 to 1.25 mg Hg/kg (dry weight). This was substantially more than was observed in the sediments of the oligotrophic lake, which had concentrations of only 0.03 to 0.12 mg Hg/kg (dry weight) (D'Itri et al. 1971).

Humic and fulvic compounds have an essential role in the transport of

Hg between sediments and water because they reduce the release of Hg^{2+} from the sediment into the water column (Alberts *et al.* 1974, Miller 1975). Because there are more humic than fulvic materials, they are more important in trace metal transport and retention in sediments (Jenne 1976) even though the fulvic acids are especially efficient at complexing Hg and keeping it in solution (Schnitzer & Kerndorff 1981), primarily due to their smaller molecular masses, greater density of functional groups, and much greater solubility relative to the humic-type materials.

Whereas soluble Hg complexes are adsorbed onto organic or inorganic particulates and removed by sedimentation in aerobic waterways, as the sediments become increasingly anaerobic the precipitated Hg compounds usually are converted into mercuric sulfide (HgS), which reduces the possibility that they will be recycled into the overlying water (Fagerstrom & Jernelov 1971). This removal mechanism is absent from aquatic ecosystems that are aerobic year round. Under anaerobic conditions most Hg that is not present as HgS is bound to organic matter. Both forms are very stable and not likely to release Hg^{2+} to the aqueous phase in any significant quantities. The amount transported within the sediments is also small for the same reason (Hakanson 1974, Miller *et al.* 1975, Kudo *et al.* 1975, Reimers *et al.* 1975, Jernelov *et al.* 1975).

Little is known about the properties of insoluble HgS, which composes a large fraction of the mercurials in the environment. It has been proposed that HgS (solubility product constant Ksp = 10^{-53}) is formed as the precipitate of a reaction between Hg^{2+} and S^{2-} ions (Fagerstrom & Jernelov 1972). The Ksp is so small in this reaction that S^{2-} ions may be appropriated from other sulfides such as ferrous sulfide (FeS) or via an interaction with the equilibria of organometallic compounds. The equilibrium concentration of Hg^{2+} resulting from the dissociation of HgS is greater than would be expected from the theoretical solubility product constant due to the effects of competing complexing agents, redox potential, pH, and temperature (Fagerstrom & Jernelov 1971).

Under aerobic conditions HgS can be oxidized to sulfate (SO_4^{2-}), which is much more soluble, so that Hg^{2+} becomes available to be methylated by microorganisms. The rate of oxidation by physicochemical processes is very slow and depends on the redox potential. Direct biological, enzymatic oxidation may lead to a faster release of Hg^{2+} ions. This suggests that the oxidation of HgS is the rate-determining step for the transformation of Hg^{2+} ions to CH_3Hg^+ (Jensen & Jernelov 1969, Fagerstrom & Jernelov 1971). Until it is oxidized, Hg bound as HgS demonstrates little methylation even under aerobic conditions. The rate of methylation in aerobic sediments has been found to be 100 to 1000 times less for HgS than for mercuric chloride ($HgCl_2$). However, a small amount of CH_3Hg^+ was found to be produced from HgS which was incubated with aquarium sediments. This suggests that some mechanism for solubilization existed (Fagerstrom & Jernelov 1971). Even under ideal conditions, the greatest conversion rate of inorganic Hg to CH_3Hg^+ has been estimated to be less than 1.5% per month (Jensen & Jernelov 1969). However, Yamada & Tonomura (1972a, 1972b) reported no CH_3Hg^+ produced from HgS either by chemical methylation with methylcobalamin or by microbial systems which had great methylating activities.

Because HgS, like other mercurials, is labile in light, it may be converted photochemically into water-soluble mercurials available for methylation. CH_3Hg^+ has been synthesized by irradiating with sunlight aerobic sediments which were contaminated with Hg (Fujiki & Tajima 1973). Pure HgS has also been photochemically solubilized and methylated in aqueous solutions (Akagi *et al.* 1977).

CH_3Hg^+ can also react with hydrogen sulfide (H_2S) and be lost from solution in the form of a volatile methylated Hg complex which is probably dimethylmercury [$(CH_3)_2Hg$] (Rowland *et al.* 1977, Craig & Bartlett 1978). This mechanism involves the disproportionation of CH_3Hg^+ by H_2S (Equation 1).

$$2CH_3Hg^+ + H_2S \longrightarrow (CH_3)_2Hg + HgS + 2H^+ \tag{1}$$

$(CH_3)_2Hg$ is volatile and can be released from water to the atmosphere, but then it is unstable because the Hg-carbon bonds are susceptible to homolytic cleavage by UV light. Therefore, it decomposes to yield methane, ethane, and Hg° (Equation 2).

$$(CH_3)_2Hg \rightarrow Hg^\circ + 2CH_3^\bullet \longrightarrow H_3C\text{-}CH_3 + CH_4 + Hg^\circ \tag{2}$$

Although much of the Hg that is not water soluble or volatile remains stabilized in the sediments where it is available to be methylated, sedimentation, especially in river systems, may be reversed and the process repeated if the sediments are resuspended. This recycling may continue until the sediments are transported out of the system or until they become a permanent part of the bottom sediments. During resuspension, the finer particles may aggregate to form larger particles (flocculation), and the larger particles may disaggregate as well. The circumstances and rates depend on the turbulence, sediment concentration, and other factors. The components of sediments are also dispersed by wave actions, currents, and bioturbation and then are redeposited on the bottom.

3.0 LEAD

Lead, a toxic metal, is found in all surface waters. The maximum permissible concentration allowed by the U.S. EPA under its drinking water standards is 50 μg Pb/L. Atmospheric input is the primary source of Pb to aquatic sediments (Edington & Robbins 1976, Kemp & Thomas 1976). Besides emissions from natural erosion and volcanic activity, internal combustion engines, especially in the past, have released five alkyllead compounds. These are: 1) tetramethyllead, $(CH_3)_4Pb$; 2) trimethyllead, $(CH_3)_3Pb^+$; 3) dimethyldiethyllead, $(CH_3-CH_2)_2(CH_3)_2Pb$; 4) methyltriethyllead, $(CH_3-CH_2)_3CH_3Pb$; and 5) tetraethyllead, $(CH_3-CH_2)_4Pb$ (DeJonghe & Adams 1982). However, chemical reactions in the atmosphere convert these compounds into $PbCO_3PbO_x$, $(PbO)_2PbCO_3 \cdot PbSO_4$, and $PbBrCl$ (Corrin & Natusch 1977). Thus, only small concentrations reach aquatic systems directly. Atmospheric input reaches the land and surface waters by wet and dry deposition, through precipitation and runoff.

The partitioning of Pb between the aqueous and solid phases in aquatic environments is affected by input from diverse sources as well as biological activities and a variety of other factors. These include particle composition, texture, organic and inorganic characteristics of the sediment, temperature, pH, redox potential, and ionic competition. More Pb is retained by solutions of high ionic strength, and Pb complexes by most of the common sulfur-, phosphorus-, oxygen-, and nitrogen-containing ligands. Soluble Pb in the water column can cause significant toxicity to organisms. It may either remain in suspension or become fixed within sediments. This occurs by a variety of physical processes, including precipitation of the mineral phases and adsorption and complexation with organic matter and inorganic mineral components (Agemian & Chau 1977). The dense and sometimes gelatinous communities of algae and associated microconsumers act as filters, removing soluble Pb from the aqueous environment (Gale et $al.$ 1973). Their enrichment is enhanced by both precipitation as metallic coatings and incorporation in crystalline structures and in organic matter. The high polarizability of Pb^{2+} enables it to form strong complexes with polarizable anions. The portion of Pb that is not water soluble sinks to the bottom and becomes part of the sediment. It tends to form compounds of low solubility with major anions in natural waters and so has little effect on the aquatic environment. Pb concentrates primarily in sediments containing large amounts of clay and organic matter.

4.0 ARSENIC

Arsenic is a toxic metalloid which can be found in almost all natural waters. The maximum permissible concentration allowed by the U.S. EPA under its drinking water standards is 50 μg As/L. Seawater contains an average of 3 to 4 μg As/L. The concentrations of As are greater in areas that drain As-rich soils or receive industrial discharges. Concentrations of As in freshwater vary widely depending on the soil type in the drainage basin. They tend to be greater in areas of geothermal activity.

Arsenic is stable in four oxidation states ($+5$, $+3$, 0, -3) depending on the pH and redox potential (eH) of the solution (Lemmo *et al.* 1983). As metal, As° rarely occurs, and As^{3-} only at extremely low E_h values. At the E_h values which are encountered in oxygenated waters, arsenic acid species such as H_3AsO_4, $H_2AsO_4^-$, $HAsO_4^{2-}$, and AsO_4^{3-} are stable. At E_h values typical of mildly reducing conditions, arsenous acid species such as H_3AsO_3, $H_2AsO_3^-$, and $HAsO_3^{2-}$ become stable. Aqueous $HAsS_2$ is the primary species at low pH in the presence of sulfide (Gavis & Ferguson 1972).

The solubility of As in freshwater depends upon many factors. These include oxygen concentration, the presence of sulfur and ferric hydroxide, biological methylation and demethylation, and chemical and physical adsorption onto particulates (Foley *et al.* 1978). These processes frequently remove As from solution, thus preventing large concentrations from remaining in surface waters. Although the efficacy of these self-cleaning processes is variable, As generally decreases to background concentrations in waters within 400 to 1300 m of a specific source of pollution (NAS 1977). Arsenic is transported to the sediments and accumulates there. Sediments in Lake Michigan and Lake Superior contain average concentrations of 15.2 and 3.8 mg As/kg, respectively (Seydel 1972). The greater concentration in sediments from Lake Michigan has been attributed to discharges due to human activities. The variability of concentrations of As among sediment samples from a number of locations has been attributed primarily to different types of sediments and their capabilities to bind As and only secondarily to local pollution sources (Seydel 1972).

5.0 METHYLATION

When metals such as Hg and Pb and metalloids such as As occur as alkylated compounds, they are lipophilic and toxic. Methylation is a key step in their biogeochemical cycling. The cycling, transformation, deposition, and availabilities of the various forms of Hg and other metals and metalloids in the aquatic ecosystem depend not only on their physico-chemical natures but also on the metabolic interactions of the transforming microbes with other microbial species. Modern research into these biological and chemical reactions began over 50 years ago with Frederick Challenger. He identified the mechanism of As methylation and also coined the term "biological methylation" (Challenger 1935). However, most of the information on this process has been collected in the last 20 years. It demonstrates that the methylation of metals and metalloids is a common phenomenon in the aquatic environment. Methylation is considered to be a method whereby microorganisms can detoxify their cells because it changes the solubility, volatility, and toxicity of the element and, in most cases, removes it from the immediate area of the methylating organism.

In biological methylation a chemical bond is formed between a metal ion and a methyl ion (Equation 3).

$$M^{+n} + CH_3^- \longrightarrow CH_3M^{+n-1} \tag{3}$$

The methyl group can theoretically be donated to the metal ion by one of three enzyme systems: 1) methylcobalamin derivatives (vitamin B_{12}); 2) S-adenosylmethionine, and 3) N^5-methyltetrahydrofolate derivatives (Wood & Fanchiang 1979). Of these systems, only methylcobalmin derivatives are capable of transferring the methyl group as a carbanion (CH_3^-) (Figure 2). Two general mechanisms for the transfer of a methyl group from the methylcobalamin moiety to a specific heavy metal have been identified. The Type I reaction involves an electrophilic attack by the metal or metalloid ion on the $Co-CH_3$ bond of the methylcobalamin moiety, which results in a heterolytic cleavage of that bond. This facilitates the transfer of a carbanion to the heavy metal ion in the more oxidized state. Alternatively, the Type II reaction involves a homolytic cleavage of the $Co-CH_3$ bond, forming a methyl radical. In this reaction an ion pair is transferred from the attacking metal ion to the corrin macromolecule, which is the part of the methylcobalamin that sequesters the central cobalt ion (Ridley et al. 1977).

In the absence of methylcobalamin, methylation may also occur through coenzyme systems such as S-adenosylmethionine or chemical reactions involving transmethylation, methyl donation from fulvic acid, or photoionization in the presence of other organic compounds. Of the metals and metalloids being considered here, only As has been determined to be methylated by both methylcobalamin and S-adenosylmethionine (Ridley et al. 1977).

5.1 Mercury

While Hg undergoes a number of chemical and biological transformations in the environment, the primary reactions involve methylation, demethylation, and reduction of Hg^{2+} to Hg° (Figure 3). Despite some disagreement over whether the following transformations are biologically, chemically, or photochemically mediated, current experimental evidence supports the theory that inorganic Hg can be methylated primarily by two mechanisms:

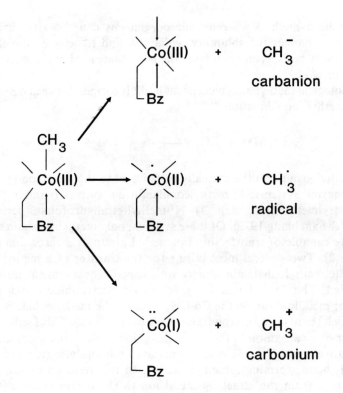

Bz = 5,6-dimethylbenzimidazole

Figure 2. Three theoretical mechanisms which may be involved in the methylation of metals and metalloids. (Reprinted from Ridley *et al.* 1977. With permission.)

1. biologically by microbes and fungi (Jensen & Jernelov 1969, Landner 1971)
2. chemically (abiotically) by at least five means:
   a. by reacting with methylcobalamin (Wood *et al.* 1968, Imura *et al.* 1971)
   b. via a transmethylation reaction if other methylated metal compounds such as biologically synthesized methyl tin or $(CH_3)_4Pb$ species are present (Jernelov 1972, Huey *et al.* 1974)
   c. through UV irradiation in the presence of organic methyl donors (Kitamura & Sumino 1972, Akagi *et al.* 1972)
   d. by reacting with humic and fulvic acid methyl donors (Rogers 1976, 1977, Nagase *et al.* 1982, 1984)
   e. in a mixture of acetaldehyde, Hg^{2+}, and NaCl (Irukayama 1968)

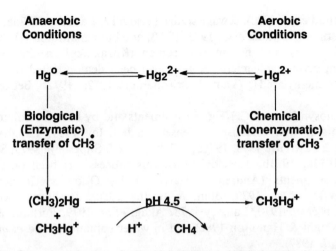

Figure 3. General mechanisms for the biological and chemical synthesis of methylmercury compounds in the environment. (Reprinted from Wood 1971. With permission.)

5.1.1 Biological Methylation

5.1.1.1 Enzymatic. The biological formation of CH_3Hg^+ was first shown to occur in anaerobic organic sediments (Jensen & Jernelov 1967). Initial studies indicated that CH_3Hg^+ was formed from Hg^{2+} by anaerobic microorganisms that depended on the availability of methylcobalamin. However, more recent studies have demonstrated that methylation of Hg occurs principally as an aerobic, microbiological process (Rissanen *et al.* 1970, Wallace *et al.* 1971, Parks *et al.* 1973, Bisogni & Lawrence 1975a, 1975b). Methylation rates for aerobic aqueous systems are greater than those for corresponding anaerobic systems (Bisogni & Lawrence 1973).

Studies of microbial methylation which have been performed with enriched pure cultures may not have direct application to the *in situ* conditions in natural environments. The physiological state and composition of microbiological communities in these cultures are likely to bear little resemblance to those in the natural environment. However, these studies do contribute to understanding the mechanisms of methylation and can demonstrate the presence of a microorganism and its potential methylating activity in a natural medium. However, they cannot predict or assess a microorganism's level of activity or relative contribution to the overall reactions *in situ* in the environment (Wollast *et al.* 1975).

Mercury has been reported to be methylated by microorganisms in many aquatic matrices. Methylation has been observed in the mucus and intestines of fish (Rudd *et al.* 1980), in surface slimes (Furutani & Rudd

1980, Rudd *et al.* 1980), sewage sludge (Bisogni & Lawrence 1975a, 1975b), rat intestines (Rowland *et al.* 1975, 1977), and human intestines (Edwards & McBride 1975), but not in cow rumen (Kozak & Forsberg 1979). In addition, certain soil microbes have also been demonstrated to have the ability to methylate Hg (Yamada & Tonamura 1972a, 1972b, Beckert *et al.* 1974).

The biosynthesis of CH_3Hg^+ in sediments and by bacteria isolated from sediments is well established (Jensen & Jernelov 1969, Langley 1973, Bisogni & Lawrence 1973, 1975a, 1975b, Vonk & Sijpesteijn 1973, Spangler *et al.* 1973a, 1973b, Jernelov 1974); this process has been observed in marine sediments (Andren & Harriss 1973, Olson & Cooper 1974, Berdicevsky *et al.* 1979, Blum & Bartha 1980), freshwater sediments (Rissanen *et al.* 1970, Langley 1973, Akagi *et al.* 1979, Furutani & Rudd 1980, Wright & Hamilton 1982) and in water columns (Rudd *et al.* 1980, Xun *et al.* 1987).

Three methylating coenzymes in biological systems are known to participate in enzymatic biomethylation: S-adenosylmethionine, N^5-methyltetrahydrofolate derivatives, and methylcobalamin (D'Itri *et al.* 1978). While all of these are capable of donating a methyl group, the S-adenosylmethionine and N^5-methyltetrahydrofolate cannot transfer methyl groups to Hg^{2+} ions because both of these coenzymes can only transfer a methyl group as CH_3^+, a carbonium ion (Shapiro & Schlenk 1965, Ridley *et al.* 1977).

Methylcobalamin appears to be the only biological methylating agent capable of transferring a methyl group to the inorganic Hg^{+2} ion. It can transfer methyl groups as a carbanion (CH_3^-) and a methyl radical ($CH_3^{\bullet}$) to produce CH_3Hg^+ and $(CH_3)_2Hg$ under both aerobic and anaerobic conditions (Wood *et al.* 1968, Hill *et al.* 1970, Adin & Espenson 1971, Neujahr & Bertilsson 1971, Schrauzer *et al.* 1971, 1973, Wood 1971, 1974, 1987, Ridley *et al.* 1977). Therefore, methylcobalamin serves as the methyl donor in the bacterial methylation of Hg and other heavy metals (Shapiro & Schlenk 1965, Wood *et al.* 1968, Neujahr & Bertilsson 1971, Wood 1974, 1975). Significant quantities of methylcobalamin may be available in the sediments because it is a common coenzyme in both aerobic and anaerobic bacteria (Wood 1972).

The rate of biological CH_3Hg^+ synthesis is determined primarily by the concentration and form of the available Hg in the aquatic ecosystem as well as the composition of the microbial species and the size and activity of the natural population capable of methylating it. This capability is not restricted to one or a few types of microorganisms but seems to be a widespread process associated with many bacteria.

The physical, chemical, and biological characteristics of aquatic ecosystems also contribute to determining the rates of Hg methylation and its subsequent bioavailability to fishes (Jackson & Woychuck 1980, Park *et al*. 1980). These characteristics include parameters such as 1) concentrations of nutrient and mineral substrate required for growth of bacteria, 2) pH, 3) temperature, 4) redox potential, 5) dissolved organic concentration, 6) inorganic and organic particulate matter, 7) sulfate/sulfide concentration, and 8) contaminants which may synergistically or antagonistically affect the microbes' metabolic or chemical processes.

Biomethylation of Hg in sediments depends on its chemical form as well as the chemical and physical characteristics of the sediment (Fagerstrom & Jernelov 1971). In order for a metal ion to be methylated by methylcobalamin, it must be in a specific oxidation state (Ridley *et al*. 1977). The Hg enters the aquatic ecosystem in both the elemental and ionic forms. Which form is preserved in the sediment depends at least partly on the pH and redox potential of the environment.

In any aquatic environment only a small portion of the total Hg exists as CH_3Hg^+, usually less than 0.1%. Methylation appears to occur primarily in the top 1 to 2 cm of the sediment because most of the microbial population responsible for CH_3Hg^+ synthesis is localized in the top 10 cm (Bianchi 1973). Consequently, only small concentrations or even the absence of CH_3Hg^+ could be expected in the deeper sediment layers. Because CH_3Hg^+ has been found in the deeper sediments, however, several possible reasons have been postulated, such as the stability of CH_3Hg^+ in anaerobic conditions (Olson & Cooper 1976) and a very slow transport to the upper sediment layers (Jernelov 1970).

In situ methylation of Hg has been reported in estuarine sediments of San Francisco Bay (Olson & Cooper 1974). At least in the initial phases of methylation, sediments 6 cm deep produced more CH_3Hg^+ than those on the surface. These higher values in the deeper sediments may indicate that in this system 1) more methylation occurs under anaerobic conditions, 2) the loss is greater from the shallower sediments, or 3) CH_3Hg^+ degrades at a greater rate in the shallow sediments. While the amount of CH_3Hg^+ was significantly above background levels, it was far below what was reported for laboratory studies (Jensen & Jernelov 1969, Fagerstrom & Jernelov 1971).

The Ottawa River bed sediments have been estimated to contain 97% or about 200 times more total Hg than the biomass of fish, invertebrates, and plants, even assuming that all of the Hg in the biomass was CH_3Hg^+ (Kudo *et al*. 1977). While the accuracy of this calculation is not high, it clearly shows that the total Hg in the sediments was higher by at least two orders of magnitude than in the biomass. In Mobile Bay, Alabama, CH_3Hg^+ in estuarine and freshwater sediments averaged 0.03% and did not exceed 0.07% of the total Hg (Andren & Harriss 1973). The background levels decreased with the depth. CH_3Hg^+ levels in the Hg-polluted sediments of

Table 1. Methyl and Total Mercury in Natural Sediments

Site	Total Hg[a] (μg/g)	CH_3Hg^+ (ng/g)	CH_3Hg^+ (%)	Reference
Minamata Bay, Japan	2010	–	–	Kitamura et al. 1959
River Seine, France	15.8	ND[b]	–	Batti et al. 1975
Monte Amiati, Italy	63.5–288	20–40	0.01–0.06	Batti et al. 1975
Mobile Bay, U.S.A.	0.21–0.60	0.06–0.19	0.03	Andren & Harriss 1973
Mississippi River, U.S.A.	0.08–0.57	ND–0.05	0.01	Andren & Harriss 1973
Everglades, U.S.A.	0.12–0.49	0.06–0.12	0.03–0.07	Andren & Harriss 1973
San Francisco Bay, U.S.A.	0.1–1.3	0.04–1.9	0.03–1.0	Olsen & Cooper 1976
Rhine River, Holland	5–17	10–110	0.2–1.0	Van Faassen 1975
Yssel River, Holland	12	70	0.6	Van Faassen 1975
Irish Sea	0–0.2	0–2.7	0–1.35	Bartlett et al. 1978
River Clyde, Scotland	0.4–4.4	0.3–5.4	0.01–0.4	Bartlett et al. 1978
River Mersey, Scotland	1.3–11.3	1.6–60.6	0.05–1.4	Bartlett et al. 1978
				Craig & Morten 1976
Kastela Bay, Yugoslavia	6	2–20	1.35	Mikac & Picer 1985

Source: After Bartlett et al. (1978).
[a]Mercury sediment concentrations exceeding 5 to 10 times the "background" levels are normally considered to be contaminated.
[b]ND = not detected.

Kastela Bay, Yugoslavia, in the central Adriatic Sea averaged about 7 μg CH_3Hg^+/kg in a range between 2 and 20 μg CH_3Hg^+/kg on a wet weight basis (Mikac & Picer 1985).

The concentrations in the sediments are usually much lower than in some species of fish. The CH_3Hg^+ concentrations in sediments rarely exceed 100 μg/kg (dry weight) (Bartlett et al. 1978, Bartlett & Craig 1981) and are generally much lower in the marine environment. In some areas CH_3Hg^+ could not be detected in sediments, but methylation was demonstrated by the high content in aquatic organisms (Gardner et al. 1978).

Fewer data are available on CH_3Hg^+ concentrations and distribution in deeper marine sediment layers (Bartlett & Craig 1981). Determining the synthesis and concentration is complicated by a wide range of factors. CH_3Hg^+ synthesis rates range from 0.1 to 240 ng/cm^2/year (Langley 1973, Olson & Cooper 1976, Shin & Krenkel 1976, Windom et al. 1976, Bartlett et al. 1978). Field studies show that the proportion of CH_3Hg^+ in the sediments ranges from 0.01 to 1.4% (Table 1). A CH_3Hg^+ production rate of 50 to 3000 ng CH_3Hg^+/m^2/day equivalent to 1.8 to 120 ng CH_3Hg^+/cm^2/year has been shown (Wright & Hamilton 1982).

Microaerophilic bacteria which methylate Hg are rather vulnerable in some respects. They can be killed if the sediment is agitated, grow only in a narrow pH range, and grow slowly even under ideal conditions (Spangler et al. 1973a, 1973b). Besides the composition, quantity, activity, and Hg-resistant character of the microbial population, the rate of CH_3Hg^+ synthesis also depends on the redox potential, the organic substrate, availability of mercuric ion, pH, and temperature (D'Itri et al. 1978). The availability of sulfur and iron, which complex with Hg and each other, can also affect the availability of Hg for methylation and uptake (Rudd et al. 1983). Macroorganisms also play an important role in the redistribution of CH_3Hg^+ in the sediment column (Boddington et al. 1979). Their burrowing can expose Hg in deeper layers to the methylating process (Jernelov 1970).

5.1.1.2 Other Biological Methylation Mechanisms.

The ability to convert inorganic Hg compounds into CH_3Hg^+ is not restricted to aerobic and anaerobic bacteria. It also occurs through a number of biological transformations not involving the methylcobalamin synthesis. For example, *Neurospora crassa*, a methylcobalamin-independent fungus, is capable of aerobically synthesizing CH_3Hg^+ from Hg^{2+} ion by one or more steps of the methionine biosynthesis pathway (Landner 1971). Many molds and photosynthetic bacteria are also capable of synthesizing ethylene from methionine; therefore, CH_3Hg^+ may also be formed by a reaction between ethylene and Hg^{2+} ion (D'Itri et al. 1978). This mechanism involves the attachment of the mercuric ion to the sulfur group on the homocysteine coenzyme. It is methylated in an "incorrect" synthesis of methionine normally formed through methylation of the sulfur group in homocysteine (Landner 1971). *Escherichia coli* also methylate Hg^{2+} ion in the absence of methylcobalamin (Vonk & Sijkesteijn 1973, Silver et al. 1976).

5.1.2 Demethylation

While some microorganisms synthesize methylated metal/metalloids in aquatic ecosystems, others can demethylate them. These methylation-demethylation interconversions may establish an ecologically dynamic system of competing reactions that can result in steady-state concentrations of various metals/metalloids and their methylated forms in the environment. However, the introduction of additional metals/metalloids released by human activities can disturb the equilibrium and the concentration of intermediate compounds in natural waters (Saxena & Howard 1977). Therefore, when the fates of these metals/metalloids are assessed, an important dimension is whether or not demethylation also occurs.

Demethylation by bacteria was demonstrated by Tonomura et al. (1968). A CH_3Hg^+-resistant bacterium of the genus *Pseudomonas* that was able to induce the demethylation of CH_3Hg^+ into volatile Hg° has been isolated (Tonomura et al. 1968). The rate of release of Hg° to the atmosphere during aerobic incubation was nearly three times the rate observed during anaerobic incubation. The microbial degradation of CH_3Hg^+ into Hg° and methane has been observed in both lake and river sediments (Spangler et al. 1973a, 1973b, Billen et al. 1974), but the process may also take place in the water column where it is affected by the turbidity and dissolved oxygen concentration (Xun et al. 1987).

It has been suggested that demethylating organisms may maintain environmental CH_3Hg^+ concentrations at a minimum (Spangler *et al.* 1973a, 1973b). Nevertheless, the evidence demonstrating substantial quantities of CH_3Hg^+ in fish suggests that the natural demethylation process does not degrade equivalent quantities of the CH_3Hg^+ produced in the sediments, water column, or fish intestines. At least the rate is not sufficient to prevent accumulation in some components of the ecosystem.

Whether or not the concentrations of CH_3Hg^+ are increasing in the sediments depends on the difference between the rates of the two processes as well as the rate of transport from the system (Billen *et al.* 1974). Whenever the methylating process is more efficient than the demethylating process, a net CH_3Hg^+ synthesis rate develops to produce a greater equilibrium CH_3Hg^+ concentration in sediments as well as a greater equilibrium rate of transport to the water column from sediments, which leads to accumulation of Hg by aquatic organisms. If the demethylation process were more efficient, the net equilibrium CH_3Hg^+ synthesis rate and the sediment concentrations would be smaller and so would the corresponding concentrations in aquatic biota. If the input of Hg and the synthesis or degradation of CH_3Hg^+ are disturbed, these processes readjust to establish a new equilibrium.

Most experiments have been conducted to determine the concentration of CH_3Hg^+ in sediments. More recently, the methylation rates have been measured but not the rates of demethylation. The results achieved by most researchers have not differentiated between the two processes. Therefore, most of the methylation rates reported probably reflect the net rate of methylation and demethylation reactions occurring simultaneously. The kinetics of these simultaneously occurring reactions, relative to external stimuli such as increased food, bacteria, temperature, redox potential, pH, or effect of xenobiotic chemicals in the sediments, are even more difficult to elucidate. All of these effects complicate interpretations of data on the rates of methylation and demethylation. Together, comparisons of methylation and demethylation ratios can indicate potential sites of accelerating CH_3Hg^+ synthesis as well as the effects of the various chemical and biological environmental factors on the microorganisms' ability to support the processes (Ramlal *et al.* 1986).

5.1.3 Acid Lakes: Effects of pH on Methylation and Demethylation

The main causes of acidification of surface waters are smelting operations, auto emissions, and fossil fuel combustion. They contribute not only sulfur and nitrogen oxides, the precursors of sulfuric and nitric acids, but many airborne heavy metals and metalloids, including Hg, As, and Pb. All of these contaminants enter the aquatic ecosystem through wet and dry deposition.

Depending on the buffering capacity of the soil, acid deposition may cause the terrestrial environment to become slightly acidified, resulting in increased leaching of cations such as Mg^{2+}, Ca^{2+}, K^+, and Al^{3+} as well as the low-molecular-weight fulvic acids. This type of environmental leaching has two consequences. First, the low-molecular-weight fulvic acids become more labile. Second, the inorganic Hg of airborne origin may also be chemically (abiotically) methylated (Rogers 1977, Nagase *et al.* 1982).

As acid rain lowers the pH of the surface waters, the likelihood of methylation is enhanced. In a watercourse, soluble Hg deposited in rain or other sources can be methylated in suspension without entering the bottom sediments, or it may be adsorbed by organic particles suspended in the water column and transported to the sediments, where it can also be methylated.

Greater availability of Hg in acid-stressed, poorly buffered lakes may explain the greater concentrations of Hg in fish taken from these lakes. Some dystrophic lakes may contain large quantities of organic matter and still not produce substantial growth of algae and plants. For instance, some reservoirs in Finland (Verta *et al.* 1986), Northern Minnesota, and Northern Wisconsin contain great concentrations of organic matter because of input of humic materials from surrounding bogs (Lillie & Mason 1983, Helwig & Heiskary 1985). Mosses, which replace the macrophytes whenever a lake becomes acidified, efficiently bind Hg. Major portions of the Hg associated with natural organic matter in acidic Swedish lakes were bound in the mosses (Jernelov & Johansson 1984). Consequently, dystrophic water bodies also contain fish with relatively great concentrations of Hg (Helwig & Heiskary 1985, Verta *et al.* 1986).

Biomethylation of the Hg^{2+} ion may be thermodynamically enhanced at low pH and at positive redox potentials. Both conditions can occur in the water column and at the sediment surface of acidified, aerobic lake systems but not in anaerobically incubated subsurface sediments (Wood 1980). The methylcobalamin-dependent methylation process is optimized at pH 4.7 (Fagerstrom & Jernelov 1972, DeSimone *et al.* 1973). The pH indirectly affects the methylation synthesis through its effect on microbial activity. The role of microorganisms in the methylation of Hg changes with increased acidity. CH_3Hg^+ is biomethylated at neutral or slightly acidic pH, whereas $(CH_3)_2Hg$ is the predominant form at alkaline values (Figure 4; Fagerstrom & Jernelov 1972).

Evaluating the effects of pH on the methylation rate is difficult because of the confounding effects of small concentrations of calcium (Ca) and low alkalinities. Waters with a low pH and small concentrations of Ca often contain fish with great Hg concentrations. One hypothesis is that the uptake of Hg by fish in slightly acidic waters is affected by Ca-mediated changes in

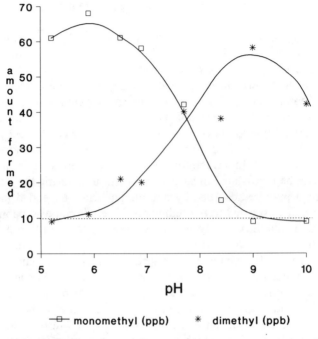

Figure 4. Formation of mono- and dimethylmercury in organic sediments. (After Fager-strom & Jernelov 1972.)

gill permeability (Rogers & Beamish 1983). It has also been theorized that any CH_3Hg^+ produced in the water column of acidified lakes is directly available for uptake by fish via the gill membranes (Xun *et al*. 1987). At pH values greater than 6.0 the effect was minimal even though effects of Ca were still important (McWilliams & Potts 1978).

Many studies of aerobic and anaerobic methylation in sediments *in situ* and under laboratory conditions have demonstrated that the rates of methylation of Hg increase linearly with greater concentrations of Hg^{2+} ion (DeSimone *et al*. 1973, Nagase *et al*. 1982, Rudd *et al*. 1983). Acidification has been observed to cause an increase in the net rate of CH_3Hg^+ production in lake water and epilimnetic sediments with an aerobic surface (Xun *et al*. 1987). However, the rates of methylation of Hg in samples of sediment incubated under anaerobic conditions were directly proportional to the pH (Ramlal *et al*. 1985). CH_3Hg^+ formed in sediment samples only in the pH range of 5.5 to 6.5 (Baker *et al*. 1983). Acidification also results in a decrease in the rate of net CH_3Hg^+ production in near-surface anaerobic sediments (Pan-Hou & Imura 1982a, 1982b). These lesser rates of methylation of Hg do not explain why fish taken from lakes with anaerobic sediments tend to contain greater concentrations of Hg. An alternative hypothesis is that more CH_3Hg^+ is synthesized in the water column under these conditions.

Methylation of Hg in the water column has seldom been described (Armstrong & Scott 1979, Furutani & Rudd 1980, Topping & Davies 1981, Xun *et al*. 1987), and only Xun *et al*. (1987) have described demethylation in the water column. In samples of lake water the rate of methylation is proportionally greater than would be predicted from a linear relationship within concentrations of Hg^{2+} ions. Perhaps the proportion of unbound and microbiologically available inorganic Hg increases when the available binding sites are saturated. This could be the result of a greater exchange of hydrogen ions with inorganic Hg adsorbed onto particulate matter. Therefore, a proportionately higher concentration of uncomplexed Hg^{2+} may cross the cell membrane more effectively at low pH (Bienvenue *et al*. 1984), which would enhance the methylation process because the rate depends on the Hg concentration (Xun *et al*. 1987).

In a few lakes where the pH of the anaerobic sediment interstitial water has been lowered, methylation in the subsurface sediments was observed to decrease. This was probably due to the reaction of Hg^{2+} ions with free sulfide to produce HgS (Furutani *et al*. 1984, Ramlal *et al*. 1985, Xun *et al*. 1987). It appears that the whole-lake effect of acidification can be determined by ascertaining the relative importance of net methylation in the water column and surface sediments. This could explain the elevated concentrations of Hg in fish in acidified lakes.

In most acid lakes the pH of interstitial water in subsurficial sediments is usually acidic because anaerobic bacteria produce hydrogen ions, except in very oligotrophic lakes (Kelly & Rudd 1984, Rudd *et al*. 1987). Thus, unlike net methylation in the surface waters and sediments, which appears to be enhanced at reduced pH, the impact of acidification of surface water on the net rates of methylation in subsurficial sediments is likely to vary from lake to lake according to the eH and pH of interstitial water in subsurficial sediments.

Acidification of lakes would also be expected to increase the solubility of Hg in the sediments. However, acidification of a softwater lake in one study did not appreciably increase the amount of Hg^{2+} released from the bottom sediments (Schindler *et al*. 1980, Jackson *et al*. 1980). In another system more CH_3Hg^+ was released at greater acidity (Miller & Akagi 1979). The net synthesis of CH_3Hg^+ in lake water has been observed to be about seven times faster at lesser pH (about 4.5) than at greater pH (about 8.5) (Johansson & Jernelov 1983). Acidification in most lakes is likely to increase the net amount of CH_3Hg^+ synthesized in the water column and in the aerobic surface sediments without affecting methylation in the subsurface sediments (Furutani & Rudd 1980, Ramlal *et al.* 1985, Rudd *et al.* 1987).

The specific rate of demethylation was observed to decrease regardless of whether the pH was raised or lowered from ambient values. Overall, any change in pH had a smaller effect on the specific rate of demethylation than on methylation. Changes in pH affected methylation to a greater extent than demethylation, with the net rate of methylation consistently increasing when the pH was lowered. The ratio of methylation to demethylation rates can also be a function of the dissolved organic carbon (DOC) concentrations. In aerobic, acidic lake systems the methylation rate varied inversely with the DOC concentrations, while the demethylation rate varied directly with the DOC (Miskimmin *et al.* 1989). The methylation rate was greater at pH 5 than 7, which is the opposite of the demethylation rate, which was greater at pH 7 than 5. Over sediments, significantly more CH_3Hg^+ remained in the overlying water containing greater concentrations of DOC than in water with lesser concentrations (Miskimmin *et al.* 1989).

5.1.4 Abiotic Methylation

5.1.4.1 Nonenzymatic Methylcobalamin. Hg in an aqueous solution can also be abiotically methylated (Neujahr & Bertilsson 1971, Imura *et al.* 1971). A nonenzymatic as well as an enzymatic transfer of methyl groups from methylcobalamin to Hg^{2+} ion has been shown to occur in cell-free extracts of the strictly anaerobic bacterium *Methanobacterium omelianskii* (Wood *et al.* 1968). Methylcobalamin was the methyl donor, and the reaction was extremely rapid and nearly quantitative under both aerobic and anaerobic conditions. Whereas Hg^{2+} ion reacts very quickly with methylcobalamin, mercurous (Hg_2^{2+}) ion does not react at all (Wood *et al.* 1968, DeSimone *et al.* 1973). The nonenzymatic methylation of Hg^{2+} ion demonstrated in the laboratory has not been shown to occur in the natural environment. Evaluating the ecological significance of this mechanism is difficult because methylcobalamin is not very stable in natural aquatic systems.

5.1.4.2 Humic Material. Inorganic Hg deposited on the soil can also be methylated (Beckert *et al.* 1974). The formation of CH_3Hg^+ is favored, and the reaction proceeds abiotically. Hg^{2+} ion was also methylated in steam-sterilized agricultural soils, but the methylating agent was not identified (Rogers 1977).

Sterilized humic and fulvic acids extracted from river sediments and leaf mold can also donate methyl groups to Hg^{2+} and, therefore, can be an additional source of CH_3Hg^+ to the aquatic ecosystem. CH_3Hg^+ has been produced from extracts of both fulvic and humic acid. Fulvic acid methylated without regard to its molecular weight, although the smaller-molecular-weight fraction (<200 MW) was more active than the greater-molecular-weight fraction. Subsequently, 2,6-di-tert-butyl-4-methyl-phenol (BHT), p-xylene, and mesitylene were identified as naturally occurring non-biological methyl donors (Nagase *et al.* 1982, 1984).

Only BHT, which produced CH_3Hg^+ at pH 7, was identified as a possible methyl donor under environmental conditions. However, because concentrations of BHT in river sediments and leaf mold were very small, Nagase *et al.* (1984) concluded that any contribution to abiotic Hg methylation by humic material in the aquatic environment would also be very small (Nagase *et al.* 1984). This conclusion was substantiated by Berman & Bartha (1986), who estimated that biotic methylation by humic compounds accounted for less than 10% of CH_3Hg^+ formed by microorganisms in sediments.

5.1.4.3 Methyl Silicon Methylation. Water-soluble nuclear magnetic resonance (NMR) reference standards 2,2-dimethyl-2-sila-pentane-5-sulfonate (DDS) and sodium 3-trimethylsilyl-propionate (TSP) have been observed to react with mercuric acetate under ambient conditions to form CH_3Hg^+ acetate (DeSimone 1972). Therefore, trimethylsilyl salts in sediments may also abiotically methylate Hg^{2+}, and under special conditions polydimethylsiloxanes may contribute to the production of CH_3Hg^+ in aquatic ecosystems (DeSimone 1972, Watanabe *et al.* 1986, Nagase *et al.* 1986). However, the probability of Hg^{2+} ions being methylated by a methyl silicon species is extremely small under prevalent conditions in aquatic environments and is not likely to contribute measurably to concentrations of CH_3Hg^+ in the environment (Frye and Chu 1988). This conclusion was based on the following: 1) $HgCl_2$ can be methylated at 80°C, a temperature far higher than that of aquatic sediments (Watanabe *et al.* 1986); 2) Cl^- ions in the sediments markedly inhibit the Hg methylation reaction (Jones & Nickless 1974, Bellama *et al.* 1985); and 3) highly insoluble methylsiloxanes are not very susceptible to electrophilic cleavage and, therefore, would be even less reactive. Overall, these mechanisms are thought to be of little environmental significance because methylation has not been observed in nature without biological activity (Jensen & Jernelov 1969, Rissanen *et al.* 1970, Bishop & Kirsch 1972, Olson & Cooper 1976, Blum & Bartha 1980).

5.1.4.4 Transmethylation Reactions. In addition to the biomethylation mechanism, several other possible mechanisms for the transalkylation or chemical alkylation of mercurials in soils and sediments have been demonstrated. CH_3Hg^+ can be formed chemically through a transmethylation reaction involving methyl tin derivatives (Craig 1980). It can also be produced chemically in the sediments via a transalkylation reaction between inorganic Hg and ethyl and methyl Pb compounds discharged into the same water body (Jernelov 1972).

5.1.4.5 Remobilization. CH_3Hg^+ does not bind as tightly with organic matter in sediments as does the Hg^{2+} ion. Therefore, CH_3Hg^+ can be desorbed from the sediment particles at a relatively faster rate (Menzer & Nelson 1986). Consequently, CH_3Hg^+ readily remobilizes from the stable and less reactive sediments into the overlying water. The rate of the CH_3Hg^+ remobilization influences bioaccumulation in aquatic organisms, although the amount of CH_3Hg^+ may be very small relative to the total concentration of Hg in the sediments. The estimated monthly releases of CH_3Hg^+ amount to only about 0.1% of the total Hg which is transported to the surficial sediments (Jernelov & Lann 1973). This means that, relatively speaking, usually little and sometimes no CH_3Hg^+ is found in sediments, even those which contain great concentrations of inorganic Hg. In experiments involving water-sediment systems receiving Hg^{2+} or Hg^o, the CH_3Hg^+ content was less than 1% of the total (Holm & Cox 1974).

5.2 Methylation of Lead

Whether environmental Pb methylation is a biological or chemical process has been the subject of some controversy. Evidence has been presented that anaerobic microbes can form volatile $(CH_3)_4Pb$ in lake sediments polluted with inorganic lead (Wong *et al.* 1975). Adding $Pb(NO_3)_2$ or $(CH_3)_3Pb^+$ greatly increased the synthesis of $(CH_3)_4Pb$ in sediments from Lakes Erie, Ontario, and St. Clair. Inorganic Pb did not easily convert to organic Pb, but $(CH_3)_3Pb^+$ readily converted to $(CH_3)_4Pb$, and volatile $(CH_3)_4Pb$ was detected above the sediment-water system (Wong *et al.* 1975). However, the absence of $(CH_3)_4Pb$ after these sediments were sterilized suggested a biological conversion (Schmidt & Huber 1976, Dumas *et al.* 1977).

In an alternative abiotic mechanism, $(CH_3)_3Pb^+$ cations convert to $(CH_3)_4Pb$ by a chemical disproportionation reaction under anaerobic conditions in the presence of sulfide (Jarvie *et al.* 1975). In this process trimethyllead acetate, $(CH_3)_3Pb(C_2H_5O_2)$, is first converted to trimethyllead sulfide, $[(CH_3)_3Pb]_2S$, in the sediment followed by decomposition to yield $(CH_3)_4Pb$ (Equation 4).

$$3[(CH_3)_3Pb]_2S \longrightarrow 4(CH_3)_4Pb + (CH_3)_2S + 2PbS \qquad (4)$$

No $(CH_3)_4Pb$ has been found in aqueous solutions of methylcobalamin and $(CH_3)_3PbCl$, $(CH_3)_2PbCl_2$, or $Pb(NO_3)_2$. Therefore, because Pb compounds apparently do not react with methylcobalamin in the environment, they may not be methylated like Hg in this regard.

CH_3I and several other methyl donors can methylate Pb^{2+} compounds to produce $(CH_3)_4Pb$. The biologically mediated methylation of Pb has been ascribed to a CH_3I synthesis pathway (Craig & Rapsomanikis 1985). CH_3I has been detected in open ocean waters at 1.2–135 ng/L, and concentrations which were 1000–fold greater have been measured in the vicinity of *Laminaria digitata* kelp. Although Pb^{2+} has not been shown to methylate via methylcobalamin, Pb can be methylated by CH_3I. CH_3Pb^{3+}, $(CH_3)_2Pb^{2+}$, $(CH_3)_3Pb^+$, and $(CH_3)_4Pb$ were produced from CH_3I and Pb° as well as Pb salts. The reaction starts with Pb° and involves an oxidative methyl transfer from CH_3I (Equations 5–10).

$$Pb: + CH_3I \longrightarrow Pb^{+}_{\cdot} + CH_3^{\cdot} + I^- \tag{5}$$

$$Pb^{+}_{\cdot} + I^- \longrightarrow PbI^{\cdot} \tag{6}$$

$$PbI^{\cdot} + CH_3^{\cdot} \longrightarrow CH_3PbI \tag{7}$$

$$CH_3IPb: + CH_3I \longrightarrow CH_3IPb^{+}_{\cdot} + CH_3^{\cdot} + I^- \tag{8}$$

$$CH_3IPb^{+}_{\cdot} + I^- \longrightarrow CH_3PbI^{\cdot} \tag{9}$$

$$CH_3PbI^{\cdot}_{2} + CH_3^{\cdot} \longrightarrow (CH_3)_2PbI_2 \tag{10}$$

The varying results suggest that the relative importance of biotic versus abiotic $(CH_3)_4Pb$ formation still has not been determined (Thayer & Brinckman 1982).

5.3 Methylation of Arsenic

Like Hg, As is susceptible to methylation by microorganisms such as bacteria and fungi (McBride & Wolfe 1971, Cox & Alexander 1973). They produce toxic di- and trimethylarsines (Saxena & Howard 1977). Early studies indicated that the methylation of As involved some "activated" methionine intermediate (Challenger *et al.* 1933). The subsequent discovery that S-adenosylmethionine was the methyl donor in the methylation of As by fungi provided biochemical evidence for the *in vivo* synthesis of trimethylarsine (Challenger 1935).

In aquatic environments, the As^{5+} oxidation state predominates. Conversion of As^{5+} to As^{3+} by phytoplankton is apparently a survival adaptation because As^{5+} is much more phytotoxic. Phytoplankton excrete arsenite and methylarsenate into the aquatic environment, where they are oxidized and demethylated (Smies 1983, Saxena & Howard 1977).

Indirect evidence has also been presented for the methylation of As by bacteria in the environment (McBride & Wolfe 1971). *Methanobacterium*

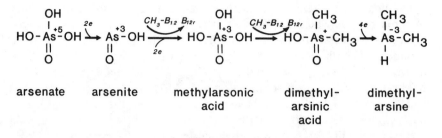

| arsenate | arsenite | methylarsonic acid | dimethyl-arsinic acid | dimethyl-arsine |

Figure 5.   The biosynthesis of dimethylarsine from arsenate by *Methanobacterium*. (After McBride & Wolfe 1971.)

can anaerobically reduce methylarsonic acid to dimethylarsine with methyl-cobalamin as the methyl donor (Figure 5). As with Hg, cell-free extracts of whole-cell preparations of this microorganism can methylate As.

Volatile organoarsenicals formed by methylation in anaerobic environments are released into the aerobic environment at the sediment/water or water/air interface (Figure 6; Ridley *et al*. 1977). These methylated As compounds slowly oxidize in air and water to produce less toxic compounds such as dimethylarsinic acid or cacodylic acid (Wood 1974). Because cacodylic acid is an intermediate in the synthesis of dimethylarsine, apparently a biological cycle for As exists in the environment similar to that of Hg (Ridley *et al*. 1977).

Arsenic (V) does not react with sulfhydryl groups, although it can be reduced to arsenite both abiotically and biotically (Lemmo *et al*. 1983). Inorganic As can be converted to an organic form in the digestive tracts of fish (Lunde 1972a, 1972b, 1977, Penrose 1974). However, adding As to tank water (Lunde 1972a, 1972b) or injecting it intramuscularly (Penrose *et al*. 1977) results in negligible concentrations of organic forms of As, which suggests that the conversion of As to organic forms is mediated by intestinal flora.

6.0 BIOACCUMULATION/CONCENTRATION IN AQUATIC ORGANISMS

Toxic elements can be introduced into an aquatic food web through algae and phytoplankton at the bottom. Fish and other aquatic organisms also assimilate the methylated compound through their gills via respiration and/or food intake. Organometallic forms of metals are more easily assimilated due to their affinity for lipids and their ease of diffusion through cell membranes. Then the methylated compound is readily incorporated into tissues. There, the organometallic compound or ion interacts with enzyme

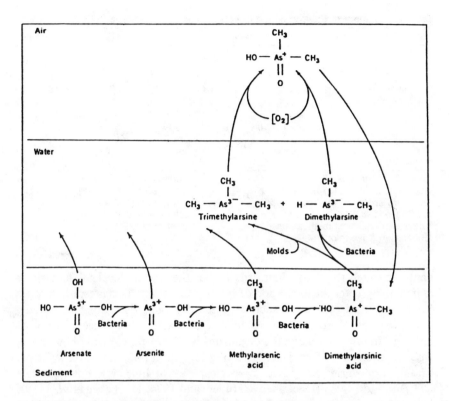

Figure 6. Proposed mechanism for the methylation of arsenic compounds in the environment. (Reprinted from Wood 1971. With permission.)

systems of the organism and often deactivates them because the toxic elements have an affinity for sulfur and sulfide linkages.

6.1 Mercury

A substantial body of evidence supports the contention that aquatic organisms directly accumulate great concentrations of CH_3Hg^+ primarily from the extremely small quantities in the water column. The CH_3Hg^+ in fish, aquatic birds, and mammals is the culmination of a complex food chain beginning with the methylation of inorganic Hg, usually in the sediments. As the first links in the aquatic food chain, phyto- and zooplankton begin the process of biomagnification by assimilating both inorganic Hg and CH_3Hg^+ compounds from the water, probably passively through surface adsorption (Armstrong 1979). The potential for Hg concentration in phytoplankton has been estimated to be approximately 1000 fold.

The organisms at the highest trophic levels bioconcentrate Hg predomi-

Table 2. Summary of Environmental Levels of Mercury in Water and Fish

Matrix	Range	Average
	Water (ng Hg/ml)	
Ocean	0.03[a]–5.0[b]	0.03
Freshwater[c]	<0.1[a]–136[d]	<0.01
Freshwater[e]	0.0001[f]–0.005[g]	<0.005
	Fish (μg Hg/g, wet weight)	
Ocean	<0.02[a]–40[h]	0.04–0.4[i]
Freshwater	0.03[a,j]–28[k]	0.05–0.2[i]

[a]Stock & Cucuel (1934).
[b]Jonasson & Boyle (1971).
[c]Standard analytical methods used.
[d]Dall'Aglio (1968).
[e]Ultraclean analytical methods used.
[f]Johansson & Iverfeldt (1989).
[g]Watras et al. (1989).
[h]Tsubaki et al. (1977).
[i]Friberg et al. (1971).
[j]Huckabee et al. (1974).
[k]Annett et al. (1975).

nantly through the food chain; the higher the trophic level of the fish or animal the greater the likelihood of biomagnification. This is especially true for longer-lived organisms such as predatory fish, fish-eating mammals, and predatory birds. In unpolluted freshwaters the concentration of CH_3Hg^+ in lower-level aquatic organisms is often below 0.1 mg Hg/kg wet weight, whereas it often exceeds 1.0 mg Hg/kg in polluted water systems. The amount varies according to species, way of life, etc., but as a rule CH_3Hg^+ is 50–80% (Huckabee & Hildebrand 1974, Huckabee et al. 1979). From 80 to more than 90% of the Hg in fish biota is in the form of CH_3Hg^+ (Westoo 1966, 1973, Bache et al. 1971, Kamps et al. 1972, Huckabee et al. 1975). Data on the uptake and excretion of Hg in invertebrates show good agreement with the accumulations in fish, although the Hg content of invertebrates is generally lower due to their shorter life cycle (deFreitas et al. 1977). A summary of concentrations of Hg in water and fish is presented in Table 2.

Biomagnification in fish has been observed to range between 5,000 and 100,000 times the levels in the surrounding water. Species that breathe faster, eat more, and live longer, such as tuna, accumulate more Hg than do other fish, but they all exhibit a continual uptake and slow depuration. That is why older, larger fish often contain greater CH_3Hg^+ body burdens than smaller fish of the same species in similar lake conditions.

In general, the rate at which Hg is methylated, rather than the concentration of total Hg in the ecosystem, is the principal determinant of how much Hg is concentrated to the fish. Great concentrations of total Hg in fish not only indicate the availability of Hg, but also indicate that it is in the form of CH_3Hg^+ that can be readily assimilated by fish. Laboratory studies have demonstrated that fish assimilate CH_3Hg^+ from water across gill surfaces and from food by digestive absorption (Hannerz 1968, Lock 1975, Olson *et al.* 1975, Phillips *et al.* 1980, Rogers & Besmish 1983). Hg accumulated from these two sources is additive (Phillips & Buhler 1978).

Whatever the route of bioaccumulation, the uptake of CH_3Hg^+ is much more efficient than for inorganic Hg. In fish the body burden derived from food is between 70 and 90% CH_3Hg^+ compared with only 5 to 15% inorganic Hg. Likewise, assimilation of inorganic Hg across the gills is between 10 and 100 times slower than CH_3Hg^+ (deFreitas 1977).

Because the diffusion rate of CH_3Hg^+ across cell membranes is very fast, it bioaccumulates rapidly in the tissues of organisms such as fish (Tillander *et al.* 1969, Jarvenpaa *et al.* 1970, Miettinen *et al.* 1970, Lockhart *et al.* 1972, Olson *et al.* 1973, Reinert *et al.* 1974, D'Itri *et al.* 1978, Olson 1986). CH_3HgCl diffuses through a lipid bilayer into cells on the order of 2×10^{-8} sec. Once inside the cell CH_3Hg^+ binds rapidly to sulfhydryl groups, thereby maintaining the concentration gradient across the membrane. This process allows aquatic organisms to accumulate great concentrations of CH_3Hg^+ rapidly even though its steady-state concentrations are relatively small (Wood 1983). The combination of the lipophilic properties and affinity for the sulfhydryl groups of amino acid compounds results in a rapid accumulation in the muscles and fat tissues until CH_3Hg^+ is metabolized and removed by the liver, kidney, and spleen. Because of its relative stability in biological systems and its unusual biochemical characteristics, CH_3Hg^+ is more slowly metabolized and eliminated than the inorganic forms of Hg. The overall result is a net bioconcentration in the organism over time.

Because CH_3Hg^+ is assimilated rapidly and eliminated slowly, its synthesis in the sediments doesn't have to be very rapid in order for it to be bioconcentrated by fish. As the rate of synthesis, release into the water, and bioconcentration exceed the rate at which the CH_3Hg^+ is metabolized, fish readily bioconcentrate great quantities which are passed on to their predators.

The mechanisms and rates of accumulation and elimination are not clear, but they appear to depend on the specific biological characteristics of each species of fish as well as the chemical, biological, and/or physical properties of the watercourse, i.e., the region of habitation (Abernathy & Cumbie 1977). They directly or indirectly affect the metabolic rates in individual fish or species, as do differences in the selection of food as the fish mature and their epithelial surface areas as well as their growth rate, length of life, size, and exposure times. The foraging habits and proximity to sediments also appear to have a significant effect on the rate of uptake by different species of fish. Bottom feeders have lesser body burdens of CH_3Hg^+ because they normally have great amounts of sediments in their intestines and the percentage of CH_3Hg^+ is small.

The half-lives of CH_3Hg^+ in fishes are a function of the species, respiration rate, and temperature. Metabolic rate and lipid content are the most important factors to determine the rate of CH_3Hg^+ elimination as well as accumulation. Radioisotopic studies have shown that the CH_3Hg^+ excretion rates vary greatly among species, but they usually follow a bi-exponential equation. The biological half-life of the fast component at the ambient water temperature usually varies from 1 to 10 days, while the slow one ranges from 200 to 1200 days depending on the species of aquatic organism (Miettinen et al. 1969). A half-life of 700 days has been reported for northern pike (Phillips & Buhler 1978) . The half-life of excretion in a small fish (15 g) has been observed to be 60 days and 350 days in a larger fish (100 g) (deFreitas 1979). This means that bioconcentration takes place during about 4 years before equilibrium is reached.

The half-life of Hg decreases with increasing temperature (Jarvenpaa et al. 1970), but higher temperatures also increase the metabolic rate and Hg uptake (MacLeod & Pessah 1973, Rogers & Beamish 1981). These results suggest seasonal and geographic temperature variability. The excretion rate at 10°C is twice as fast as at 4°C (Miettinen et al. 1969). Therefore, fish from watercourses where the temperature reaches 20° can be expected to eliminate Hg approximately twice as fast as fish in waters of about 10°C (MacLeod & Pessah 1973).

Other factors which contribute to Hg concentrations and biomagnification in fish and invertebrates include total Hg in water (dissolved and particulate), total Hg in the sediments, redox potential, pH, alkalinity, sulfur concentration, the organic content of the sediments and particulates, particle size and agitation of the sediments, and the total Hg concentration in benthic invertebrates. A series of experiments conducted in Canada showed that Hg concentrations in fish which were relocated from a contaminated lake to a decontaminated lake decreased about 30% in 1 year (Lockhart et al. 1972). However, because the sizes and ages of the fish also are influencing factors, it is difficult to determine the relative importance of the duration of exposure and increase in mass of tissue related to Hg accumulation during exposure. For example, part of the observed decrease could have been caused by dilution during growth. In another, more extensive study of the distribution and accumulation of Hg in the Ottawa River, the data indicated that the concentrations in fish were not less when the concentration in the sediment was less (Miller 1977). Apparently, this was because of a slower turnover rate of Hg in fish compared with the sediments. This difference could influence the relationship between the levels of Hg in fish and sediments, especially in systems where the source has been eliminated.

The total mercury concentration in fish is a function of 1) the mercury content of the sediments, 2) distance from the mercury source, 3) the lake retention time, 4) the lake watershed drainage area, and 5) the lake surface area (Håkanson et al. 1988). In addition, many lake chemistry characteristics such as water pH, hardness, alkalinity, and conductivity can also influence the Hg concentration in fish. Unproductive waters, those with small masses of algae and little plant growth, tend to have small ionic content, low pH, and high fish Hg concentrations, whereas productive waters tend to have a great ionic content, greater pH, and low fish Hg concentrations. Concentrations of Hg in fish from eutrophic lakes are often less than the concentrations of Hg that would be predicted. Lakes with large macrophyte/plankton/algae biomasses usually have a relatively smaller Hg content per unit of biomass. They effectively dilute the Hg content of the food base for other organisms. The adsorbed Hg is directed to the sediments when the plankton/algae die, which further decreases the exposure of aquatic organisms in the upper trophic levels. This is one reason that organisms in eutrophic waters with large biomasses often have lower Hg levels than comparable organisms in oligotrophic waters (D'Itri et al. 1971, Jernelov et al. 1975, Hultberg & Hasselrot 1981).

Numerous studies have demonstrated that water bodies containing fish with great concentrations of Hg share five characteristics: 1) low pH (Jernelov et al. 1975, Håkanson 1980, Stokes et al. 1983, Weiner 1983, 1986, Wren & MacCrimmon 1983, Helwig & Heiskary 1985), 2) low alkalinity (Scheider et al. 1979, Akielaszek & Haines 1981, Helwig & Heiskary 1985), 3) low calcium (Helwig & Heiskary 1985), 4) low productivity (D'Itri et al. 1971), and 5) high dissolved organic content (Helwig & Heiskary 1985, Verta et al. 1986).

A number of researchers have examined the relationship between Hg loading and pH in the aquatic environment and the subsequent Hg concentrations in fish. The effects of pH on bioaccumulation and biomagnification are very important but mostly indirect. In Sweden, the United States, and Canada, the Hg content in fish was correlated with the acidity of the lake. In Sweden, predatory fish taken from selected small, slightly acidic lakes with no point source of Hg still contained elevated levels (Landner & Larsson 1972, Håkanson et al. 1988). These finding were substantiated by other researchers studying similar aquatic systems (Brouzes et al. 1977, Suns et al. 1980, Jernelov 1980). The pH of lake systems may affect the association with particulate and dissolved organic matter, altering the availability of Hg to fish and microorganisms.

In lakes with low pH values, concentrations of Hg in fish are generally greater than in less acidic lakes. Aquatic organisms, especially fish in acidic lakes, are often stressed due to inadequate food supply, osmotic salt regulation, and greater than normal concentrations of aluminum and heavy metal ions in the water. Consequently, the plankton-feeding fish tend to disappear first because their food base, the phytoplankton and zooplankton, dies off. Thus the remaining large, predatory fish species are forced to feed on predatory fish at higher trophic levels. They are usually older and have higher body burdens of CH_3Hg^+. As these older and larger fish die off last, the result is a fish population gradually concentrating a higher average concentration of Hg. Therefore, as the pH of a system decreases, the CH_3Hg^+ concentrations increase in the remaining fish (Jernelov & Johansson 1984).

Northern Wisconsin contains many lakes with the above limnological characteristics (Lillie & Mason 1983). Part of the Hg may come from natural sources because few Wisconsin lakes suffer from point-source Hg contamination; however, increases in atmospherically borne Hg have also substantially contributed to the loadings in the lakes over the past century. Especially in those with a low buffering capacity, acid deposition no doubt increased the availability of Hg to fish as the pH was lowered and the alkalinity reduced. One year after a lake in Northern Wisconsin was artificially acidified from a pH of 6.0 to 5.5, yellow perch contained significantly greater concentrations of Hg than previously (Wiener 1986). Older walleyes from naturally acidic lakes contained more Hg than similarly aged fish from circumneutral lakes (Wiener 1983).

In surficial aerobic sediments a decreased pH can reduce the adsorption of Hg to particulate matter, making it more available for methylation and/ or uptake (Verta et al. 1986). The reduction of the biomass in an acidified lake leads to greater availability of CH_3Hg^+ for each organism. Also, a reduction in the growth rate of fish leads to a greater rate of uptake per unit of body weight. Low pH conditions also can increase mucus formation in fish, which could indirectly result in additional methylation of Hg (Varanasi et al. 1975).

Fish in more productive lakes have been found to contain lesser amounts of Hg for two reasons. First, the fish are well fed and actively growing; therefore, the assimilated Hg is, in effect, diluted relative to the growing body mass. Second, the anoxic conditions often found in the sediments of eutrophic lakes facilitate the formation of HgS (D'Itri et al. 1971, Fagerstrom & Jernelov 1972). This Hg cannot be effectively released except by microbes under aerobic conditions. In addition, because of the great organic content of productive lakes, the Hg that has complexed with particulate matter settles into the sediments. If they are anaerobic, HgS can form.

Oligotrophic lakes, especially if they are slightly acidic with well-oxygenated sediments, provide an aquatic ecosystem with less dissolved and particulate organic matter and also less dissolved inorganic bicarbonate ions with which the Hg can complex. This causes more unbound Hg to be methylated and available to fish and other aquatic organisms (Scheider et al. 1979, Akielaszek & Haines 1981). The acidic conditions in the oligotrophic lakes allow any $(CH_3)_2Hg$, which normally would be volatilized, to be converted to CH_3Hg^+ and be available for bioaccumulation. The specific methylation rate in softwater lakes increases not only in the sediments but also in the water column as the pH decreases (Rudd & Turner 1983, Wren & MacCrimmon 1983, Verta *et al.* 1986, Xun *et al.* 1987).

6.2 Lead

Although organolead compounds are generally more toxic to aquatic organisms than inorganic Pb compounds (Wong *et al.* 1978, Grandjean & Nielson 1979), few laboratory studies have been conducted to describe these effects in detail. For one thing, maintaining a relatively constant concentration of the volatile Pb compounds and measuring them is very difficult (Wong *et al.* 1981). A summary of concentrations of Pb in water, sediments, and biota is presented in Table 3. Tetraalkyllead compounds are sparingly soluble in water and have a very great vapor pressure, so they may be only momentarily present in natural waters (Wong *et al.* 1975). Also, accumulation of $(CH_3)_4Pb$ into sediments appears not to constitute a significant source of contamination in aquatic plants and animals.

Regardless of the great vapor pressure of $(CH_3)_4Pb$ and its expected off-gassing, laboratory experiments and ambient concentrations in fish suggest

Table 3. Summary of Environmental Levels of Lead in Water, Sediments, and Biota

Matrix	Range	Average	Reference
Air (ng/m^3)			
Urban, particulate Pb	500–14,500	–	Harrison & Perry 1977
Rural, particulate Pb	10–660	–	Harrison & Laxen 1978
Urban, $(CH_3)_4Pb$	<6–206	–	Harrison & Perry 1977
Rural, $(CH_3)_4Pb$	0.5–230	–	Harrison & Laxen 1978
Water (ng Pb/ml)			
Ocean[a]	<0.02–>1.0	~0.02	Nriagu 1978
Freshwater[a]	<0.1–>100	<0.1	Nriagu 1978
Sediment porewater	<0.1–36	–	Duchard *et al.* 1973
Sediments (μg Pb/g)			
Uncontaminated	<20–160	~15–25	Nriagu 1978
Contaminated	930–37,000	–	Nriagu 1978
Biota (μg Pb/g)			
Plankton (Lake Ontario)	2.5–43[b]	–	Wong *et al.* 1978
Zooplankton (Lake Ontario)	1.3–5.8[b]	–	Wong *et al.* 1978
Fish (Great Lakes, uncontaminated area)	0.03–0.33[c]	–	Wong *et al.* 1978, 1981
Fish (Great Lakes, contaminated areas)	<0.1–2.0[c]	–	Wong *et al.* 1978, 1981
Fish (Great Lakes, hexane extractables)[d]	0.002–0.020	–	Wong *et al.* 1978, 1981
Marine biota	–	0.5	Nriagu 1978
Freshwater biota	–	2.5	Nriagu 1978

[a]Ultraclean analytical methods may reduce these concentrations by 10 to 100 times.
[b]Dry weight.
[c]Wet weight.
[d]The alkyllead fraction is usually less than 10% of the total lead body burden.

that alkylated forms of Pb may pose an environmental risk similar to that of Hg and As. Based on their configuration, $(CH_3)_4Pb$ compounds would be expected to concentrate in the lipid tissues of aquatic organisms; this has been reported for some cases (Sirota & Uthe 1977, Chau 1980, Chau et al. 1980). Rainbow trout accumulated $(CH_3)_4Pb$ rapidly from water. The highest concentration was observed in the lipid layer of the intestine (Wong et al. 1981). It has been suggested that either biological methylation occurs in surface waters or these organisms selectively concentrate the alkyl forms (Craig & Wood 1981). Fish bioconcentrated approximately 100 and 700 times the $(CH_3)_4Pb$ in the first and seventh days, respectively, on a whole fish basis. Even so, there is no evidence of biomagnification from aquatic vegetation to the edible portions of fish and shellfish.

6.3 Arsenic

When As is washed into lakes and streams, mainly in the form of arsenate, algae and other microorganisms adsorb or absorb it and often accumulate As to concentrations which are much greater than the surrounding waters. The biochemical pathway of As in phytoplankton is complex, involving at least eight steps. Arsenate is reduced to arsenite and methylated. Then the resulting arsine reacts with products of intermediate metabolism to form complexes (Wrench & Addison 1981).

Arsenic has an electronic structure like that of phosphorus and reacts similarly. Arsenic in fish and algae has been shown to be as both water-soluble and lipid-soluble organic As compounds (Lunde 1972a, 1972b, Woolson et al. 1974). It has also been found in crustaceans (Irgolic et al. 1977). Arsenic concentrations in both marine and freshwater fishes vary widely depending on the species, location (coastal vs. off-shore), and amount in the water. In unpolluted areas, concentrations range from less than 0.2 to 2 mg As/kg, while as great as 50 mg As/kg has been reported in As-contaminated waters. Although trimethylarsine has been shown to oxidize to cacodylic acid, trimethyl arsenic containing phospholipid was isolated from various fishes, shellfish, and marine algae (Wood 1971). A summary of environmental concentrations of As in water and biota is presented in Table 4.

Although aquatic organisms can bioaccumulate organoarsenicals from their surroundings, these compounds are rapidly excreted from fish (Penrose 1974). Higher trophic levels apparently do not contribute much to the accumulation, as there is no clear evidence of biomagnification in the aquatic food web (Gavis & Ferguson 1972, Seydel 1972, Lunde 1972a, 1972b, Isensee et al. 1973, Woolson 1975, Foley et al. 1978, Zingaro 1983). Fish and other freshwater organisms bioaccumulate dimethylarsine and cacodylic acid from water, but they do not appear to biomagnify them

Table 4. Summary of Environmental Concentrations of Arsenic in Water and Biota

Matrix	As species	Range	Average	References
		Air (ng As/m^3)		
Air	$(As)_T$[a]	0–10	–	Braman 1975
		Ocean (ng As/ml)		
Water	$(As)_T$	<0.1–6	~2	Woolson 1975
	As^{3+}	0.12–0.62	–	Braman & Foreback 1973; Andreae 1977, 1978
	As^{5+}	0.35–42.5	–	Braman & Foreback 1973; Andreae 1977, 1978
	MAA[b]	<0.002–0.13	–	Braman & Foreback 1973; Andreae 1977, 1978
	DMA[c]	<0.002–1.0	–	Braman & Foreback 1973; Andreae 1977, 1978
		Freshwater (ng As/ml)		
Water	$(As)_T$	<0.1 to 243,000[b]	~0.5	Braman 1975, Woolson 1975
	As^{3+}	<0.02–2.74	–	Braman & Foreback 1973
	As^{5+}	<0.16–0.96	–	Braman & Foreback 1973
	MAA	<0.02–0.22	–	Braman & Foreback 1973
	DMA	<0.02–0.62	–	Braman & Foreback 1973

Matrix	As species	Range	BCF[e]	References
		Marine Biota (μg As/g wet wt.)		
Shrimp	$(As)_T$	1–48	9–13,000	Braman 1975, Woolson 1975, Lunde 1973a, Coulson et al. 1935
Fish	$(As)_T$	0.1–50	38–12,150	Woolson 1975, Lunde 1973a, Windon et at. 1973, Cox 1925
Algae	$(As)_T$	0.1–95	50–47,500	Lunde 1973a, 1972a, 1972b, Jones 1922, Vinogradov 1953
Seawood	$(As)_T$	0.7–142	350–71,000	Schroeder & Balassa 1966, Lunde 1973b, Wilson & Fieldes 1942
		Freshwater Biota (μg As/g wet wt)		
Fish	$(As)_T$	0.035–0.3	3–30	Woolson 1975, Pratt et al. 1972, Wilderhaus 1966
Algae	$(As)_T$	2–550	4–7,000	Lunde 1973b, Schuth et al. 1974, Isensee et al. 1973, Reay 1972, Woolson et al. 1974.

Table 4. continued

Matrix	As species	Range	BCF[e]	References
		Freshwater Biota (μg As/g wet wt) continued		
Submergent weeds	(As)$_T$	20–971	800–20,000	Reay 1972
Emergent weeds	(As)$_T$	8–12	100	Reay 1972
Duckweed	(As)$_T$	1–3	1–3	Schuth et al. 1974
Lakweeds	(As)$_T$	11–1450	110–14,500	Lancaster et al. 1971

[a](As)$_T$ = total arsenic concentration.
[b]DMA = dimethylarsinic acid.
[c]MAA = methylarsenic acid.
[d]Searles Lake California, a highly saline lake.
[e]BCF = biological concentration factor.

(Isensee et al. 1973). Fish from a lake treated with sodium arsenite from 1955 through 1963 accumulated As from the water but did not biomagnify it between trophic levels (Foley et al. 1978).

Nonetheless, as with Hg, the rate of elimination of As was much slower than the rate of uptake. The sites of greatest concentration were the gills and digestive glands, which accounted for 70 to 80% of the total body burden. The gills eliminated As relatively rapidly, whereas elimination from the digestive glands was slower, indicating that the As is tightly bound to these tissues. Overall, however, the research to date indicates that any inorganic and organic forms are rapidly excreted, thus preventing food web biomagnification (Penrose et al. 1977).

7.0 CONCLUSIONS

The sources of Hg, Pb, and As to natural waters are diverse. Geologic weathering is the primary source of Hg and As, whereas much of the Pb has been released by human activities. Historical reports indicate that, prior to human interference, the concentrations of methylated metals and metalloids in the aquatic environment usually remained fairly small.

Although discharges of all three elements from human-related activities have decreased in the past two decades, indirect releases from fossil fuel combustion, auto emissions, smelting, and agricultural applications still add to the environmental cycle. Consequently, methylated Hg and Pb have been found in fish living in remote lakes with no direct input. As more has been learned about the natural methylation of Hg by bacteria in both anaerobic and aerobic sediments, the evidence now indicates that most forms of Hg released into the aquatic environment can be transformed into Hg^{2+} which is then available for methylation.

Laboratory studies and field research suggest that As and Pb can also be methylated. Inorganic Pb seldom converts into tetraalkyllead in natural waters, but $[(CH_3)_3Pb]_2S$ undergoes a disproportionation reaction to form $(CH_3)_4Pb$. As it is only slightly soluble in water and has a high vapor pressure, tetraalkyllead may be readily eliminated from the aquatic environment. Similarly, natural releases of As such as arsenic acid may produce methylated compounds.

Methylation is a detoxification mechanism for bacteria when a volatile compound is synthesized. In anaerobic environments, organisms are more likely to transmethylate Pb and As rather than to synthesize methane, but compounds like di- and trimethylarsine are very volatile and release quickly to overlying waters where they are oxidized to the less toxic cacodylic acid.

The volatility of many naturally formed compounds leads to their rapid discharge from the environment, but both methylated arsenicals and tetraalkyllead have been detected in aquatic organisms, and both bioaccumulate in fish to some extent. Tetraalkyllead has been shown to be toxic to aquatic organisms. The toxicity of organic arsenicals has also been demonstrated, but the extent and circumstances have not been evaluated.

Because organo-As and -Pb compounds appear to be eliminated readily, the present evidence indicates little or no biomagnification of these methylated compounds through the food web. However, their synthesis appears to play a significant part in their biogeochemical cycling. In addition, as transmethylation from such methylated metals and metalloids as Pb and As may contribute to the methylation of Hg, they also contribute indirectly to that problem. Because CH_3Hg^+ has such a long half-life, it continues to pose the greatest threat to the environment and organisms, including human beings, in whom it accumulates relentlessly and readily crosses the blood-brain and placental barriers.

The scope of current research must be broadened to understand these and other reactions in nature. Many of the microbial methylation studies were performed *in vitro* with enriched, pure cultures and are not directly applicable to the natural environment. The *in vitro* studies contributed vital information about the processes of methylation, transmethylation, bioaccumulation, bioconcentration, and biomagnification of Hg. However, because the physiological state and composition of these microbiological communities bear little resemblance to the ones that occur naturally, these studies have limited application except to demonstrate likely biological and chemical reactions. These are not a suitable basis on which to assess the quantitative or qualitative value of *in situ* activity.

Of course, adding to the information about the biological and chemical reactions in nature as well as in the laboratory must continue to be of major importance. In light of atmospheric cycling and ongoing releases of the various elements from natural as well as anthropogenic sources, it is appropriate to place greater emphasis on atmospheric research in the future as well as on devising means to decrease anthropogenic emissions, to reclaim the quality of the air as well as the waterways.

8.0 SUMMARY

The geochemical cycling of metals and metalloids like As, Pb, and Hg through the environment has assumed major proportions with the additions from anthropogenic sources to those from natural processes. Despite decreases in domestic, industrial, and agricultural uses and releases of these elements into the environment over the last two decades, large quantitites are stored in aquatic sediments and continue to be transported to the water and atmosphere and back. This cycle is now being delineated along with the processes of biomethylation, accumulation, concentration, and magnification through the food web.

Although the mechanism for the biomethylation of As was reported by Frederick Challenger in the 1930s, the latest wave of research was generated only after new concerns were raised in the 1950s. When an acetaldehyde factory at Minamata, Japan produced CH_3Hg^+ as a by-product, it was discharged into the bay, bioconcentrated by fish, and concentrated in human beings to the extent of causing disability and death. Then Swedish scientists established that biomethylation of the element readily occurred in the natural environment as well. Because microbes were affirmed to methylate Hg in anaerobic waters, the process has also been shown to be common among microbial communities detoxifying their aerobic environments, as well as by some fungi.

While much of the CH_3Hg^+ produced in sediments is released back to the waterways, where it may be bioaccumulated in aquatic organisms, some of it is demethylated by other microbes, so the quantity actually available to the aquatic biota may reflect only the net difference between the amounts methylated and demethylated. Other factors also affect the process. Some of these are: the species of microorganism and its degree of resistance to Hg, the amounts and kinds of organic matter available for sedimentation, and the ambient water conditions such as pH, temperature, and redox potential. Generally, lakes with greater acidity and less organic matter and aquatic plant growth produce fish with higher levels of CH_3Hg^+, perhaps because more CH_3Hg^+ is produced in the sediments and, upon release to the overlying water, is more available for bioaccumulation.

Hg can also be methylated chemically (abiotically) at least under laboratory conditions. This has been demonstrated for reactions with methylcobalamin by UV irradiation in the presence of organic methyl donors, transmethylation reactions, and in reactions with humic and fulvic acid methyl donors, as well as in a mixture of acetaldehyde, Hg^{2+}, and NaCl.

The geochemical cycles of As and Pb have not been as thoroughly researched, but their organic compounds do not appear to biomagnify in the food web like Hg because of their greater tendency to volatilize in the environment and to be more readily metabolized and excreted from aquatic organisms.

REFERENCES

Abernathy, A.R. & P.M. Cumbie, 1977. Mercury accumulation by largemouth bass (*Micropterus salmoides*) in recently impounded reservoirs. *Bull. Environ. Contam. Toxicol.* 17: 595–602.

Adin, A. & W. Espenson, 1971. Kinetics for methyl-transfer to mercury. *Chem. Commun.* 13: 653–654.

Agemian, H. & A.S.Y. Chau, 1977. A study of different analytical extraction methods for non-detrital heavy metals in aquatic sediments. *Arch. Environ. Contam. and Toxicol.* 6: 69–82.

Akagi, H., M. Fujita & Y. Sakagami, 1972. Studies on photochemical alkylation of inorganic mercury. I. *Eiseikagaku* 18: 309–314 (in Japanese).

Akagi, H., D.R. Miller & A. Kudo, 1977. Photochemical transformation of mercury. In *Ottawa River Project*. Final Report, National Research Council of Canada, Ottawa, Ontario: 16.1–16.31.

Akagi, H., D.C. Mortimer & D.R. Miller, 1979. Mercury methylation and partition in aquatic systems. *Bull. Environ. Contam. Toxicol.* 23: 372–376.

Akielaszek, J.J. & T.A. Haines, 1981. Hg in the muscle tissue of fish from three northern Maine lakes. *Bull. Environ. Contam. Toxicol.* 27: 201–208.

Alberts, J.J., J.E. Schindler, R.W. Miller & D.E. Nutter, 1974. Elemental mercury evolution mediated by humic acid. *Science* 184: 895–896.

Andreae, M.O., 1977. Determination of arsenic species in natural waters. *Anal. Chem.* 49: 820–824.

Andreae, M.O., 1978. Distribution and speciation of arsenic in natural waters and some marine algae. *Deep-Sea Res.* 25: 391–402.

Andren, A.W., & R.C. Harriss, 1973. Methylmercury in estuarine sediments. *Nature* 245: 256–257.

Annett, C.S., F.M. D'Itri, J.R. Ford & H.H. Prince, 1975. Mercury in fish and waterfowl from Ball Lake, Ontario. *J. Environ. Qual.* 4: 219–222.

Armstrong, F.A.J., 1979. Mercury in the aquatic environment. In *Effects of Mercury in the Canadian Environment*. NRCC No. 16739. National Research Council of Canada, Ottawa, Ontario: 84–100.

Armstrong, F.A.J. & D.P. Scott, 1979. Decrease of mercury content of fishes in Ball Lake since imposition of controls of mercury discharges. *J. Fish. Res. Bd. Can.* 36: 670–672.

Aston, S.R., D. Bruty, R. Chester & R.C. Padgham, 1973. Mercury in lake sediments: A possible indicator of technological growth. *Nature* 241: 450–451.

Bache, C.A., W.H. Gutenmann & D.J. Lisk, 1971. Residues of total mercury and methylmercury salts in lake trout as a function of age. *Science* 171: 951–952.

Baker, M.D., W.E. Inniss, C.I. Mayfield, P.T.S. Wang & Y.K. Chau, 1983. Effect of pH on the methylation of mercury and arsenic by sediment microorganisms. *Environ. Technol. Lett.* 4: 89–100.

Bartlett, P.D. & P.J. Craig, 1981. Total mercury and methylmercury levels in British estaurine sediments. II. *Water Res.* 15: 37–47.

Bartlett, P.D., P.J. Craig & S.F. Morton, 1978. Total mercury and methyl mercury levels in British estuarine and marine sediments. *Sci. Total Environ.* 10: 245–251.

Batti, R., R. Magnaval & E. Lanzola, 1975. Methylmercury in river sediments. *Chemosphere* 1: 13–14.

Beckert, W.F., A.A. Moghissi, F.H.F. Au, E.W. Bretthauer & J.C. McFarlane, 1974. Methylmercury: Evidence for its formation in a terrestrial environment. *Nature* 249: 674–675.

Bellama, J.M., J.L. Jewett & J.D. Nies, 1985. Rates of methylation of mercury (II) species in water by organotin and organosilicon compounds. In A.E. Martell and K.J. Irgolic (Eds.), *Environmental Inorganic Chemistry*. VCH Publishers, Deerfield Beach, FL: 239–247.

Berdicevsky, I., H. Shoyerman & S. Yannai, 1979. Formation of methylmercury in the marine sediments under *in vitro* conditions. *Environ. Res.* 20: 325–334.

Berman, M. & R. Bartha, 1986. Control of the methylation process in an mercury-polluted aquatic sediment. *Bull. Environ. Contamin. Toxicol.* 36: 401–404.

Bianchi, A.J.M., 1973. Variations in the bacterial concentration of littoral waters and sediments. *Mar. Biol.* 22: 23–29.

Bienvenue, E., A. Boudou, J.P. Desmazes, C. Gavach, D. Georgescauld, J. Sandeaux, R. Sandeaux & P. Seta, 1984. Transport of mercury compounds across bimolecular lipid membranes: Effect of lipid composition, pH and chloride concentration. *Chem. Biol. Interact.* 48: 91–101.

Billen, G., C. Joiris & R. Wollast, 1974. A bacterial methylmercury-mineralizing activity in river sediments. *Water Res.* 8: 219–225.

Bishop, P.L. & E.J. Kirsch, 1972. *Biological Generation of Methylmercury in Anaerobic Pond Sediment.* Eng. Bull. Series 1, 14–1 part 2, Purdue University, West Lafayette, IN: 628–638.

Bisogni, J.J. & A.W. Lawrence, 1973. *Kinetics of Microbially Mediated Methylation of Mercury in Aerobic and Anaerobic Aquatic Environments.* Report to OWRR, Department of the Interior. Technical Report No. 63. Cornell University Water Resources and Marine Sciences Center, Ithaca, NY.

Bisogni, J.J. & A.W. Lawrence, 1975a. Kinetics of mercury methylation in aerobic and anaerobic aquatic environments. *J. Water Pollut. Control Fed.* 47: 135–152.

Bisogni, J.J. & A.W. Lawrence, 1975b. Metabolic cycles for toxic elements in the environment: A study of kinetics and mechanism (J.M. Wood). In P.A. Krenkel (Ed.), *Heavy Metals in the Aquatic Environment*. Pergamon Press, Oxford, England: 113–115.

Blum, J.E. & R. Bartha, 1980. Effect of salinity on methylation of mercury. *Bull. Environ. Contam. Toxicol.* 25: 404–408.

Boddington, M.J., A.S.W. deFreitas & D.R. Miller, 1979. The effect of benthic invertebrates on the clearance of mercury from sediments. *Ecotox. Environ. Safety* 3: 236–244.

Braman, R.S., 1975. Arsenic in the environment. In E.A. Woolson (Ed.), *Arsenical Pesticides*, ACS Symposium Series, Number 7. American Chemical Society, Washington, D.C.: 108–123.

Braman, R.S. & C.C. Foreback, 1973. Methylated forms of arsenic in the environment. *Science* 182: 1247–1249.

Brouzes, R.J.P., R.A.N. McLean & G.H. Tomlinson, 1977. Mercury—The link between pH of natural waters and the mercury content of fish. Paper presented at the meeting of the U.S. National Academy of Sciences—National Research Council Panel on Mercury, Washington, D.C.

Challenger, F., 1935. Biological methylation of compounds of arsenic and selenium. *Chem. Ind. (London)* 54: 657–662.

Challenger, F., C. Higginbottom & L. Ellis, 1933. Formation of organo-metalloidal compounds by microorganisms. *J. Chem. Soc.* 1933: 95–101.

Chau, Y.K., 1980. Biological methylation of tin compounds in the aquatic environment. In *Proc. 3rd Int. Conf. Organometalloidal Coordination Chemistry of Germanium, Tin, Lead*. University of Dortmund, Dortmund, West Germany, July 1980.

Chau, Y.K., P.T.S. Wong, O. Kramar, G.A. Bengert, R.B. Cruz, J.D. Kinrode, J. Lye & J.C. Van Loon, 1980. Occurrences of tetraalkylead compounds in the aquatic environment. *Bull. Environ. Contam. Toxicol.* 24: 265–269.

Corrin, M.L. & D.F.S. Natusch, 1977. Physical and chemical characteristics of environmental lead. In W.R. Boggess & B.G. Wixson (Eds.), *Lead in the Environment*. National Science Foundation, Washington, D.C.: 7–31.

Colwell, R.R. & J.D. Nelson, 1975. *Metabolism of Mercury Compounds in Microorganisms*. U.S. Environmental Protection Agency, Office of Research and Development, Report No. EPA-600/3-75-007, Environmental Research Laboratory, Narragansett, RI: 84 pp.

Coulson, E.J., R.E. Remington & K.M. Lynch, 1935. Metabolism in the rat of the naturally occurring arsenic of shrimp as compared with arsenic trioxide. *J. Nutr.* 10: 255–270.

Cox, H.E., 1925. On certain new methods for the determination of small quantities of arsenic, and its occurrence in urine and fish. *Analyst* 50: 3–13.

Cox, D.P. & M. Alexander, 1973. Production of trimethylarsine gas from various arsenic compounds by three sewage fungi. *Bull. Environ. Contamin. Toxicol.* 9: 84–91.

Craig, P.J., 1980. Metal cycles and biological methylation. In *Handbook of Environmental Chemistry*, Vol. 1A. Springer-Verlag, Berlin: 169–227.

Craig, P.J. & S.F. Morton, 1978. Kinetics and mechanisms of the reaction between methylcobalamin and mercuric chloride. *J. Organometallic Chem.* 145: 79–89.

Craig P.J. & S.F. Morton, 1976. Mercury in Mersey estuary sediments and the analytical procedure for total mercury. *Nature* 261: 125–126.

Craig, P.J. & J.M. Wood, 1981. Biological methylation of lead: An assessment of the present position. In D.R. Lynam, L.E. Piatanida & J.F. Cole (Eds.), *Environmental Lead*. Academic Press, New York: 333 pp.

Craig, P.J. & S. Rapsomankis, 1985. Methylation of tin and lead in the environment: Oxidative methyl transfer as a model for the environmental reactions. *Environ. Sci. Technol.* 19: 726–730.

Cranston, R.E. & D.E. Buckley, 1972. Mercury pathways in a river and estuary. *Environ. Sci. Technol.* 6: 274–278.

Dall'Aglio, M., 1968. The abundance of mercury in 30 natural water samples from Tuscany and Latium (central Italy). In L.H. Abren (Ed.), *Origin and Distribution of the Elements*. Pergamon Press, New York: 1065–1081.

deFreitas, A.S.W., Q.N. LaHam, M.A.J. Gidney, B.A. MacKenzie, J.G. Trepanies, D. Rogers, M.A. Sharpe, S. Qadri, A.E. McKinnon, P. Clay & E. Javorski, 1977. Mercury uptake and retention by fish. In *Ottawa River Project*. Final report, National Research Council of Canada, Ottawa, Ontario: 30.1–30.61.

DeJonghe, W.R.A. & F.C. Adams, 1982. Measurements of organic lead in air — a review. *Talanta* 29: 1057–1067.

DeSimone, R.E., 1972. Methylation of mercury by common nuclear magnetic resonance reference compounds. *Chem. Soc. London J. Chem. Commun.* 13: 780–781.

DeSimone, R.E., M.W. Penley, L. Charbonneau, S.G. Smith, J.M. Wood, H.A.O. Hill, J.M. Pratt, S. Ridsdale & R.J.P. Williams, 1973. The kinetics and mechanisms of cobalamin-dependent methyl and ethyl transfer to mercuric ion. *Biochim. Biophys. Acta* 304: 851–863.

D'Itri, F.M., 1972. *The Environmental Mercury Problem*. CRC Press, Cleveland, OH: 194 pp.

D'Itri, P.A. & F.M. D'Itri, 1977. *Mercury Contamination: A Human Tragedy*, John Wiley & Sons, New York: 311 pp.

D'Itri, F.M., C.S. Annett & A.W. Fast, 1971. Comparison of mercury levels in an oligotrophic & a eutrophic lake. *Mar. Technol. Soc. J.* 5: 10–14.

D'Itri, F.M., A.W. Andren, R.A. Doherty & J.M. Wood, 1978. *An Assessment of Mercury in the Environment*. National Academy of Sciences, Washington, D.C.: 185 pp.

Duchard, P., S.E. Calvert & N.B. Price, 1973. Distribution of trace metals in the pore waters of shallow water marine sediments. *Limnol. Oceanogr.* 18: 605–611.

Dumas, J.P., L. Pazdernik, S. Bellonick, D. Bouchard & G. Vaillancourt, 1977. Methylation of lead in aquatic medium. *Water Pollut. Res. Can.* 12: 91–100.

Edington, D.H. & J.A. Robbins, 1976. Records of lead deposition in Lake Michigan sediments since 1800. *Environ. Sci. Technol.* 10: 266–273.

Edwards, T. & B.C. McBride, 1975. Biosynthesis and degradation of methylmercury in human faeces. *Nature* 253: 462–464.

Fagerstrom, T. & A. Jernelov, 1971. Formation of methyl mercury from pure mercuric sulphide in anaerobic organic sediment. *Water Res.* 5: 121–122.

Fagerstrom, T. & A. Jernelov, 1972. Some aspects of the quantitative ecology of mercury. *Water Res.* 6: 1193–1202.

Faust, S.D. & O.M. Aly, 1981. *Chemistry of Natural Waters*. Ann Arbor Science Publishers, Inc., Ann Arbor, MI: 400 pp.

Foley, R.E., J.R. Spotila, J.P. Giesy & C.H. Hall, 1978. Arsenic concentrations in water and fish from Chautauqua Lake, New York. *Environ. Biol. Fish.* 3(4): 361–367.

Friberg, L., F. Berglund, M. Berlin, G. Birke, R. Cederlof, U. vonEuler, B. Holmsledt, E. Jonsson, K.G. Luning, C. Ramel, S. Skerfving, A. Swensson & S. Tejning, 1971. Methylmercury in fish. *Nord. Hyg. Tidskr.* Suppl. 4: 364 pp.

Frye, C.L. & H.K. Chu, 1988. The case against mercury (II) methylation by aquatic environmental methysiloxanes. *Environ. Toxicol. Chem.* 7: 95–98.

Fujiki, M. & S. Tajima, 1973. Pollution of Minamata Bay and the neighboring sea by factory wastewater containing mercury. In F. Couiston (Ed.), *New Methods in Environmental Chemistry and Toxicology. Collection of Papers from the Research Conference on Ecolological Chemistry*, International Academic Print Co. Ltd., Tokyo: 217–229.

Furutani, A. & J.W.M. Rudd, 1980. Measurement of mercury methylation in lake water and sediment samples. *Appl. Environ. Microbiol.* 40: 770–776.

Furutani, A., J.W. Rudd & C.A. Kelly, 1984. A method for measurement of the response of sediment microbial communities to environmental change. *Can. J. Microbiol.* 30: 1408–1414.

Gale, N.L., B.G. Wixson, M.G. Hardie & J.C. Jennett, 1973. Aquatic organisms and heavy metals in Missouri's New Lead Belt. *Am. Water Res. Bull.* 9: 673–688.

Gardner, W.S., D.R. Kendall, R.R. Odom, H.L. Windom & J.A. Stephens, 1978. The distribution of methyl mercury in a contaminated salt marsh ecosystem. *Environ. Pollut.* 15: 243–251.

Gavis, J. & J. Ferguson, 1972. The cycling of mercury through the environment. *Water Res.* 6: 989–1008.

Gilderhaus, P.A., 1966. Some effects of sublethal concentrations of sodium arsenic on bluegills and the aquatic environment. *Trans. Am. Fish. Soc.* 95: 289–296.

Grandjean, P. & T. Nielson, 1979. Organolead compounds: Environmental health aspects. *Residue Rev.* 72: 97–148.

Håkanson, L., 1974. Mercury in some Swedish lake sediments. *Ambio* 3: 37–43.

Håkanson, L., 1980. The quantitative impact of pH, bioproduction and Hg contamination on the Hg content of fish (pike). *Environ. Pollut. Ser. B* 1: 285–304.

Håkanson, L., A. Nilsson & T. Andersson, 1988. Mercury in fish in Swedish lakes. *Environ. Pollut.* 49: 145–162.

Hannerz, L., 1968. *Experimental Investigations on the Accumulation of Mercury in Water Organisms.* Institute of Freshwater Research, Rep. No. 48, Drottningholm, Sweden: 120–176.

Harrison, R.M. & R. Perry, 1977. The analysis of tetraalkyl lead compounds and their significance as urban air pollutants. *Atmos. Environ.* 11: 847–852.

Harrison, R.M. & D.P.H. Laxen, 1978. Natural source of tetraalkyllead in air. *Nature* 275: 738–739.

Head, P.C., 1976. Organic processes in estuaries. In J.D. Burton and P.S. Liss (Eds.), *Estuarine Chemistry.* Academic Press, New York: 53–91.

Helwig, D.D. & S.A. Heiskary, 1985. *Fish Mercury in NE Minnesota Lakes.* Minnesota Pollution Control Agency, St. Paul, MN: 81 pp.

Hill, H.A.O., J.M. Pratt, S. Ridsdale, F.R. Williams & R.J.P. Williams, 1970.

Kinetics of substitution of co-ordinated carbanions in cobalt (III) corrinoids. *Chem. Commun.* 6: 341–342.

Holm, H.N. & M.F. Cox, 1974. *Mercury in Aquatic Systems: Methylation Oxidation-Reduction, and Bioaccumulation.* Ecological Series Report EPA 6601 3-74-021. U.S. Environmental Protection Agency, Corvallis, OR: 38 pp.

Huckabee, J.W., J.W. Elwood & S.G. Hildebrand, 1979. Accumulation of mercury in freshwater biota. In J.O. Nriagu (Ed.), *The Biogeochemistry of Mercury in the Environment.* Elsevier/North Holland Biomedical Press, New York: 277–302.

Huckabee, J.W. & S.G. Hildebrand, 1974. Background concentrations of mercury and methylmercury in unpolluted freshwater environments. In *1st Congr. Int. Mercurio* (Tomo II). Barcelona. Fabrica Nacional de Moneda y Timbre, Madrid, Spain: 219–223.

Huckabee, J.W., R.A. Goldstein, S.A. Janzen & S.E. Woock, 1975. Methylmercury in a freshwater food chain. In T.C. Hutchinson (Ed.), *Proc. Int. Conf. Heavy Metals in the Environment,* Vol. 2. National Research Council of Canada, Ottawa and Institute for Environmental Studies, University of Toronto, Toronto, Ontario, Canada. October 27-31, 1975: 199–213.

Huey, C., F.E. Brinckman, S. Grim & W.P. Iverson, 1974. The role of bacterial methylation. In *Proc. Int. Conf. Transport of Persistent Chemicals in Aquatic Ecosystems,* Vol. 2. National Research Council of Canada, Ottawa, Canada: 73–78.

Hultberg, H. & B. Hasselrot, 1981. *Kvicksilver i ekosystemet.* Information 04, K.H.M., Stockholm, Sweden: 33–35.

Imura, N., E. Sukewaga, S. Pan, K. Nagae, J. Kim, T. Kwan & T. Ukita, 1971. Chemical methylation of inorganic mercury with methylcobalamin, a vitamin B_{12} analog. *Science* 172: 1248–1249.

Irgolic, K.J., E.A. Woolson, R.A. Stockton, N.R. Newman, N.R. Bottino, R.A. Zingaro, P.C. Kearney, R.A. Pyles, S. Maeda, W.J. McShane & E.R. Cox, 1977. Characterization of arsenic compounds formed by *Daphnia magna* and *Tetraselmis chuii* from inorganic arsenic. *Environ. Health Perspect.* 19: 61–66.

Irukayama, K., 1968. Minamata Disease as a public nuisance. In M. Kutsuma (Ed.), *Minamata Disease: Study Group of Minamata Disease.* Kumamoto University, Kumamoto, Japan: 302–324.

Isensee, A.R., P.C. Kearney, E.A. Woolson, G.E. Jomes & V.P. Williams, 1973. Distribution of alkyl arsenicals in model ecosystems. *Environ. Sci. Technol.* 7: 841–845.

Jackson, T.A. & R.N. Woychuck, 1980. The geochemistry and distribution of mercury in the Wabigoon River System. In T.A. Jackson (Ed.), *Mercury Pollution in the Wabigoon-English River System of Northwestern Ontario, and Possible Remedial Measures: A Progress Report.* Department of the Environment, Winnipeg, Manitoba: 1–28.

Jackson, T.A., G. Kipphut, R. Hesslein & D.W. Schindler, 1980. Experimental study of trace metal chemistry in soft-water lakes at different pH level. *Can. J. Fish. Aquat. Sci.* 37: 387–402.

Jarvenpaa, T., M. Tillander & J.K. Miettinen, 1970. Methylmercury: Halftime of elimination in flounder, pike and eel. *Suom. Kemistil. B* 43: 439–442.

Jarvie, A.W.P., R.N Markall & H.R. Potter, 1975. Chemical alkylation of lead. *Nature* 255: 217–218.

Jenne, E.A., 1976. Trace element sorption by sediments and soils - sites and processes. In W. Chappell & K. Peterson (Eds.), *Symposium on Molybdenum*, Vol. 2. Marcel Dekker, New York: 425–553.

Jensen, S. & A. Jernelov, 1967. Biosyntes av metylkvicksilver. I. *Nordforsk Biocidinf.* 10:4.

Jensen, S. & A. Jernelov, 1969. Biological methylation of mercury in aquatic organisms. *Nature* 223: 753–754.

Jernelov, A., 1970. Release of methyl mercury from sediments with layers containing inorganic mercury at different depths. *Limnol. Oceanogr.* 15: 958–960.

Jernelov, A., 1972. Unpublished report prepared for the Government of Ontario for analysis of St. Clair River Sediments. Institutet for Vatten-Och Luftvardsforskning, Stockholm, Sweden.

Jernelov, A., 1974. Heavy metals, metalloids and synthetic organics. In E.D. Goldberg (Ed.), *The Sea: Ideas and Observations on Program in the Study of the Seas*. Vol. 5, Marine Chemistry. John Wiley & Sons, New York: 799–815.

Jernelov, A., 1980. The effects of acidity on the uptake of mercury in fish. *Environ. Sci. Res.* 17: 211–222.

Jernelov, A. & H. Lann, 1973. Studies in Sweden on the feasibility of some methods for restoration of mercury-contaminated bodies of water. *Environ. Sci. Technol.* 7: 713.

Jernelov, A. & K. Johansson, 1984. Effects of acidity on the turnover of mercury in lakes and cadmium in soil. In G. Muller (Ed.), *4th International Conference on Heavy Metals in the Environment*. CEP Consultants Ltd., Edinburgh: 727–740.

Jernelov, A., L. Landner & T. Larsson, 1975. Swedish perspectives on mercury pollution. *J. Water Pollut. Control Fed.* 47: 810–822.

Jernelov, A. & K. Johansson, 1983. Effects of acidity on the turnover of mercury in lakes and cadmium in soil. In *4th International Conference on Heavy Metals in the Environment*, Vol. 2. CEP Consultants, Edinburgh: 727–740.

Johansson, K. & A. Iverfeldt, 1989. Mercury run off from Swedish forest areas. Oral presentation during Session topic 17 at the 1989 Society International Limnology Conference, Munich, West Germany.

Jonasson, I.R. & R.W. Boyle, 1971. Geochemistry of mercury. In *Mercury in Man's Environment*, Royal Society of Canada, Ottawa, Ontario: 5–21.

Jones, A.J., 1922. The arsenic content of some of the marine algae. *Pharm. J.* 109: 86–87, 104.

Jones, P. & G. Nickless, 1974. Determination of inorganic mercury by gas-liquid chromotagraphy. *J. Chromatogr.* 89: 201–208.

Kamps, L.R., R. Carr & H. Miller, 1972. Total mercury-monomethylmercury content of several species of fish. *Bull. Environ. Contam. Toxicol.* 8: 273–279.

Kelly, C.A. & J.W.M. Rudd, 1984. Epilimnetic sulfate reduction and its relationship to lake acidification. *Biogeochemistry* 1: 63–77

Kemp, A.L.W. & R.L. Thomas, 1976. Cultural impact on the geochemistry of the sediments of Lakes Ontario, Erie and Huron. *Geosci. Can.* 3: 191–207.

Kitamura, S. & K. Sumino, 1972. Photochemical synthesis of methylmercury. *Jpn. J. Hyg.* 27: 123–129 (in Japanese).

Kitamura, S., Y. Hirano, Y. Noguchi, T. Kojima, T. Kakita & H. Kumamoto, 1959. Epidemiological studies on "Minamata Disease" (Supplementary Report No. 2), *J. Kumamoto Med. Soc.* 33 (Suppl. 3): 559–571 (in Japanese).

Kitamura, S., T. Kakita, S. Kojo & T. Kajima, 1960. Epidemiological studies on Minamata Disease. *J. Kumamoto Med. Soc.* 34 (Suppl. 3): 477–480 (in Japanese).

Kozak S. & C.W. Forsberg, 1979. Transformation of mercuric chloride and methylmercury by rumen microflora. *Appl. Environ. Microbiol.* 38: 626–636.

Krenkel, P.A., 1973. *Mercury Environmental Considerations, Part I.* CRC Press, Cleveland, OH: 303–373.

Kudo, A., D.C. Mortimer & J. Hart, 1975. Factors influencing desorption of mercury from bed sediments. *Can. J. Earth Sci.* 12: 1036–1040.

Kudo, A., H. Akagi, D.C. Mortimer & D.R. Miller, 1977. Equilibrium concentrations of methylmercury in Ottawa River sediments. *Nature* 270: 419–420.

Kudo, A., H. Nagase & Y. Ose, 1982. Proportion of methylmercury to the total amount of mercury in river waters in Canada and Japan. *Water Res.* 16: 1011–1015.

Lancaster, R.J., M.R. Coup & J.W. Hughes, 1971. Toxicity of arsenic present in lakeweed. *N.Z. Vet. J.* 19: 141–145.

Landner, L., 1971. Biochemical model for the biological methylation of mercury suggested from methylation studies *in vivo* with *Neurospora crassa*. *Nature* 230: 452–453.

Landner, L. & P.O. Larsson, 1972. IVL Report B 115. Swedish Institute for Water and Air Pollution Research, Stockholm, Sweden.

Langley, D.G., 1973. Mercury methylation in an aquatic environment. *J. Water Pollut. Control Fed.* 45: 44–51.

Lemmo, N.W., S.D. Faust, T. Belton & R. Tucker, 1983. Assessment of the chemical and biological significance of arsenical compounds in a heavily contaminated watershed. I. The fate and speciation of arsenical compounds in aquatic environments—a literature review. *J. Environ. Sci. Health* A18 (3): 335–387.

Lillie, R.A. & J.W. Mason, 1983. *Limnological characteristics of Wisconsin Lakes.* Tech. Bull. No. 138, Wisconsin Department of Natural Resources, Madison, WI: 116.

Lock, R.A.C., 1975. Uptake of methylmercury by aquatic organisms from water and food. In J.H. Koeman & J.J.T.W.A. Strik (Eds.), *Sublethal Effects of Toxic Chemicals on Aquatic Animals.* Elsevier Scientific Publishing Company, New York: 61–79.

Lockhart, W.L., J.F. Uthe, A.R. Kenney & P.M. Mehrle, 1972. Methylmercury in northern pike (*Esox lucius*): distribution, elimination, and some biochemical characteristics of contaminated fish. *J. Fish. Res. Bd. Can.* 30: 485–492.

Lunde, G., 1972a. The absorption and metabolism of arsenic in fish. *Fiskeridir. Norw. Skri. Ser. Teknol. Unders.* 5: 12–18.

Lunde, G., 1972b. The analysis of arsenic in the lipid phase from marine and limnetic algae. *Acta Chem. Scand.* 26: 2642–2644.

Lunde, G., 1973a. The synthesis of fat and water soluble arsenoorganic compounds in marine and limnetic algae. *Acta Chem. Scand*. 27: 1586-1594.

Lunde, G., 1973b. Separation and analysis of organic-bound and inorganic arsenic in marine organisms. *J. Sci. Food Agric*. 24: 1021-1027.

Lunde, G., 1977. Occurrence and transformation of arsenic in the marine environment. *Environ. Health Perspect*. 19: 47-52.

MacLeod, J.C. & E. Pessah, 1973. Temperature effects on mercury accumulation, toxicity and metabolic rate in rainbow trout, *Salmo gairdneri*. *J. Fish. Res. Bd. Can*. 30: 485-492.

McBride, B.C. & R.S. Wolfe, 1971. Biosynthesis of dimethylarsine by methanobacterium. *Biochemistry* 10: 4312-4317.

McWilliams, P.C. & W.T.W. Potts, 1978. The effects of pH and calcium concentrations on gill potentials in the brown trout, *Salmo trutta*. *J. Comp. Physiol*. 126: 277-286.

Menzer, R.O. & J.O. Nelson, 1986. Water and soil pollutants. In C.D. Klaassen, M.O. Amdur and J. Doull (Eds.), *Casarett and Doull's Toxicology: The Basic Science of Poisons*, 3rd ed. Macmillan Publishing Company, New York: 825-853.

Miettinen, J.K., M. Tillander, K. Rissanen, V. Miettinen & Y. Okanomo, 1969. Distribution and excretion rate of phenyl-and methylmercury nitrate in fish, mussels, molluscs and crayfish. In *Proceedings of the 9th Japanese Conference on Radioisotopes*, Japan Industrial Forum, Inc., Tokyo.

Miettinen, V., E. Blankenstein, K. Rissanen, M. Tillander, J.K. Miettinen & M. Valtonen, 1970. Preliminary study on the distribution and effects of two chemical forms of methylmercury in pike and rainbow trout. In Mario Ruivo (Ed.), *Marine Pollution and Sea Life*. FAO-Fishing News (Books) Ltd., London: 298-303.

Mikac, N. & M. Picer, 1985. Mercury distribution in a polluted marine area. Concentrations of methylmercury in sediments and some marine organisms. *Sci. Total Environ*. 43: 27-39.

Miller, R.W., 1975. The role of humic acids in the uptake and release of mercury by freshwater sediments. *Verh. Int. Ver. Theor. Angew. Limnol*. 19: 2082-2086.

Miller, D.R., 1977. Distribution and transport of pollutants. In *Flowing Water Ecosystems*. Final Report, Ottawa River Project, University of Ottawa and the National Research Council of Canada, Ottawa, Canada.

Miller, D.R. & H. Akagi, 1979. pH affects mercury distribution, not methylation. *Ecotoxicol. Environ. Safety* 3: 36-38.

Miller, R.W., J.E. Schindler & J.J. Alberts, 1975. Mobilization of mercury from freshwater sediments by humic acid. In *Mineral Cycling in Southeastern Ecosystems*. Technical Information Center, Augusta, GA: 445-451.

Miskimmin, B.M., C.A. Kelly & J.W.M. Rudd, 1989. Mercury methylation and demethylation in lakes as influenced by dissolved organic carbon (DOC). In *Abstr. 1989 Annu. Meet. American Society of Limnology and Oceanography*, University of Alaska, Fairbanks, AK, June 18-22: 40.

Nagase, H., Y. Ose & T. Sato, 1986. Possibility of mercury methylation of silicones in the environment. Abstr. (Division of Geochemistry) 192nd Natl. Meet. American Chemical Society, Anaheim, CA, September 11, 1986, Paper 126.

Nagase, H., Y. Ose, T. Sato & T. Ishikawa, 1982. Methylation of mercury by humic substances in an aquatic environment. *Sci. Total Environ.* 24: 133–142.

Nagase, H., Y. Ose, T. Sato & T. Ishikawa, 1984. Mercury methylation by compounds in humic materials. *Sci. Total Environ.* 32: 147–156.

NAS, 1977. *Arsenic.* Report of the Subcommittee on Arsenic, Committee on Medical and Biologic Effects of Environmental Pollutants, National Research Council. U.S. Government Publishing and Printing Office, Washington, D.C.: 332 pp.

Neujahr, H. & L. Bertilsson, 1971. Methylation of mercury compounds by methylcobalamin. *Biochemistry* 10: 2805–2808.

Nriagu, J.O., 1978. Properties and the biogeochemical cycle of lead. In J. O. Nriagu (Ed.), *The Biogeochemistry of Lead in the Environment.* Part A. *Ecological Cycles.* Vol. 1A. Elsevier/North-Holland Biomedical Press, New York: 1–284.

Ogner, G. & M. Schnitzer, 1970. Humic substances: fluvic acid-dialkyl phthalate complexes and their role in pollution. *Science* 170: 317–318.

Olson, B.H. & R.C. Cooper, 1974. *In situ* methylation of mercury in estuarine sediment. *Nature* 252: 682–683.

Olson, B.H. & R.C. Cooper, 1976. Comparison of aerobic and anaerobic methylation of mercuric chloride by San Francisco Bay sediments. *Water Res.* 10: 113–116.

Olson, K.R., H.L. Bergman & P.O. Fromm, 1973. Uptake of methylmercuric chloride by trout: A study of uptake pathways into the whole fish and uptake by erythrocytes *in vitro. J. Fish. Res. Bd. Can.* 30: 1293–1299.

Olson, G.F., D.I. Mount, V.M. Snarski & T.W. Throslund, 1975. Mercury residues in fathead minnows, *Pimephales promelas* Rafinesque, chronically exposed to methylmercury in water. *Bull. Environ. Contam. Toxicol.* 14: 129–134.

Olson, G.J., 1986. Microbial intervention in trace element containing industrial process streams and waste products. In M. Bernhard, F.E. Brinckman & P. Sadler (Eds.), *The Importance of Chemical Speciation in Environmental Processes*, Dahlem Konferenzen, Springer-Verlag, Berlin, West Germany: 493–512.

Pan-Hou, H.S. & N. Imura, 1982a. Involvement of mercury methylation in microbial mercury detoxification. *Arch. Microbiol.* 131: 176–177.

Pan-Hou, H.S. & N. Imura, 1982b. Physiological role of mercury methylation in *Clostridium cochlearium. Bull. Environ. Contam. Toxicol.* 29: 290–297.

Park, J.W., J.D. Hollinger & P.M. Almost, 1980. The transport and dynamics of the total and methyl mercury in the Wabigoon River and Clay Lake. In T.A. Jackson (Ed.), *Mercury Pollution in the Wabigoon-English River System of Northwestern Ontario, and Possible Remedial Measures: A Progress Report.* Department of the Environment, Winnipeg, Manitoba: 1–32.

Parks, G.A., F.W. Dickson, J.O. Leckie, P.I. McCarty, P. Berendson & K.L. Pering, 1973. *Part of Trace Elements in Water: Origin, Fate and Control.* Progress Rep., Stanford University, Stanford, CA, March 1972 to February 1973: 247.

Penrose, W.R., 1974. Arsenic in the marine and aquatic environments. Analysis, occurrence, and significance. *Crit. Rev. Environ. Contam.* 4: 465–482.

Penrose, W.R., H.B.S. Conacher, R. Black, J.C. Meranger, W. Miles, H.M. Cunningham & W.R. Squires, 1977. Implications of inorganic/organic interconversion on fluxes of arsenic in marine food webs. *Environ. Health Perspect.* 19: 53–59.

Phillips, G.R. & D.R. Buhler, 1978. The relative contributions of methylmercury from food or water to rainbow trout (*Salmo gairdneri*) in a controlled laboratory experiment. *Trans. Am. Fish. Soc.* 107: 853–861.

Phillips, G.R., T.E. Lenhart & R.W. Gregory, 1980. Relation between trophic position and mercury accumulation among fishes from the Tongue River Reservoir, Montana. *Environ. Res.* 22: 73–80.

Portmann, J.E. & J.P. Riley, 1964. Determination of arsenic in sea water, marine plants and silicate and carbonate sediments. *Anal. Chim. Acta* 31: 509–519.

Pratt, D.R., J.S. Bradshaw & B. West, 1972. Arsenic and selenium analysis in fish. *Proc. Utah Acad. Sci. Arts Lett.* 49: 23–26.

Ramlal, P.S., J.W.M. Rudd, A. Furutani & L. Xun, 1985. The effect of pH on methyl mercury production and decomposition in lake sediments. *Can. J. Fish. Aquat. Sci.* 42: 685–692.

Ramlal, P.S., J.W.M. Rudd & R.E. Hecky, 1986. Measurement of specific rates of mercury methylation and degradation and their use in determining factors controlling net rates of mercury methylation. *Appl. Environ. Microbiol.* 51: 110–114.

Reay, P.F., 1972. The accumulation of arsenic from arsenic-rich waters by aquatic plants. *J. Appl. Ecol.* 9: 557–565.

Reimers, R.S., P.A. Krenkel, M. Eagle & G. Tragitt, 1975. Sorption phenomenon in the organics of bottom sediments. In P.A. Krenkel (Ed.), *Heavy Metals in the Aquatic Environment*. Pergamon Press, Oxford, England: 117–136.

Reinert, R.E., L.J. Stone & W.A. Willford, 1974. Effect of temperature on accumulation of methylmercuric chloride and p,p' DDT by rainbow trout, *Salmo gairdneri. J. Fish. Res. Bd. Can.* 31: 1649–1652.

Ridley, W.P., L.J. Dizikes & J.M. Wood, 1977. Biomethylation of toxic elements in the environment. *Science.* 197: 329–332.

Rissanen, K. & J.K. Miettinen, 1972. Use of mercury compounds in agriculture and its implications. In *Mercury Contamination in Man and His Environment*. Technical Reports Series No. 137. International Atomic Energy Agency, Vienna, Austria: 5–34.

Rissanen, K., J. Erkama & J.K. Miettinen, 1970. Experiments on microbiological methylation of mercury (2 +) ion by mud and sludge under aerobic and anaerobic conditions. In *FAO Technical Conference on Marine Pollution and Its Effects on Living Resources and Fishing*, Rome, Italy, December 9–18, 1970.

Rodgers, D.W. & F.W.H. Beamish, 1981. Uptake of waterborne methylmercury by rainbow trout (*Salmo gairdneri*) in relation to oxygen consumption and methylmercury concentration. *Can. J. Fish. Aquat. Sci.* 38: 1309–1315.

Rodgers, D.W. & F.W.H. Beamish, 1983. Water quality modifies uptake of waterborne methylmercury by rainbow trout, *Salmo gairdneri*. *Can. J. Fish. Aquat. Sci.* 40: 824–828.

Rogers, R.D., 1976. Methylation of mercury in soil. *J. Environ. Qual.* 5: 454–458.

Rogers, R.D., 1977. A biological methylation of mercury in soil. *J. Environ. Qual.* 6: 463–467.

Rowland, I.R., M.J. Davies & P. Grasso, 1975. The methylation of mercury by the gastro-intestinal contents of the rat. *Biochem. Soc. Trans.* 3: 502–504.

Rowland, I.R., M.I. Davies & P. Grasso, 1977. Biosynthesis of methylmercury compounds by the intestinal flora of the rat. *Arch. Environ. Health* 32: 24–28.

Rudd, J.W.M. & M.A. Turner, 1983. The English-Wabigoon river system. II. Suppression of mercury and selenium bioaccumulation by suspended and bottom sediments. *Can. J. Fish Aquat. Sci.* 40: 2218–2227.

Rudd, J.W.M., A. Furutani & M.A. Turner, 1980. Mercury methylation by fish intestinal contents. *Appl. Environ. Microbiol.* 40: 777–782.

Rudd, J.W.M., M.A. Turner, A. Furutani, A.L. Swick & B.E. Townsend, 1983. A synthesis of recent research on the English-Wabigoon system with a view towards mercury amelioration. *Can. J. Fish. Aquat. Sci.* 40: 2206–2217.

Rudd, J.W.M., C.A. Kelly, V. St. Louis, R.H. Hesslein, A. Furutani & M. Holoka, 1987. Microbial consumption of nitric and sulfuric acids in acidified north temperate lakes. *Limnol. Oceanogr.* 31: 1267–1280.

Saxena, J. & P. Howard, 1977. Environmental transformations of alkylated and inorganic forms of certain metals. *Adv. Appl. Microbiol.* 21: 185–226.

Scheider, W.A., D.S. Jefferies & P.J. Dillon, 1979. Effects of acidic precipitation in precambrian freshwaters in southern Ontario. *J. Great Lakes Res.* 5: 45–51.

Schindler, D.W., R.H. Hesslein, R. Wagemann & W.S. Broecker, 1980. Effects of acidification on mobilization of heavy metals and radionuclides from the sediments of a freshwater lake. *Can. J. Fish. Aquat. Sci.* 37: 373–377.

Schmidt, U. & F. Huber, 1976. Methylation of organolead and lead (II) compounds to $(CH_3)_4Pb$ by microorganisms. *Nature* 259: 157–158.

Schnitzer, M. & H. Kerndorff, 1981. Reactions of fulvic acid with metal ions. *Water Air Soil Pollut.* 15: 97–108.

Schrauzer, G.N., J. Weber, R. Holland, T. Beckham & R. Ho, 1971. Alkyl group transfer from cobalt to mercury: Reaction of alkylcobalamins, alkylcobaloximes, and of related compounds with mercuric acetate. *Tetrahedron Lett.* 3: 275–277.

Schrauzer, G.N., J.A. Seck, R.J. Holland, T.M. Beckham, E.M. Rubin & J.W. Sibert, 1973. Reductive dealklation of alkycobaloximes, alkylcobalamins, and related compounds: Simulation of corrin dependent reductase and methyl group transfer reactions. *Bioinorg. Chem.* 2: 93–124.

Schroeder, H.A. and J. J. Balassa, 1966. Abnormal trace metals in man: Arsenic. *J. Chronic Dis.* 19: 85–106

Schuth, C.K., A.R. Isensee, E.A. Woolson & P.C. Kearney, 1974. Distribution of carbon-14 and arsenic derived from carbon-14 labeled cacodylic acid in aquatic ecosystem. *J. Agric. Food Chem.* 22: 999–1003.

Seydel, I.S., 1972. Distribution & circulation of arsenic through water, organisms and sediments of Lake Michigan. *Arch. Hydrobiol.* 71: 17–30.

Shapiro, S.K. & F. Schlenk, 1965. *Transmethylation and Methionine Biosynthesis.* University of Chicago Press, Chicago, IL: 261.

Shin, E.B. & P.A. Krenkel, 1976. Mercury uptake by fish and biomethylation mechanisms. *J. Water Pollut. Control Fed.* 48: 473–501.

Silver, S., J. Schottel & A. Weiss, 1976. Bacterial resistance to toxic metals determined by extrachromosomal R factors. In S.M. Sharpley & A. M. Kaplan (Eds.), *Proceedings of the Third International Biodegradation Symposium.* Applied Science Publishers Ltd., London, England: 899–917.

Sirota, G.R. & J.F. Uthe, 1977. Determination of tetraalkyllead compounds in biological material. *Anal. Chem.* 49(6): 823–825.

Smies, M., 1983. Biological aspects of trace element speciation in the aquatic environment. In G. Leppard (Ed.), *Trace Element Speciation in Surface Waters.* Plenum Press, New York: 177–193.

Spangler, W.J., J.L. Spigarelli, J.M. Rose, R.S. Flippen & H.H. Miller, 1973a. Degradation of methylmercury by bacteria isolated from environmental samples. *Appl. Microbiol.* 25: 488–493.

Spangler, W.J., J.L. Spigarelli, J.M. Rose & H.M. Miller, 1973b. Methylmercury: Bacterial degradation in lake sediments. *Science.* 180: 192–193.

Stock, A. & F. Cucuel, 1934. Die verbreitung des quecksilbers. *Naturwissenschaften* 22: 390–393.

Stokes, P.M., S.I. Dreier, M.O. Farkas & R.A.N. McLean, 1983. Mercury accumulation by filamentous algae: A promising biological monitoring system for methylmercury in acid stressed lakes. *Environ. Pollut.* 5: 255–271.

Suns, K., C. Curry & D. Russell, 1980. *The Effects of Water Quality and Morphometic Parameters on Mercury Uptake by Yearling Yellow Perch.* Technical Report LTS 80-1. Ontario Ministry of the Environment, Toronto, Ontario Canada.

Tillander, M., J.K. Miettinen, K. Rissanen, V. Miettinen & E. Minkkinen, 1969. Excretion of phenyl-and methylmercuric nitrate following oral or intramuscular administration into fish, mussels, snails, and crayfish. *Nord. Hyg. Tidskr.* 50 (2): 181–183 (in Swedish).

Thayer, J.S. & F.E. Brinkman, 1982. The biological methylation of metals and metalloids. *Adv. Organometal. Chem.* 20: 313–356.

Thomas, R.L., 1972. The distribution of mercury in the sediments of Lake Ontario. *Can. J. Earth Sci.* 9: 636–651.

Thomas, R.L., 1973. The distribution of mercury in the surficial sediments of Lake Huron. *Can. J. Earth Sci.* 10: 194–204.

Tonomura, K., K. Maeda, F. Futai, T. Nakagami & M. Yamada, 1968. Stimulative vaporization of phenylmercuric acetate by mercury-resistant bacteria. *Nature* 217: 644–646.

Topping, G. & I.M. Davies, 1981. Methylmercury production in the marine water column. *Nature* 290: 243–245.

Tsubaki, T., K. Shirakawa, K. Hirota, K. Kondo, T. Sato & K. Kanabayashi, 1977. Case history of Niigata. In Tsubaki, T. & K. Irukayama (Eds.), *Minamata Disease: Methylmercury Poisoning in Minamata and Niigata, Japan.* Elsevier Scientific Publishing Co., New York: 317.

Van Faassen, H.G., 1976. Methylation of mercury compounds in soil, sediment and sewage sludge samples. *Plant Soil* 44: 505–509.

Varanasi, V., P.A. Robisch & D.A. Malins, 1975. Structural alterations in fish epidermal mucus produced by waterborne lead and mercury. *Nature* 258: 431–432.

Verta, M., S. Rekolaianen & K. Kinnunen, 1986. Causes of increased fish mercury levels in Finnish reservoirs. *Publ. Water Res. Inst. (Helsinki)* 65: 44–58.

Vinogradov, A.P., 1953. *The Elementary Chemical Composition of Marine Organisms*. Memoir 2. Sears Foundation for Marine Research, Yale University, New Haven, CT.

Vonk, J.W. & A.K. Sijpesteijn, 1973. Studies on the methylation of mercuric chloride by pure cultures of bacteria and fungi. *Antonie van Leeuwenhoek J. Microbiol. Serol.* 39: 505–513.

Wallace, R.A., W. Fulkerson, W.D. Schultz & W.S. Lyon, 1971. *Mercury in the Environment*. ORNL-NSF-EP-1. Oak Ridge National Laboratory, Oak Ridge, TN: 61 pp.

Watanabe, N., H. Nagase, T. Nakamura, E. Watanabe & Y. Ose, 1986. Chemical methylation of mercury (II) salts by polydimethylsiloxanes in aqueous solutions. *Ecotoxicol. Environ. Safety* 11: 174–178.

Watras, C.J., N.S. Bloom, W.F. Fitzgerald, J.P. Hurley, D.P. Krabbenhoft, R.G. Rada & J.G. Wiener, 1989. Mercury in temperate lakes: A mechanistic field study. Oral presentation during Session topic 17 at the 1989 Society of International Limnology Conference, Munich, West Germany.

Westoo, G., 1966. Determination of methylmercury compounds in foodstuffs. I. Methylmercury compounds in fish, identification and determination. *Acta Chem. Scand.* 20: 2131–2137.

Westoo, G., 1973. Methylmercury as percentage of total mercury in flesh and viscera of salmon and sea trout of various ages. *Science* 181: 567–568.

Wiener, J.G., 1983. *Comparative Analyses of Fish Populations in Naturally Acidic and Circumneutral Lakes in Northern Wisconsin*. FWS/OBS-80/40.16. U.S. Fish and Wildlife Service, Eastern Energy and Land Use Team, Kearneysville, WV: 107 pp.

Wiener, J.G., 1986. *Effect of Experimental Acidification on Mercury Accumulation by Yearling Yellow Perch in Little Rock Lake, Wisconsin*. Report to U.S. Environmental Protection Agency, Corvallis, OR: 20 pp.

Wilson, S.H. & M. Fieldes, 1942. Studies in spectrographic analysis. II. Minor elements in sea weed (*Macrocystic pyrifera*). *N.Z. J. Sci. Technol.* 23B: 47–48.

Windom, H.L., R. Stickney, R. Smith, D. White & F. Taylor, 1973. Arsenic, cadmium, copper, mercury and zinc in some species of North Atlantic finfish. *J. Fish. Res. Bd. Can.* 30: 275–279.

Windom, H.L., W. Gardner, J. Stephens & F. Taylor, 1976. The role of methylmercury production in the transfer of mercury in a salt marsh ecosystem. *Estuar. Coastal Mar. Sci.* 4: 579–583.

Wollast, R., G. Billen & F.T. Mackenzie, 1975. Behavior of mercury in natural systems and its global cycle. In A.D. McIntyre & C.F. Mills (Eds.), *Ecological Toxicology Research: Effects of Heavy Metal and Organohalogen Compounds*.

Proceedings of a NATO Science Committee Conference. Plenum Press, New York: 145–166.

Wong, P.T.S., Y.K. Chau & P.L. Luxon, 1975. Methylation of lead in the environment. *Nature* 253: 263–264.

Wong, P.T.S., B.A. Silverberg, Y.K. Chau & P.V. Hodson, 1978. Lead in the aquatic biota. In J.O. Nriagu (Ed.), *Biogeochemistry of Lead. Part B. Biological Effects*, Vol. 1B. Elsevier/North-Holland Biomedical Press, New York: 279–342.

Wong, P.T.S, Y.K. Chau, O. Kramar & G.A. Bengert, 1981. Accumulation and depuration of tetramethyllead by rainbow trout. *Water Res.* 15: 621–625.

Wood, J.M., 1971. Toxic elements in the environment. In J.M. Pitts and R.L. Metcalf (Eds.), *Advances in Environmental Science and Technology*, Vol. 2. Wiley Interscience, New York: 39–56.

Wood, J.M., 1972. Personal communication. Gray Freshwater Biological Institute, University of Minnesota, Navarre, MN.

Wood, J.M., 1974. Biological cycles for toxic elements in the environment. *Science* 183: 1049–1052.

Wood, J.M., 1975. Biological cycles for elements in the environment. *Naturwissenschaften* 62 (8): 357–364.

Wood, J.M., 1980. The role of pH and oxidation reduction potential in the mobilization of heavy metals. In T.Y. Toribaca, M.W. Miller & D.E. Morrow (Eds.), *Polluted Rain.* Plenum Press, New York: 223–238.

Wood, J.M., 1983. Selected biochemical reactions of environmental significance. *Chem. Scripts* 21: 155–162.

Wood, J.M., 1987. Biological processes involved in the cycling of elements between soil or sediments and the aqueous environment. *Hydrobiologia* 149: 31–42.

Wood, J.M. & Y.T. Fanchiang, 1979. Mechanisms for B_{12}-dependent methylation. In B. Zagalak & W. Friedrich (Eds.), *Proc. 3rd Vitamin B_{12} Proc. Eur. Symp.*, Walter de Gruyter & Co., Berlin, West Germany: 539–556.

Wood, J.M, F.S. Kennedy & C.G. Rosen, 1968. Synthesis of methylmercury compounds by extracts of a methanogenic bacterium. *Nature* 220: 173–174.

Woolson, E.A., 1975. Bioaccumualtion of arsenicals. In E.A. Woolson (Ed.), *Arsenical Pesticides.* ACS Symp. Ser. No. 7. American Chemical Society, Washington, D.C.: 97–107.

Woolson, E.A., P.C. Kearney, A.R. Isensee, W.G. McShane, K.J. Irgolic and R.A. Zingaro, 1974. 168th ACS Meeting, Pesticide Chemistry Division, Paper #8.

Wren, C.D. & H.R. MacCrimmon, 1983. Mercury levels in the sunfish, *Lepomis gibbosus,* relative to pH and other environmental variables of Precambrian Shield Lakes. *Can. J. Fish. Aquat. Sci.* 40: 1737–1744.

Wrench, J.J. & R.F. Addison, 1981. Reduction, methylation, and incorporation of arsenic into lipids by the marine phytoplankton *Dunaliella tertiolecta. Can. J. Fish. Aquat. Sci.* 38: 518–523.

Wright, D.R. & R.D. Hamilton, 1982. Release of mercury from sediments: Effects of mercury concentration low temperature and nutrient addition. *Can. J. Fish. Aquat. Sci.* 39: 1459–1466.

Xun, L., N.E.R. Campbell & J.W.M. Rudd, 1987. Measurements of specific rates of net methylmercury production in the water column and surface sediments to acidified and circumneutral lakes. *Can. J. Fish. Aquat. Sci.* 44: 750–757.

Yamada, M. & K. Tonomura, 1972a. Microbial methylation of mercury in hydrogen sulfide evolving environments. *J. Ferment. Technol.* 50: 901–909.

Yamada, M. & K. Tonomura, 1972b. Formation of methylmercury compounds from inorganic mercury by *Clostridium cochlearium. J. Ferment. Technol.* 50: 156–159.

Zingaro, R., 1983. Biochemistry of arsenic: recent developments. In W. Lederer & R. Fensterheim (Eds.), *Arsenic.* Van Nostrand Reinhold Company, New York: 327–347.

CHAPTER 7

Biogenic Gas Production and Mobilization of In-Place Sediment Contaminants by Gas Ebullition

Donald D. Adams, Nicholas J. Fendinger, and Dwight E. Glotfelty

1.0 INTRODUCTION

A major portion of the mineralization of deposited organic matter in sediments results in the accumulation of gaseous microbial waste products, such as CH_4, N_2, and CO_2, as well as trace concentrations of other gases. The most important bacterial pathways responsible for gas production are denitrification, which yields N_2, methanogenesis, which produces CH_4, and CO_2 from numerous fermentation reactions including sulfate reduction. These microbiological processes have recently been reviewed in the literature (Postgate 1984, Winfrey 1984, Capone & Kiene 1988, Kuenen & Robertson 1988, Oremland 1988, Seitzinger 1988, Zumft et al. 1988) and therefore will not be discussed in detail. Production and venting of these gases from sediments represent substantial losses (termed "sinks") of nitrogen and carbon from aquatic ecosystems (Molongoski & Klug 1980, Kelly & Chynoweth 1981, Martens & Klump 1984, Adams & Fendinger 1986, Fendinger & Adams 1987, Seitzinger 1988). Therefore, bubble ebullition may also be an important mechanism which affects the extent of eutrophication in waters receiving substantial nutrient loading.

Much of the organic matter reaching the sediments of aquatic systems is decomposed in such a way that various inorganic electron acceptors are reduced. As pointed out by Capone and Kiene (1988), organic matter mineralization is accompanied by a sequence of microbial respiration reactions at different depths in the sediments, with the most energetically favored electron acceptors used first in the order of: $O_2 > NO_3^- > Mn^{4+} > Fe^{3+} > SO_4^{2-} > CO_2$. Even though the depth of oxygen penetration into sediments is diffusion limited (Revsbech & Jørgensen 1986, Cappenberg 1988, Sweerts & Cappenberg 1988), oxygen as the terminal electron acceptor usually accounts for the major amount of organic carbon oxidation at the sediment-water interface (Adams & van Eck 1988, Sweerts & Cappenberg 1988). Below the zone of oxygen penetration and in most sediments containing less than 1 mM sulfate, methanogenesis displaces sulfate reduction as the dominant pathway for terminal carbon metabolism (Capone & Kiene 1988, Kelly *et al*. 1988).

The formation of sediment gases may also mobilize organic compounds from sediments. This pathway can be considered a three-step process. First, gas bubbles are produced in the sediments as described earlier. This gas typically contains 45 to 95% CH_4, 5 to 80% N_2, and trace quantities of CO_2 and H_2 (Kuznetsov 1968, Howard *et al*. 1971, Chen *et al*. 1972, Ward & Frea 1979, Chanton *et al*. 1988). In addition to gases, the sediment interstitial water contains dissolved organic compounds, some of which are contaminants released by human activities. As a second step, it is expected that dissolved organic contaminants will partition into sediment gas bubbles. The third step would be the transport of gases, along with their vapor-phase constituents, from the sediments to overlying waters either by diffusion or bubble ebullition. In shallow water environments which receive large amounts of nutrients, bubble ebullition is probably the dominant mechanism for advection of gases from sediments (Fendinger 1981). For example, diffusion has been observed to account for only 13 to 36% of total methane transfer from a variety of different Florida wetland environments (Barber *et al*. 1988). The remainder was bubble loss. By this pathway in-place organic compounds can be transported from sediments directly into the atmosphere (Thomas *et al*. 1987). Transit through the water column could represent some loss as a result of partitioning from gas to water phases and into particulate organic matter. However, it is expected that this would be negligible because of transit time. Even though Cicerone and Oremland (1988) reported that bacterial oxidation would have little influence on the methane content of rising bubbles, Snodgrass (1977) reported 52 to 56% CH_4 loss as a result of oxidation, but no change in N_2 concentration, between surface and bottom gas collection samplers at two stations which were at 10 and 20 m in Hamilton Harbour, Ontario, Canada. Therefore, methane diffusion from rising gas bubbles can be an important contributor to hypolimnion deoxygenation. However, diffusive transport from gas to water phases of other bubble components is as yet an unknown factor.

The objective of our laboratory study was an evaluation of the second step of this pathway, that is, the partitioning of dissolved organic compounds from sediment interstitial water (porewater) into the gas phase of bubbles. The behavior of other porewater contaminants in polluted harbor sediments could be predicted from the partitioning of ^{14}C-labelled lindane.

Furthermore, if the partitioning of in-place organic contaminants into gas bubbles were known, it would then be possible to estimate the sediment-to-atmosphere transport rate of these pollutants by measuring gas bubble ebullition from various environments.

2.0 METHODS AND MATERIALS

The collection of sediment samples for extraction of interstitial water from sites in open Lake Erie and Hamilton Harbour was divided into two different time sequences: 1979–1980 for the measurements of porewater-dissolved organic contaminants by gas chromatography, followed by mass spectrometry (GC/MS) (Adams *et al.* 1984), and 1987–1988 for large-scale removal of sediment porewater for laboratory gas/water ^{14}C-labelled lindane partition studies reported here.

2.1 Study Areas

Lake Erie, the southernmost of the Laurentian Great Lakes, is located between 42°45′ and 42°50′ north latitude and 78°55′ and 83°30′ west longitude. The lake has a total area of 25,300 km^2, total volume of 470 km^3, length of 386 km, mean width of 17 km, and is divided into three major, distinct subbasins. The western basin is the shallowest, with a mean depth of 11 m. The central basin is the largest in terms of area (16,317 km^2) and has a mean depth of 25 m, while the eastern basin is the deepest, with a maximum depth of 64 m (Sly 1976).

Sediments were collected from the central and eastern basins of Lake Erie. The central basin is dominated by fine-grained muds, particularly in the central depositional areas. A west to east increase in grain size is due to increasing wave energy caused by the long fetch of the basin and prevailing winds (Thomas *et al.* 1976). Eastern basin sediments exhibit a mean off-shore decrease in grain size as a result of reduced hydraulic energy with water depth. Coarse silt material found in the deepest portion of the basin is derived directly from erosion of Long Point on the northern shore of the lake. Sedimentation rates for the central basin range from 1.5 to 4.6 mm/year, while rates of 1.3 to 14 mm/year were reported for the eastern basin (Mortimer 1987). Some of the greatest sedimentation rates for stations in the open Great Lakes have been observed in the eastern basin (Robbins *et al.* 1978). Concentrations of some organic contaminants in surface water of Lake Erie, which were measured in 1986, have been reported by Rathke and McRae (1989). Of the 17 compounds listed in their IJC Report, only two, α-benzene hexachloride (α-BHC) and polychlorinated biphenyls (PCBs), exhibited lakewide averages greater than 1 ng/L. Concentrations of organic compounds in surface water at our four sampling locations (stations 84a, A-1, and C-11 in the middle of the central basin and 23 in the eastern basin) averaged 3 ng/L for α-BHC, 0.4–0.7 ng/L for total PCBs, 3 ng/L for lindane, and 0.4–0.5 ng/L for p-dichlorobenzene. Concentrations of synthetic organic compounds in the sediments at these locations were not reported even though data for other "areas of concern" and Lake Erie harbors and tributaries have been tabulated by Rathke and McRae (1989).

With an average depth of 13 m and a surface area of 21.5 km², Hamilton Harbour is considered one of the most polluted sites in the Great Lakes (Barica 1989) and has been designated as one of the "areas of concern" by the International Joint Commission (IJC 1989). It is connected to the western end of Lake Ontario by a 9.5 m deep by 107 m wide ship canal which affords a theoretical hydraulic retention time of 73–107 days (OME 1987). The harbor receives 4.3 m³/sec of treated municipal waste, 3.5 m³/sec of storm water overflow from an urbanized watershed, and 27 m³/sec of water withdrawn and recirculated back to the harbor by south-shore steel mills (Barica 1989, Poulton 1989). Besides exceeding Ontario provincial water quality objectives for ammonia, zinc, total phosphorus, turbidity, iron, phenols, cyanides, copper, nickel, chromium, and coliforms (Barica 1989), contamination from numerous organic pollutants such as polynuclear aromatic hydrocarbons (PNAs), PCBs, hexachlorobenzenes, and hexachlorocyclohexane isomers has also been observed (Poulton 1987).

2.2 Sediment Collection and Processing

During the 1979–80 sampling period, sediments were collected from Hamilton Harbour (station 258) and the central basin of Lake Erie (C-11) for GC/MS identification of dissolved organic components in porewater (Adams et al. 1984). The sediments were divided into two segments, 0–25 cm and 25 cm and deeper (normally 25–50 cm), and labelled top and bottom. Station C-11 bottom samples were collected by multiple coring while top sediments were obtained with a "shipek" dredge. The sediments were quickly transferred to quart mason jars which had been cleaned with chromic acid and heated in a muffle furnace at 450°C for 4 h. The containers were covered with precombusted aluminum foil and sealed with a lid and retaining ring, leaving no headspace over the sediments. Samples were packed in ice and kept at approximately 4°C until processing in the laboratory. Additional cores were also obtained for porewater DOC measurements at stations A-1 in the central basin of Lake Erie, within 2 miles of station C-11, and EB near NOAA station 28 in the eastern basin of Lake Erie.

Sediment porewater was removed by centrifugation at 9000 rpm and 4°C. Porewater samples were stored in a nitrogen atmosphere to avoid oxidation, then passed through a glass fiber filter and adjusted to pH 12 with 12 M NaOH. Even though sorption of organic substances to the glass fiber filters was not tested, routine DOC measurements before and after filtration did not indicate such losses. Filtration was necessary, however, to remove resuspended colloidal material during manipulation of the sample after centrifugation. Organic compounds were extracted with pesticide-grade methylene chloride (Burdick and Jackson, Muskegon, MI). The porewater was then adjusted to pH 2 with 12 M HCl and extracted again with a second portion of methylene chloride. "Base-neutral"- and "acid"-extractable methylene chloride fractions were dried with sodium sulfate, reduced in volume using a Kuderna-Danish apparatus, and refrigerated until analysis. All glassware was washed in dilute soap water followed by thorough cleaning with chromic acid, rinsing with distilled, deionized water, and heating to 450°C for 4 h. Aluminum foil was heated in a muffle furnace before wrapping glassware during storage. Sodium sulfate was heated at 450°C for 4 h, and sodium hydroxide and diluted hydrochloric acid were prepared with low-carbon water. The same glassware was used for method blanks by employing low-carbon water which was prepared as outlined in Strickland and Parsons (1972).

Purgeable organic analysis, following the Bellar and Lichtenberg (1974) technique, utilized GC/MS (Hewlett-Packard Model 5983 MS with data system) for compound separation and identification. Samples of porewater were spiked with internal standards [$BrClCH_2$ and $Cl(CH_2)_4Cl$], sparged onto Tenax®/silica gel, then resorbed onto a 183 cm × 0.32 cm 0.2% Carbowax® 1500 on 80/100 mesh Carbopak C column held at –40°C. This was the same program used for an EPA validation study of purgeable priority pollutants. The base/neutral extracts were analyzed using a 183 cm × 2 mm i.d. SP-225ODB glass column programmed from 50°C to 260°C at 8°C/min. An internal standard, biphenyl-d_{10}, was used in each sample with the method blank additionally containing dialkyl adipate and dioctyl phthalate. Samples were quantified using base/neutral priority pollutants from Supelco (Bellefonte, PA). The analytical procedure for the acid fraction was designed to identify compounds based upon the EPA phenolic priority pollutant methodology. Methylene chloride extracts were injected onto a 2 mm i.d. × 183 cm SP-124ODA glass column with 50°C to 250°C temperature programming at 8°C/min. In all cases the GC was held at 250°C for 10–15 min for elution of additional organic compounds.

During the 1987–88 sample collection period, sediments were obtained using box cores at station 84a in the central basin of Lake Erie, approximately 12 miles NNE of stations A-1 and C-11, from which samples were collected during the 1979–80 period, and station 23 in the eastern basin. Sediments were sectioned into 0 to 10 cm top and 10 to 20–30 cm bottom sections. Sediments from station 258 in Hamilton Harbour were from the same location as during 1979–80. Multiple gravity coring was conducted to collect enough sediments. Cores were sectioned immediately into 0–20 cm and 20–40 cm portions. After sectioning, sediments were stored in 3-L glass canning jars with glass lids (Arc, France) which had been rinsed with deionized water. Jars were iced or refrigerated until interstitial water was removed by centrifugation at 7000 rpm and 4°C. Porewater was filtered through precombusted (6 h at 400°C) Whatman® GF/C glass fiber filters. The dissolved organic carbon content of the porewater was determined with an OI Analytical (College Station, TX) model 700 organic carbon analyzer.

2.3 Reagents

Organic-free water used in this study was produced by the Hydro (Rockville, MD) model 4C2–18 ultrapure water system with a critical applications column. Labelled (^{14}C) γ-hexachlorocyclohexane (lindane) was obtained from Mallinckrodt (St. Louis, MO) and was purified by preparative thin-layer chromatography (TLC) on silica gel plates developed with methylene chloride. The final purity ($>98\%$) of the material was determined by TLC and liquid scintillation counting.

Waters (Milford, MA) C-18 Sep Paks were used to extract ^{14}C-labelled lindane from water and air collected from the wetted-wall column (WWC) system, which is explained in the next section. Sep Pak cartridges used for water extractions were conditioned prior to use by eluting with 5 mL of methanol, followed by 5 mL of organic-free water.

A simulated sediment gas mixture composed of 80% methane and 20% nitrogen gas was prepared by Air Products Corp. (Tamaqua, PA). This simulated mixture was similar in composition to sediment porewater gases from Lake Erie and Hamilton Harbour (Snodgrass 1977, Adams *et al.* 1984, Adams & Fendinger 1986, Fendinger & Adams 1987).

A constant vapor source of ^{14}C-lindane was produced from a generator column constructed from a 5 cm $\times$ 0.75 cm i.d. glass tube packed with ultrafine glass wool. Labelled lindane dissolved in benzene was added to the packed glass wool. The benzene was evaporated under a stream of nitrogen gas prior to use.

Porewater samples from Hamilton Harbour and Lake Erie sediments were amended with ^{14}C-lindane by applying 0.04–0.06 μCi of labelled lindane dissolved in benzene to the walls of 125-mL glass bottles. The benzene was evaporated under a stream of nitrogen before addition of porewater. The amended porewater samples were allowed to equilibrate at least 36 h before gas/water partitioning measurements were made.

2.4 Gas/Water Partitioning Measurements

A WWC (Figure 1) was used to measure partitioning of ^{14}C-labelled lindane between sediment porewater and simulated sedimentary biogenic gas. Gas/liquid partitioning of an organic solute in the WWC occurs between a thin liquid film on the inside of a vertical column (2.2 cm i.d. × 32 cm) and a concurrent flow of gas. A complete description of the WWC apparatus is found in Fendinger and Glotfelty (1988). Gas/water partitioning measurements were made by introducing ^{14}C-labelled lindane either as a vapor in the gas phase (gas-to-water partitioning) or as an aqueous solute (water-to-gas partitioning). Both modes of operation are shown in Figure 1. Water and gas collected from the WWC were passed through C-18 Sep Pak extraction columns, which were found to be 98% efficient in extracting ^{14}C-lindane from the gas and aqueous phases. Each gas/water partitioning determination consisted of at least two measurements of 30–40 min equilibration period before the first extraction cartridges were placed on the wetted-wall column. Also, each gas/water partitioning measurement consisted of one set of cartridges (one air and one water) placed on the column for 20 to 40 min.

Each of the gas/water partitioning measurements for porewater samples was made with gas flow rates of 200 mL/min and a water flow rate of approximately 0.5 mL/min. Additional gas-to-water partition measurements in organic-free water were made with air flow rates that ranged from 30 to 200 mL/min.

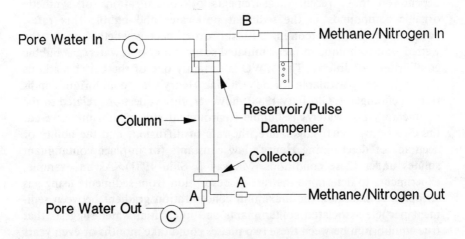

Figure 1. Wetted-wall column apparatus showing reservoir pulse dampener, collector and column components. Additional components include: A, extraction columns; B, optional vapor source generator column for gas-to-water measurements; C, valveless metering pump.

Sep Pak cartridges used for water extraction were dried for 15–30 min. The extraction cartridges were cut open and the contents emptied into a scintillation vial. After addition of 15 mL of Handifluor® (Mallinckrodt, St. Louis, MO) scintillation cocktail, the activity was determined with a Searl Analytic Inc. (Elk Grove Village, IL) Mark III Model 6880 Liquid Scintillation System. Dimensionless gas/water partition coefficients were calculated from the activity in air and water extracts and volume of air and water passed through the extraction columns from the wetted-wall column system (Equation 1):

$$\text{Gas/water partition coefficient} = \frac{(\text{activity air/volume air})}{(\text{activity water/volume water})} \quad (1)$$

3.0 RESULTS AND DISCUSSION

Partitioning of a dissolved organic compound between porewater and a prepared gas mixture which simulated sedimentary conditions was determined. Even though concentrations of organic compounds in sediment interstitial water were measured 6–8 years ago, the objectives of this investigation would not be invalid. Calculations of modern sediment-to-air fluxes of in-place contaminants from gas bubble ebullition rates would obviously depend on more recent measurements of concentrations of synthetic organic compounds in the sediment porewater and bubble flux rates. Another obvious shortcoming with this experiment is whether or not the wetted-wall technique in fact simulated sedimentary porewater/gas bubble equilibrium conditions. The WWC is certainly one of the better methods that is presently available for determining Henry's law equilibrium conditions (Fendinger & Glotfelty 1988); however, three questions related to the sedimentary porewater/gas bubble environment still remain unanswered: the direction of diffusing solute, the rate of diffusion, and the ability or accuracy of determining Henry's law constants for in-place contaminant solutes under these conditions (Mackay & Shiu 1981). As an example, experiments to determine the rate of desorption from sediments using gas stripping to maintain the maximum concentration gradient between sediment particle-associated contaminants and interstitial water suggests that true equilibrium between these two phases could take months or even years depending on the composition of the sediment, porewater, and solute (Karickhoff 1980, Karickhoff & Morris 1985, Oliver 1985, Coats & Elzerman 1986, Witkowski et al. 1988).

3.1 Synthetic Organic Compounds in Sediment Interstitial Water

Concentrations of purgeable synthetic organic compounds in interstitial water from location C-11 in the central basin of Lake Erie and 258 near the center of Hamilton Harbour were small (Table 1) compared to extractable organic compounds from the same sample. The only exceptions were a few halogenated methanes which probably represented laboratory contaminants. Smaller molecular weight organic compounds were either rapidly utilized by bacteria to form methane, carbon dioxide, and cellular material or were transformed to greater molecular weight humic material via a pathway through the fulvic acids (Krom & Sholkovitz 1977).

Purgeable priority pollutants were not measured in excess of the reportable limit, which was approximately 1 $\mu g/L$, except those shown in Table 1. Hexane and ethylbenzene were identified in samples from both Lake Erie and Hamilton Harbour. Other volatile compounds found in sediments from Hamilton Harbour were xylene, indan or methylstyrene, and C_3- and C_4-

Table 1. Purgeable Organic Compounds ($\mu g/L$) Determined in Sediment Interstitial Water from Lake Erie and Hamilton Harbour[a]

Minutes[c]	Component	Lake Erie Station C-11[b]		Hamilton Harbour Station 258[b]	
		Surface	Bottom	Surface	Bottom
10	Dichloromethane[d]	60		163	
10.2	Methylene chloride	33	11	ND	ND
15.1	Chloroform	Sample contaminated	Sample contaminated	2.8	14
21.2	Benzene[e]	0.9	ND	0.8	1.0
22.8	Hexane[d,f] (1)	ca. 30		ca. 30	
27.3	Toluene[e]	1.0	1.7	ND	ND
30.0	Ethylbenzene	ND	ND	ND	0.9
32.9	Xylene[f] (2)	ND	ND	ND	~1
33.2	Xylene[f] (2)	ND	ND	~1	~2
35.5	Indan or methylstyrene[f] (3)	ND	ND	~4	~33
36.1	C_3–benzene[f] (4)	ND	ND	ND	~2
40+	C_4–benzene[f] (5)	ND	ND	ND	?

Source: Adams et al. (1984).

[a]Concentrations were corrected for background. ND = not detected where detection limit in general was 1-5 $\mu g/L$.

[b]Surface = 0–25 cm sediment section; bottom = 25–50 cm or deeper sediment section.

[c]Approximate retention time.

[d]Data transposed from other purge and trap samples from similar locations.

[e]Possibly a contaminant of the analytical system; background level was approximately ½ of the value before correction.

[f]Tentative identification and quantification in absence of authentic standard; quantitation based on major ion, assuming same relative response as for the average of the major ion responses in toluene and ethylbenzene: (1) major masses – 57, 41, 56, 86; (2) major masses – 91, 106; (3) major masses – 117, 118, 115; (4) major masses – 105, 120, 50; and (5) major masses – 119, 134.

benzenes (Adams *et al.* 1984). Mass spectra at several time intervals during elution of an unresolved GC "hump" revealed aromatics, alkenes, and cycloalkanes but no alkanes.

Substantial numbers of base/neutral components were observed in the extractable fractions of sediment interstitial water samples from Hamilton Harbour (Table 2). The composition of these samples was diverse and included alkanes, aromatics, PNAs (polynuclear aromatic hydrocarbons), and probably some heterocyclic compounds. The concentrations of polynuclear aromatic hydrocarbons ranged from approximately 1 μg/L to 13 μg/L but were generally in the range of 1–4 μg/L. Concentrations of alkanes and some of the apparent heterocyclic components, which occurred at concentrations as great as 30 μg/L, were estimated to be approximately ten times greater than the PNAs. The series of compounds, marked (X) in Table 2, eluting at 16.9, 18.0, 18.6, 19.1, and 19.5 min have mass spectra compatible with dialkylthiophene or alkylamino-carbazole species. The great degree of similarity among the mass spectra strongly suggests a series of homologs eluting roughly 1 min apart, with isomeric forms of the two later homologs eluting at slightly different times. Examination of the general background of the unresolved "hump" early in the chromatogram at 17.9 min suggests that the mixture of compounds is very likely aliphatic and alicyclic or alkenyl in composition. Some specific compounds identified at small μg/L concentrations were acenaphthene, anthracene/phenanthrene, chrysene/benzo(a)anthracene, methylnaphthalene, and pyrene. Large-molecular-weight PNAs or chlorinated compounds were not observed in the extracts. The presence of small-molecular-weight PNAs, aromatic and heterocyclic organic contaminants dissolved in the interstitial water, particularly in the surface sediments, reinforces the results of sediment bioassays demonstrating significant mortality of mayfly larvae in Hamilton Harbour (IJC 1989).

Concentrations of base/neutral-extractable organic compounds in porewater samples obtained from the central basin of Lake Erie were much smaller than observed in Hamilton Harbour sediments. Concentrations were slightly greater in the bottom 25–50 cm section of the core's interstitial water than in the top 0–25 cm portion. Concentrations of 1 to 4 μg/L of saturated hydrocarbons, with C_{28} to C_{36} chain lengths, were also measured in the open lake station sediment porewater. These compounds were probably of natural origin.

Few, if any, substances were observed in the acidified methylene chloride extracts of sediment interstitial water (Table 3). The analytical procedure employed favored phenols, so the concentrations of extractable organic compounds were not more than a few μg/L in the original interstitial water. Phenolic priority pollutants were not detected (≥ 1 μg/L); however, the extraction procedure discriminates against carboxylic acids, especially dicarboxylic acids. Only four compounds were identified in these extracts:

Table 2. Analysis of Base-Neutral Mass Spectral Data from the Hamilton Harbour Station 258[a]

| Retention time (min.) | | | Concentration |
Major Peaks[b]	Minor Peaks[c]	Compound (Major Masses in Parentheses)	μg/L in Original Porewater[d]
	12.3	Methylnaphthelene[e]	1.6[f]
15.2		Alkane (57, 71, 85)	8[g]
	15.6	Acenaphthene	3.7
15.9		Alkane (71, 57, 85)	18[g]
16.4		Unknown (97, 84, 69)	
16.9		Unknown (X) (181, 166, 165, 196)	12[h]
17.2		Alkane (71, 57, 85)	16[g]
18.0		Unknown (X) (195, 165, 180)	16[h]
18.4		Alkane (71, 85, 57)	6[g]
18.6		Unknown (X) (195, 180, 210)	6[h]
18.9		Unknown (143, 91, 142)	6[h]
19.1		Unknown (X) (209, 224, 179)	
19.5		Unknown (X) (209, 224, 71)	
20.2		Anthracene/phenanthrene[i]	13
20.6		Unknown (71, 85, 57)	6[g]
21.1		Alkane (71, 57, 85)	8[g]
	21.5	Methyl-anthracene/phenanthrene[e,i]	3[f]
22.2		Alkane (71,57,85)	14[g]
	23.0	C_2-anthracene/phenanthrene[e,i]	1[f]
23.1		Alkane (71, 85, 57)	15[g]
	23.9	Fluoranthene	2.6
24.0		Unknown (71, 83, 69)	
24.2		Alkane (71, 57, 85)	14[g]
	24.7	Pyrene	2.2
25.3		Alkane (57, 71, 85)	12[g]
26.2		Alkane plus? (71, 57, 85)	9[g]
27.1		Alkane plus? (71, 57, 85)	9[g]
28.1		Alkane plus? (71, 57, 85)	6[g]
	28.3	Chrysene/benzo(a)anthracene[i,j]	0.9
28.3		Unknown (104, 91, 184)	
28.9		Unknown (71, 57, 85)	6[g]
29.5		Unknown (195, 196, 220)	5[h]
29.9		Unknown (57, 71, 85)	5[g]
30.6		Alkane (57, 71, 85)	5[g]
31.5		Unknown (191, 97, 95)	

Source: Adams et al. (1984).
[a]Concentrations were corrected for background. X: see text for discussion of this series of peaks.
[b]Peaks easily observable by inspection of chromatogram.
[c]Very small peaks or components not apparent as a separate peak.
[d]Concentrations not followed by a footnote were calculated relative to the Supelco B/N standards.
[e]Tentative identification based on mass spectra; not confirmed with authentic sample.
[f]Concentration of alkylated species were calculated relative to response of nonalkylated parent compound in the Supelco standards.
[g]Concentration estimated based on area of m/e 71 in the unknown relative to m/e 164 in the biphenyl-d_{10} internal standard.
[h]Concentration estimated based on area of major mass in the unknown relative to m/e 164 on the biphenyl-d_{10} internal standard.
[i]Not resolved by chromatography or mass spectra; either or both may be present.
[j]Mass spectrum too weak for positive identification, but retention time in GC identical to standard compound.

Table 3. Analysis of Acid-Extractable (Phenolic) Fractions of Sediment Interstitial Water from Lake Erie and Hamilton Harbour[a]

Min.[b]	Identity[c]	Lake Erie Station C-11			Hamilton Harbour 258		
		Top	Top	Bottom	Bottom	Blank	Bottom
10.2	p-Cresol	1–5	1	a	a	N.D.	1
12.5	Unknown[d]	≤1	a	a	a	a	a
13.7	Unknown[d]	1–5	1–2	0.5–1	≤1	≤1	1–2
17.1	Phthalate[e]	1–5	1–3	0.5–2	1–5	a	10–25
18.9	Dioctyl adipate	N.D.	N.D.	10–25	N.D.	N.D.	N.D.
21.3	bis (2-Ethylhexyl) phthalate	N.D.	N.D.	7–15	N.D.	N.D.	N.D.

Source: Adams *et al.* (1984).
[a]Concentrations (μg/L) listed below were estimated for original interstitial water from injected extracts. a: mass spectra too weak for accurate identification. N.D.: not detected.
[b]Approximate retention time.
[c]Identity based on mass spectra. This was not confirmed by authentic standards.
[d]Mass spectra too weak for identification.
[e]Found also in the Supelco standard at the 1–5 μg/L level; undoubtedly an artifact at that level.

p-cresol, an unspecified phthalate (likely a contaminant), dioctyl adipate, and bis-(2-ethylhexyl)-phthalate. Because the last two compounds were identified in only one sample from Lake Erie and not in Hamilton Harbour, their actual presence in the porewater must be considered tentative.

3.2 Gas/Water Partitioning of Lindane as a Representative In-Place Porewater Organic Contaminant

The diffusion of in-place organic pollutants into sediment gas bubbles was simulated in the laboratory using lindane and a gas mixture representing typical sedimentary conditions (80% CH_4, 20% N_2). This simulation was divided into two experiments with the wetted-wall column: partitioning of lindane using the CH_4:N_2 gas mixture with organic-free water, and a similar experiment using the same gas mixture but with porewaters collected from Lake Erie and Hamilton Harbour sediments.

3.2.1 Gas/Water Partitioning of Lindane in Organic-Free Water

Rates of gas-to-water and water-to-gas partitioning of lindane in organic-free water are presented in Figure 2. A water-to-gas partitioning coefficient for lindane with a CH_4:N_2 gas flow rate of 200 mL/min was 6.8 $\pm$ 0.7 $\times$ 10^{-5} (n = 4). Gas-to-water partitioning coefficients for lindane decreased as a function of CH_4:N_2 flow rate, from 3.4 $\pm$ 0.2 $\times$ 10^{-4} (n = 6) at 200 mL/ min to 1.7 $\times$ 10^{-4} (n = 2) at 65 mL/min and 27 mL/min. In a previous WWC study (Fendinger & Glotfelty 1988), a water-to-air (compressed air as a gas source) partition coefficient of 8.2 $\pm$ 0.9 $\times$ 10^{-5} (n = 5) for lindane, which was independent of air flow rate (37 to 223 mL/min) and column

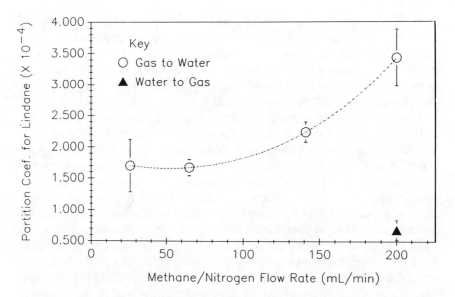

Figure 2.   Gas-to-water and water-to-gas partitioning coefficients for lindane in organic-free water. Error bars are ±1 sigma.

length (10 to 57 cm), was determined. The air-to-water partition coefficient for lindane with an air flow rate of 50 mL/min was $9.5 \pm 1.2 \times 10^{-5}$ (n = 3). This was statistically similar to the water-to-air partition coefficient. Based on agreement between air-to-water and water-to-air coefficients it was concluded that air/water equilibration was achieved in the WWC under the selected conditions (Fendinger & Glotfelty 1988). Gas/water equilibration was probably not achieved in the $CH_4:N_2$−low-carbon water system because: 1) gas-to-water and water-to-gas partition coefficients do not agree and 2) the partitioning of lindane in an air/water system is possibly dependent upon gas flow rates. Gas/water equilibration with the WWC is dependent on adequate gas/water contact time for diffusion-controlled mixing of organic solutes in each phase. The non-equilibrium conditions in the $CH_4:N_2$−organic-free water WWC system may be caused by diffusion-limited transport in the gas phase. This indicates that different gas diffusion constants may exist for lindane in a $CH_4:N_2$ gas mixture than in air.

3.2.2 Gas/Water Partitioning of Lindane in Sediment Porewater.

Mean water-to-gas partition coefficients for sediment porewater ranged from 5.76×10^{-5} for Lake Erie's eastern basin (10–30 cm sediment depth section) to 1.08×10^{-4} for Hamilton Harbour's bottom (20–40 cm) section (Figure 3). Water-to-gas partition coefficients were statistically similar for surficial and deep sediments at all stations. There were also no statistical differences between water-to-gas partition coefficients for porewater at the Lake Erie stations and similar coefficients determined with organic-free water. Water-to-gas partition coefficients for sediment porewater collected from Hamilton Harbour in 1988 were higher (ca. 1 to 4 $\times$ 10^{-5}) than coefficients measured for the Lake Erie stations.

Agreement between water-to-gas partition coefficients for sediment porewater from Lake Erie stations and organic-free water was unexpected, because the association of lindane with dissolved organic material in porewater should decrease the partition coefficient. Smaller partition coefficients in sediment interstitial water compared to low-carbon water would be expected because lindane associated with or bonded to dissolved organic material would not be free to exchange with the gaseous phase. Because water-to-gas partition coefficients for organic-free water were similar to Lake Erie sediment porewater or lower than Hamilton Harbour (HH'88) 1988 porewater, there was probably little association of free lindane with dissolved organic matter during the 36-h equilibration period before partition measurements were made. Organic matter dissolved in the sediment porewater that could associate with lindane, measured as DOC, ranged in

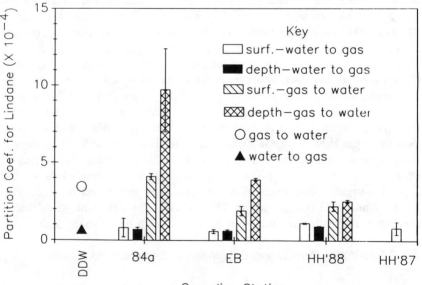

Figure 3.   Gas-to-water and water-to-gas partitioning coefficients for surface and deep sediments at stations in Hamilton Harbour (HH) and Lake Erie (84a in central basin, EB in eastern basin). Error bars are ± 1 sigma.

concentration from 6.6 mg/L at station 84a to 38.1 mg/L at HH'88 (Table 4).

Gas-to-water partition coefficients for sediment porewater were at least twice the measured coefficients for water-to-gas partitioning. The greater values for gas-to-water partitioning with sediment porewater probably resulted from the same mechanisms that cause the greater values measured for gas-to-water partitioning in organic-free water. Because all gas-to-water partitioning measurements were made under the same WWC operating conditions, comparisons of gas-to-water measurements for different stations and sediment depths can be made.

There was generally no agreement between gas-to-water partition coefficients for sediment porewater and organic-free water. Gas-to-water partition coefficients for sediment porewater from deep sediments were twice the coefficients for surface sediment porewater at the Lake Erie stations. The difference in gas-to-water partitioning between surface and deep sediments may be related to organic carbon composition at different sediment depths as a result of diagenic alteration of porewater components. Sediments collected from Hamilton Harbour in 1988 were apparently more homogeneous

Table 4. Sediment Porewater Dissolved Organic Carbon (DOC) Concentrations for 1979–1980 and 1988 Data

Date	Station-Location[a]	Core Section (cm)	DOC (mg/L)[b]
1979	A-1 L.E. Central Basin	0–25	9.3 ± 2.8 (11)
		25–50	13.4 ± 1.4 (4)
	258 Hamilton Harbour	0–25	43.1 ± 17.1 (18)
		25–50	40.2 ± 15.4 (13)
1980	A-1 L.E. Central Basin	0–25	14.4 ± 3.0 (8)
		25–50	17.8 ± 3.7 (5)
	EB L.E. Eastern Basin	0–25[a]	18.1 ± 8.4 (9)
		25–50	24.2 (2)
		0–25	23.0 ± 4.9 (15)
		25–50	19.2 ± 3.7 (6)
	258 Hamilton Harbour	0–25[a]	36.7 ± 6.7 (17)
		25–50	42.9 ± 4.7 (5)
		0–25	36.8 ± 10.0 (8)
		25–50	48.3 ± 4.3 (4)
1988	84a L.E. Central Basin	0–10	6.55[c]
		10–20	7.44
	EB L.E. Eastern Basin	0–10	20.7[c]
		10–20	19.6
	258 Hamilton Harbour	0–20	19.3[c]
		20–40	38.1

Sources: Deis (1981), Adams et al. (1984).

[a]For the eastern basin (EB) of Lake Erie (L.E.), two cores were collected at the same station at the same time; those from Hamilton Harbour were collected during May (first data set) and August 1980. Lake Erie central basin stations (A-1 and 84a) were within 12 miles of each other.

[b]Statistical average and standard deviation (number of data).

[c]Sediment porewater used for partitioning experiments.

because there was no difference between gas-to-water partition coefficients for surface and deep sediment porewater samples. However, a direct relationship between dissolved organic carbon concentration in the sediment porewater and the gas-to-water partition coefficient is not evident (Table 4 and Figure 3). Unlike water-to-air partition measurements where there is a longer equilibration period of a pesticide solute with other organic solutes, gas-to-water partition measurements have a rapid equilibration equivalent to the residence time of the two phases in the WWC. Therefore, if components of the DOC influence the partitioning behavior of lindane in the gas-to-water measurements it must be during transport across the liquid film. For sediment porewater at Lake Erie station 84a the flux of lindane vapor into the porewater was impeded, resulting in greater concentrations in the vapor phase and, thus, greater gas-to-water partition coefficients than measured in organic-free water. Sediment porewater from the Hamilton Harbour station and surface sediments at the Lake Erie eastern basin station, both sampled in 1988, caused enhanced transport or possible binding with porewater organic solutes, which resulted in lesser gas/water partition coefficients than measured in organic-free water.

3.3 Sediment Contaminant Transport by Bubble Ebullition

A graphic representation of sediment contaminant transport by gas bubble ebullition is shown in Figure 4. The transport of sediment contaminants will depend on the rate of bubble ebullition and the partitioning of organic solutes between sediment porewater and in-place gas bubbles. The behavior of ^{14}C-labelled lindane in the sediment gas/porewater WWC system was used to predict the behavior of other organic compounds measured in sediment interstitial water, as given in Tables 1–3.

Gas ebullition from sediment is mainly dependent on the rate of gas production (methanogenesis and denitrification) and water depth. Hesslein (1976) and Fendinger (1981) theorized that gas bubble loss from sediments occurred when the sum of the partial pressures of the gases in a sediment gas bubble exceeded the atmospheric plus hydrostatic pressures above the sediments. Based on these criteria and porewater concentrations of sediment gases, Fendinger (1981) speculated that gas loss would occur from Hamilton Harbour but not Lake Erie central and eastern basin stations. Other measurements of gas bubble ebullition for nearby locations indicated rates as great as 2640 mL/m^2/day for Cleveland Harbor (Ward & Frea 1979) to 224 mL/m^2/day in Hamilton Harbour (Chau *et al.* 1977). Howard *et al.* (1971) also reported bubble loss up to 2600 mL/m^2/day from Lake Erie western basin sediments. Given the range of ebullition rates for nearby sediments, a conservative estimate for gas loss from near-shore harbor sediments of 1000

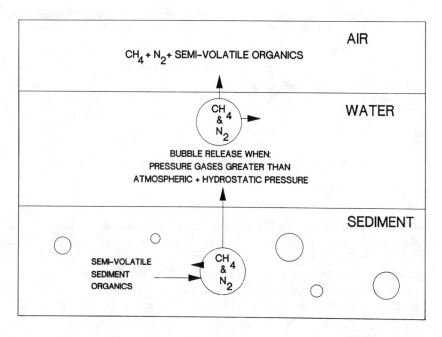

Figure 4. Possible pathway for sediment contaminant transport by gas bubble ebullition.

mL/m²/day would be reasonable. Estimated bubble loss from Hamilton Harbour sediments was calculated at 8×10^9 L of gas annually.

Sediment porewater/gas bubble equilibrium is probably best estimated by water-to-gas partition coefficients measured with the WWC. There was no evidence that porewater DOC concentrations influenced partitioning of lindane from the water phase in the WWC. It was assumed that published values for Henry law constants (HLCs) could be used to estimate partitioning of organic compounds other than lindane into sediment gas bubbles. Approximate gas/water partitioning for other compounds could then be calculated. However, other compounds may interact with porewater DOC to a greater extent than was observed for lindane and may not be readily transported into the gas phase. Regardless, use of HLC data compiled by Mackay and Shui (1981) along with gas ebullition rates from Hamilton Harbour provides an estimated upper limit of the flux of organic contaminants via sediment gas bubbles. Compounds with the greatest HLC and greatest sediment porewater concentrations (dichloromethane, chloroform, and fluoranthene) were estimated to be lost from the sediments at the greatest rates (Table 5).

The final phase of transport of synthetic hydrocarbons from sediments would be gas ejection from the sediments, movement through the water

Table 5. Compounds Identified in the Sediment Porewater of Hamilton Harbour along with Henry's Law Constants and Estimated Annual Fluxes from Hamilton Harbour Sediments by Way of Gas Bubble Ebullition[a]

Compound	Henry's Law Constants[b]	Sediment Porewater Concentration (μg/L)[c]	Estimated Annual Flux (kg)
Dichloromethane	1.0×10^{-1}	163	124
Chloroform	1.5×10^{-1}	8	9
Methylnaphthalene	1.8×10^{-2}	1.6	<1
Acenaphthene	9.7×10^{-3}	3.7	<1
Fluoranthene	8.8×10^{-3}	2.6	17
Pyrene	4.8×10^{-4}	2.2	<<1

[a]Fluxes are estimated upper limits using lindane a a model compound (see text).
[b]From Mackay and Shiu (1981).
[c]From Adams et al. (1984).

column, and release into the atmosphere. Contaminant behavior in this phase is probably best modeled by gas-to-water partition coefficients made with the WWC. Lindane introduced as vapor phase into the $CH_4:N_2-$ sediment porewater system would remain within the gas phase. Therefore, it was speculated that only a small proportion of organic compounds which partition into sediment gas bubbles would be lost to the overlying water column during the transit period. The remaining material is advected into the atmosphere.

Compounds that may also be lost from the sediments through gas ebullition could include mono-, di-, and trichlorobiphenyl congeners, toxaphene, and other semivolatile environmentally persistent organic compounds. To better estimate the flux of sediment organic contaminants by this pathway, measurements of gas ebullition rates and porewater concentrations of contaminants should be made concurrently. Uncertainty concerning the influence of sediment porewater DOC on the partitioning of hydrophobic compounds other than lindane also limits the accuracy of our current estimates. However, based on the estimated flux of these few compounds from Hamilton Harbour sediments, it is apparent that the transport of sediment-associated organic compounds by way of gas bubbles may be an important pathway and should be considered in future toxic chemical management plans and models for the Great Lakes basin. It is a mechanism by which in-place pollutants may be recycled within the basin as well as being transported outside of the basin by way of the atmosphere.

ACKNOWLEDGMENTS

The staff of Ohio State University's Center for Lake Erie Research (1979–1980 period) and C.C.I.W. of Environment Canada (for both the 1980 and 1988 periods) are gratefully acknowledged for ship time and field help. Analytical help was provided by D.A. Deis and D.J. Wagel. Dr. G.G. Hess, Department of Chemistry, Wright State University and the GC/MS facilities at Monsanto Research Corporation in Dayton, OH generously helped in the identification of porewater contaminants. Financial support was provided by OWRT A-059 Ohio, the EPA Large Lakes Laboratory Contract R806757-01-1, and a NOAA contract to D.D.A. in 1988 to participate in an LSS Limnos cruise. Carolyn Turner is thanked for typing the manuscript.

REFERENCES

Adams, D.D. & N.J. Fendinger, 1986. Early diagenesis of organic matter in the Recent sediments of Lake Erie and Hamilton Harbor. I. Carbon gas geochemistry. In P.G. Sly (Ed.), *Sediments and Water Interactions*. Springer-Verlag, Inc., New York: 305-318.

Adams, D.D. & G.Th.M. van Eck, 1988. Biogeochemical cycling of organic carbon in the sediments of the Grote Rug reservoir. *Arch. Hydrobiol. Beih. Ergebn. Limnol.* 31: 319-330.

Adams, D.D., G.G. Hess, N.J. Fendinger, D.A. Deis, D.J. Wagel, D.M. Parrish & J.A. Henry, 1984. *Chemical Study of the Interstitial Water Dissolved Organic Matter and Gases in Lake Erie, Cleveland Harbor, and Hamilton Harbour Bottom Sediments — Composition and Fluxes to Overlying Waters*. OWRT A-059, Office Water Res. Technol., Ohio State University, Columbus, OH: 187 pp.

Barber, T.R., R.A. Burke & W.M. Sackett, 1988. Diffusive flux of methane from warm wetlands. *Global Biogeochem. Cycles* 2: 411-425.

Barica, J., 1989. Unique limnological phenomena affecting water quality of Hamilton Harbour, Lake Ontario. *J. Great Lakes Res.* 15: 519-530.

Bellar, T.A. & J.J. Lichtenberg, 1974. Determination of volatile organics at microgram-per-liter levels by gas chromatography. *Am. Water Works Assoc.* Dec.: 739-744.

Capone, D.G. & R.P. Kiene, 1988. Comparison of microbial dynamics in marine and freshwater sediments: Contrasts in anaerobic carbon catabolism. *Limnol. Oceanogr.* 33: 725-749.

Cappenberg, T.E., 1988. Quantification of aerobic and anaerobic carbon mineralization at the sediment-water interface. *Arch. Hydrobiol. Beih. Ergebn. Limnol.* 31: 307-317.

Chanton, J.P., G.G. Pauly, C.S. Martens, N.E. Blair, & J.W.H. Dacey, 1988. Carbon isotope composition of methane in Florida everglades soils and fractionation during its transport to the atmosphere. *Global Biogeochem. Cycles* 2: 245-252.

Chau, Y.K., W.J. Snodgrass & P.T.S. Wong, 1977. A sampler for collecting evolved gases from sediment. *Water Res.* 11: 807-809.

Chen, R.L., D.R. Keeney, J.G. Konrad, A.J. Holding & D.A. Graetz, 1972. Gas production in sediments of Lake Mendota. *J. Environ. Qual.* 1: 155-158.

Cicerone, R.J. & R.S. Oremland, 1988. Biogeochemical aspects of atmospheric methane. *Global Biogeochem. Cycles* 2: 299–327.

Coats, J.T. & A.W. Elzerman, 1986. Desorption kinetics for selected PCB congeners from river sediments. *J. Contam. Hydrol.* 1: 191–210.

Deis, D.A., 1981. Distribution of Dissolved and Particulate Organic Carbon in Sediments of Lake Erie and Two Polluted Harbors. M.Sc. Thesis, Wright State University, Dayton, OH.: 119 pp.

Fendinger, N.J., 1981. Distribution and Related Fluxes of Dissolved Pore Water Gases (CH_4, N_2, ΣCO_2) in the Sediments of Lake Erie and Two Polluted Harbors. M.Sc. Thesis, Wright State University, Dayton, OH: 107 pp.

Fendinger, N.J. & D.D. Adams, 1987. Nitrogen gas supersaturation in the Recent sediments of Lake Erie and two polluted harbors. *Water Res.* 21: 1371–1374.

Fendinger, N.J. & D.E. Glotfelty, 1988. A laboratory method for the experimental determination of air-water Henry's law constants for several pesticides. *Environ. Sci. Technol.* 22: 1289–1293.

Hesslein, R.H., 1976. The Fluxes of CH_4, ΣCO_2, and NH_3-N from Sediments and their Consequent Distribution in a Small Lake. Ph.D. Dissertation, Columbia University, New York: 186 pp.

Howard, D.L., J.I. Frea & R.M. Pfister, 1971. The potential for methane-carbon cycling in Lake Erie. In *Proc. 14th Conf. Int. Assoc. Great Lakes Research*. Ann Arbor, MI: 236–240.

International Joint Commission, 1989. *1989 Report on Great Lakes Water Quality, Appendix A. Progress in Developing and Implementing Remedial Action Plans for Areas of Concern in the Great Lakes Basin.* IJC, Windsor, Ontario, Canada: 196 pp.

Karickhoff, S.W., 1980. Sorption kinetics of hydrophobic pollutants in natural sediments. In R.A. Baker (Ed.), *Contaminants and Sediments*, Vol. 2. Ann Arbor Science Publishers, Ann Arbor, MI: 193–206.

Karickhoff, S.W. & K.R. Morris, 1985. Sorption dynamics of hydrophobic pollutants in sediment suspensions. *Environ. Toxicol. Chem.* 4: 469–479.

Kelly, C.A. & D.P. Chynoweth, 1981. The contributions of temperature and of the input of organic matter in controlling rates of sediment methanogenesis. *Limnol. Oceanogr.* 26: 891–897.

Kelley, C.A., J.W.M. Rudd & D.W. Schindler, 1988. Carbon and electron flow via methanogenesis, SO_4^{2-}, NO_3^-, Fe^{3+}, and Mn^{4+} reduction in the anoxic hypolimnia of three lakes. *Hydrobiol. Beih. Ergebn. Limnol.* 31: 333–344.

Krom, M.D. & E.R. Sholkovitz, 1977. Nature and reactions of dissolved organic matter in the interstitial waters of marine sediments. *Geochim. Cosmochim. Acta* 41: 1565–1573.

Kuenen, J.G. & L.A. Robertson, 1988. Ecology of nitrification and denitrification. In J.A. Cole & S. Ferguson (Eds.), *The Nitrogen and Sulphur Cycles*. Cambridge University Press, Port Chester, NY: 161–218.

Kuznetsov, S.I., 1968. Recent studies on the role of microorganisms in the cycling of substances in lakes. *Limnol. Oceanogr.* 13: 211–224.

Mackay, D. & W.Y. Shiu, 1981. A critical review of Henry's law constants for chemicals of environmental interest. *J. Phys. Chem. Ref. Data* 10: 1175–1199.

Martens, C.S. & J.V. Klump, 1984. Biogeochemical cycling in an organic-rich coastal marine basin. 4. An organic carbon budget for sediments dominated by sulfate reduction and methanogenesis. *Geochim. Cosmochim. Acta* 48: 1987-2004.

Molongoski, J.J. & M.J. Klug, 1980. Anaerobic metabolism of particulate organic matter in the sediments of a hyper-eutrophic lake. *Freshwater Biol.* 10: 507-518.

Mortimer, C.H., 1987. Fifty years of physical investigations and related limnological studies on Lake Erie, 1928-1977. *J. Great Lakes Res.* 13: 407-435.

Oliver, B.G., 1985. Desorption of chlorinated hydrocarbons from spiked and anthropogenically contaminated sediments. *Chemosphere* 14: 1087-1106.

Ontario Ministry of the Environment, 1987. *Remedial Action Plan for Hamilton Harbour—Draft Summary Report.* OME, Toronto, Ontario and Environment Canada, Ottawa, Ontario, Canada.

Oremland, R.S., 1988. The biogeochemistry of methanogenic bacteria. In A. Zehnder (Ed.), *The Biology of Anaerobic Microorganisms.* John Wiley and Sons, Inc., New York: 405-447.

Postgate, J.R., 1984. *The Sulfate-Reducing Bacteria*, 2nd ed. Cambridge University Press, Port Chester, NY.

Poulton, D.J., 1987. Trace contaminant status of Hamilton Harbour. *J. Great Lakes Res.* 13: 193-201.

Poulton, D.J., 1989. Statistical zonation of sediment samples using ratio matching and cluster analysis. *Environ. Monit. Assessment* 12: 379-404.

Rathke, D.E. & G. McRae (Eds.), 1989. *1987 Report on Great Lakes Water Quality, Appendix B. Great Lakes Surveillance*, Vol. 1. International Joint Commission, Windsor, Ontario, Canada: 287 pp.

Revsbech, N.P. & B.B. Jørgensen, 1986. Microelectrodes: Their use in microbial ecology. In K.C. Marshall (Ed.), *Advances in Microbial Ecology*, Vol. 9. Plenum Press, New York: 293-352.

Robbins, J.A., D.N. Edgington & A.L.W. Kemp, 1978. Comparative ^{210}Pb, ^{137}Cs and pollen geochronologies of sediments from Lakes Ontario and Erie. *Quaternary Res.* 10: 256-278.

Seitzinger, S.P., 1988. Denitrification in freshwater and coastal marine ecosystems: Ecological and geochemical significance. *Limnol. Oceanogr.* 33: 702-724.

Sly, P.G., 1976. Lake Erie and its basin. *J. Fish. Res. Bd. Can.* 33: 355-370.

Snodgrass, W.J., 1977. Gas production by the sediments: An oxygen sink. In *Hamilton Harbour Study 1975.* Ontario Ministry of the Environment, Toronto, Ontario, Canada: D1-D31.

Strickland, J.D.H. & T.R. Parsons, 1972. *A Practical Handbook of Seawater Analysis.* Bull. 167 (2nd ed.). Fisheries Research Board of Canada, Ottawa, Ontario, Canada: 310 pp.

Sweerts, J.-P.R.A. & T.E. Cappenberg, 1988. Use of microelectrodes for measuring oxygen profiles in lake sediments. *Arch. Hydrobiol. Beih. Ergebn. Limnol.* 31: 365-371.

Thomas, R.L., J.-M. Jaquet, A.L.W. Kemp & C.F.M. Lewis, 1976. Surficial sediments of Lake Erie. *J. Fish. Res. Bd. Can.* 33: 385-403.

Thomas, R., R. Evans, A. Hamilton, M. Munawar, T. Reynoldson & H. Sadar

(Eds.), 1987. *Ecological Effects of In Situ Sediment Contaminants*, Developments in Hydrobiology 39. Dr. W. Junk Publishing, Dordrecht, The Netherlands: 272 pp.

Ward, T.E. & J.I. Frea, 1979. Estimation of microbial activities in lake sediments by measurement of sediment gas evolved. In C.D. Litchfield & P.L. Seyfried (Eds.), *Methodology for Biomass Determinations and Microbial Activities in Sediments*. Spec. Tech. Publ. 673, ASTM, Philadelphia, PA: 156–166.

Winfrey, M.R., 1984. Microbial production of methane. In R.M. Atlas (Ed.), *Petroleum Microbiology*. Macmillan Press, Riverside, NJ: 153–219.

Witkowski, P.J., P.R. Jaffe & R.A. Ferrara, 1988. Sorption and desorption dynamics of Aroclor 1242 to natural sediment. *J. Contaminant Hydrol.* 2: 249–269.

Zumft, W.G., A. Viebrock & H. Korner, 1988. Biochemical and physiological aspects of denitrification. In J.A. Cole & S. Ferguson (Eds.), *The Nitrogen and Sulphur Cycles*. Cambridge University Press, Port Chester, NY: 245–279.

CHAPTER 8

Bioavailability of Sediment-Associated Contaminants to Benthic Invertebrates

Peter F. Landrum and John A. Robbins

1.0 INTRODUCTION

Sediment contamination is one of the major end results of pollutant discharges into freshwater and marine aquatic environments. These contaminant discharges have resulted in highly contaminated sediments throughout the world, although the extent of the problem is not yet adequately delineated. In North America, highly contaminated sites have been reported for both coastal marine (NOAA 1988) and freshwater systems such as the Great Lakes (IJC 1987, 1988). For many toxic materials, sediments represent both the primary repository and in many cases the principle source of contamination to the food chain.

Field monitoring studies have shown that changes in portions of the ecosystem, particularly changes in the benthic community structure, are correlated with increased sediment-associated contaminants within the Great Lakes (Nalepa & Landrum 1988). However, developing cause/effect relationships is difficult because other factors such as changes in habitat, nutrient loads, and the amount of suspended materials often occur simultaneously or sequentially with the introduction of pollutants and may themselves alter the functioning of a specific ecosystem. Evidence for the effects of sediment contaminants within ecosystems can be found in monitoring studies that show the presence of the same contaminants in organisms and sediments collected from the same area (NOAA 1987). Furthermore, tumors in bottom-dwelling fish occur with greater than expected frequency in areas where sediments are contaminated with carcinogenic contaminants (Couch & Harshbarger 1985). Perhaps the most direct evidence of the effects of sediment contaminants lies in laboratory bioassays, where sediments collected from contaminated areas have produced toxicity in a wide range of laboratory bioassay organisms (Swartz 1987, Lamberson & Swartz 1988).

In spite of the evidence that sediment-associated contaminants are entering and affecting biological systems, the processes responsible for the transfer of contaminants from sediments to biota and the physicochemical and environmental factors modifying these processes remain ill-defined. Since sediments are such important potential sources for ecosystem modification and food chain transfer, it is important to characterize and understand the nature of the interaction between this complex medium and its living inhabitants. Our objectives are to review the current state of knowledge on the bioavailability of sediment-associated contaminants and to propose a model to help interpret our current understanding and define areas where data about important factors are missing. The focus of the discussion will be on nonpolar organic contaminants because the variables that influence the contaminant transfer from sediments to biota are best known for these classes of compounds. However, most of the principles described can eventually be applied to other contaminants, polar organics and heavy metals, as specific factors that influence sorption and accumulation become better defined for these materials.

2.0 DEFINING BIOAVAILABILITY

Bioavailability of sediment-associated contaminants can be defined as "the fraction of the total contaminant in the interstitial water and on the sediment particles that is available for bioaccumulation", whereas bioaccumulation is "the accumulation of contaminant via all routes available to the organism". Chemical measures of contaminant concentration in sediment do not always reflect the bioavailable fraction of sediment-associated contaminants (Landrum 1989, Landrum *et al.* 1989); therefore, a simple measure of the sediment residue is insufficient to define the contaminant concentration to which biota are exposed. Further, because feeding by benthic organisms is generally limited to the fine grain material (the material that sorbs most contaminants) the potential exposure would be much greater than would be anticipated from the bulk sediment concentration when ingestion is a significant route of exposure. Therefore, a measure is needed to define the fraction of total contaminant available for biological accumulation.

Two approaches to assess overall bioavailability of sediment-associated organic contaminants are prominent in the literature: (1) comparison of the organism- and sediment-contaminant concentrations, and (2) determination of the uptake clearance of sediment-associated contaminants. The first approach, a steady-state approach, compares the organism contaminant concentration to the contaminant concentration in the sediment. An accumulation factor (AF) or preference factor (PF) is calculated as the ratio of the contaminant concentration in the organism normalized to the lipid content divided by the contaminant concentration in sediment normalized to the organic carbon content (McFarland 1984, Lake *et al.* 1987, Foster & Wright 1988, McElroy & Means 1988, Pereira *et al.* 1988). Such normalizations to lipid and organic carbon make the assumption that organic contaminants partition reversibly between the sediment particles and organisms with interstitial water as the exchange medium. Further, the organic contaminants are presumed to partition predominantly to sediment organic carbon and organism lipids and come rapidly to a steady-state condition.

For comparability between studies, the methods to measure lipid content and organic carbon content must be standardized. However, total organic carbon has been measured both as carbon and as total volatile solids content. This methodological difference could result in a factor of four or more difference between reported organic carbon normalized concentrations.

Similarly, various extraction methods have been used to determine total organism lipid content. Most often, the method is a gravimetric determination of the residue from the same extraction used for the extraction of contaminants. The extractions for contaminants are not necessarily the appropriate methods for extracting lipids; thus, a standard lipid-extraction technique should be used to avoid variation between laboratories in the relative values for lipid normalization. It is possible to take a very small aliquot of a sample for total lipid determination using a proven microgravimetric technique (Gardner *et al.* 1985) and reserve the remainder of the sample for contaminant analysis. A standard approach needs to be developed so that comparisons among results obtained in different laboratories are meaningful.

In addition to potential procedural difficulties, other assumptions are required for the PF or AF to be valid. The utility of these factors rests on the assumption that sediment is the only source for the bioaccumulation. Thus, contributions from the overlying water would not be recognized. This assumption may or may not be appropriate, based on the organism and the actual sources involved. If overlying water is a significant contributing contaminant source, the AF would increase significantly over that resulting when sediment-associated contaminant is the only source. In the field, unexpected point discharges or other direct contamination of the overlying water could act as a significant co-source, invalidating the comparison between sediment and organism concentrations.

Further, the AF is likely constant only if contaminant concentrations in organisms are at steady-state with the contaminant concentrations in the sediment. Whether steady-state conditions occur widely in natural systems is not clear, but studies with *Pontoporeia hoyi* and *Hexagenia limbata* suggest that for some benthic organisms steady-state concentrations only exist during specific times of the year (Eadie *et al.* 1988, Landrum & Poore 1988). Nonetheless, this thermodynamic-based method permits comparisons between extractable residues and the amount of material bioaccumulated. Used in conjunction with laboratory exposures where the source could be controlled, AFs would produce useful comparisons between bioaccumulation and extractable residue.

A second, commonly used approach to define the extent of bioavailable contaminant determines the uptake clearance of sediment-associated contaminants for a specific organism. The uptake clearance is the conditional constant that relates the contaminant flux into the organism to contaminant concentration in the sediment and has units of g sediment g^{-1} organism hr^{-1}. This approach has been employed for both laboratory and field collected sediments (Landrum *et al.* 1985b, Tatem 1986, Foster *et al.* 1987, Shaw & Connell 1987, Landrum & Poore 1988, Landrum 1989). When measuring kinetics of accumulation from sediment in the laboratory, the contribution from overlying water can be kept at or near zero concentration under flowing-water conditions. The uptake clearance can be used as a measure of the bioavailable fraction and can be compared with the amount of extractable residue. If the elimination kinetics are known, the steady-state contaminant concentration can be calculated as a ratio of the uptake clearance divided by the elimination rate constant. From the calculated steady-state concentration an AF could also be calculated, based on the organism lipid content and the sediment organic carbon content. Kinetic relationships have the advantage that multiple sources can be accommodated in the experimental designs and the fraction of accumulated body burden from each source can be estimated. Further, the incorporation of multiple sources in kinetic calculations permits modeling of field situations where contaminant accumulation results from multiple sources. However, kinetic measurements are much more expensive and time-consuming than the measurements required for an assumed steady-state.

Bioavailability of sediment-associated contaminants can be defined in a general sense; however, no chemical measures currently exist to predict bioavailability of sediment-bound residues. The only approaches available for estimating the bioavailable fraction are comparisons between the accumulation and the sediment chemistry. A better method for defining bioavailability is needed.

3.0 FACTORS AFFECTING BIOAVAILABILITY

Several reviews have outlined major factors affecting the rate or extent of sediment contaminant bioaccumulation, particularly for organic contaminants (Neff 1984, Adams 1987, Knezovich *et al.* 1987, Reynoldson 1987, Rodgers *et al.* 1987). These factors can be divided into three general areas: (1) the characteristics of the contaminants, (2) the composition and characteristics of the sediment, and (3) the behavior and physiological characteristics of the organisms. In general, the range of bioaccumulation factors (BAF, the concentration of the contaminant in the organism divided by the concentration in the sediment) for sediment-associated contaminants is 0.1 to about 20 (Neff 1984). Further, previous studies have generally suggested that the accumulation of sediment-associated contaminants occurs via sediment interstitial water.

3.1 Compound Characteristics

For nonpolar organic compounds, the major characteristic which controls bioaccumulation from sediments is the hydrophobicity of the compound, as represented by the octanol-water partition coefficient (K_{ow}). The extent of contaminant accumulation generally decreases with increasing log K_{ow} or decreasing water solubility because of increased partitioning to sediment particles. Other factors that increase the sorption or complexation to sediment particles, such as hydrogen bonding, compound ionization, or chemical reactivity, may also reduce bioavailability (Neff 1984, Adams 1987, Knezovich *et al.* 1987, Reynoldson 1987, Rodgers *et al.* 1987).

Several relationships between K_{ow} and bioavailability have been reported. Some reports show log-linear relationships between K_{ow} and the accumulation of contaminants from sediments (Knezovitch & Harrison 1988, Gobas *et al.* 1989). However, oligochaetes exhibit a maximum BAF at approximately a log K_{ow} of 6 when exposed to chlorinated hydrocarbons covering a range of K_{ow} (Oliver 1984, 1987). The apparent maximum in the BAF seems to result from a combination of pharmacological characteristics, such as the uptake and elimination rates of compounds, and sorption properties of the compounds. Compounds with small K_{ow} are less strongly sorbed and more bioavailable, presumably from both more rapid desorption to interstitial water and increased assimilation efficiency, but are also more rapidly eliminated, which results in relatively low BAF values. Compounds with a very large K_{ow} are more slowly desorbed from sediment and, therefore, less bioavailable, presumably as a result of both decreased desorption rates and lower assimilation efficiency from ingested particles, which also results in a small BAF in spite of the generally high bioconcentration of such compounds from water. Similar relationships were observed for the accumulation of chlorinated hydrocarbons by *Pontoporeia hoyi* (Landrum *et al.* 1989) and for a variety of marine organisms exposed to compounds under field conditions (Pereira *et al.* 1988).

The uptake clearance of polycyclic aromatic hydrocarbons (PAH) from a single type of sediment was observed to be inversely proportional to the K_{ow} for *P. hoyi* (Landrum 1989). However, the uptake clearance cannot be described by a simple relationship with K_{ow} when the clearances of several classes of compounds are compared. The uptake clearances of chlorinated compounds compared to PAH with similar K_{ow} values are much greater (Landrum 1989, Landrum *et al.* 1989). Thus, the relationship between K_{ow} and accumulation of organic contaminants is not simple, and other factors can significantly influence the accumulation of sediment-associated contaminants.

3.2 Sediment Characteristics

Important sediment properties that enhance sorption, thereby reducing bioavailability, include the organic carbon content, particle size distribution, clay type and content, cation exchange capacity, and pH. Organic carbon is a major factor because it can both sorb organic contaminants and complex metals (Neff 1984, Adams 1987, Knezovich *et al.* 1987, Reynoldson 1987, Rodgers *et al.* 1987, Swindoll & Applehans 1987, Foster & Wright 1988, McElroy & Means 1988). The influence of organic carbon is greater for compounds of greater K_{ow} that sorb primarily to the organic fraction of sediment. Although normalization to organic carbon may help account for the extent of sorption to sediments and, particularly, the fine grain sedimentary material, clays and other fine materials may contribute significantly to sorption, thus reducing contaminant bioavailability (Neff 1984, Adams 1987, Knezovich *et al.* 1987, Reynoldson 1987, Rodgers *et al.* 1987, Swindoll & Applehans 1987).

Under such circumstances, normalization to organic carbon may be invalid and misleading. Sorption of hexachlorobiphenyl to clay, for example, reduces the hexachlorobiphenyl bioavailability (Swindoll & Applehans 1987). Similarly, normalization by organic carbon did not account for all the variability in measured uptake clearances of PAH and PCB congeners by *P. hoyi* from sediments, particularly where the separately dosed fine fraction was compared with sediments dosed in bulk. *P. hoyi* uptake clearances, normalized for organic carbon and measured from sediments with similar bulk organic carbon concentrations, were greatest when the fine fraction of the sediments had been labeled separately (Landrum, unpublished data). Thus, the particle size and not organic carbon content alone is likely an important characteristic which influences bioaccumulation. More efforts are required to account for the sorption to the fine fraction of sediment and to develop the appropriate normalizing factors.

The duration of contact between sediment particles and the contaminant alters both chemical extractability and bioavailability. The increased equilibration time between the contaminant and sediment in laboratory studies markedly reduces the bioavailability of organic contaminants to benthos (McElroy & Means 1988, Landrum 1989). These changes in bioavailability with increasing contact time are not always reflected in changes in the amount of chemically extractable residue from the sediments. Further, the extent of reduction in chemical residue concentration will not necessarily reflect the extent of reduction in bioavailability. The rate at which contaminants are observed to disappear from a bioavailable pool is approximately inversely proportional to K_{ow} within a single class of contaminants (Landrum 1989). Possible mechanisms driving this reduction include (1) sorption into less bioavailable sediment compartments, (2) removal of the ingestable fraction through packaging of fine material into fecal pellets, and (3) changes in organism behavior.

3.3 Partitioning

Understanding partitioning between sediment particles and interstitial water is important for understanding contaminant bioavailability. In their review of early laboratory studies, Elzerman and Coats (1987) showed that the partitioning of organic contaminants usually fits a Freundlich isotherm. Thus, the partition coefficient can be determined as the slope of the regression line between the contaminant concentration on particles and the concentration of the dissolved contaminant measured at equilibrium for different total contaminant concentrations. In virtually all cases, the experimental designs for determining partitioning followed contaminant adsorption as an approach to equilibrium partitioning. In other words, a contaminant was introduced into the aqueous phase and allowed to adsorb to suspended particles. The concentration in the water and particles is measured until an apparent equilibrium is obtained and a partition coefficient is then calculated. Partition coefficients measured for the same compound with different sediments showed a variability that could be reduced by normalization to sediment organic carbon (Karickhoff et al. 1979). The partition coefficient for sediment suspensions often increased as the solids concentration decreased. These increasing partition coefficients generally reach an apparent plateau when the solids concentration is sufficiently small (O'Connor & Connolly 1980, Voice & Weber 1985, DiToro 1985). One explanation for this effect was the presence of a third phase consisting of nonfilterable colloids or dissolved organic matter (DOM) (Voice & Weber 1985, Gschwend & Wu 1985). Such a third phase enhances the apparent solubility of nonpolar organic molecules by maintaining them in solution in a bound form. Binding of contaminants by this third phase also reduces the contaminant bioavailability in the water column (Landrum et al. 1985a, 1987, McCarthy & Jimenez 1985, McCarty et al. 1985, Black & McCarthy

1988). DOM sorption will also affect bioavailability from interstitial waters that contain large DOM concentrations.

A greater complexity in the partitioning phenomena was observed when the experimental design followed the desorption of contaminants as an approach to equilibrium partitioning. Compounds sorbed to sediments apparently reside in reversible and resistant pools, and the fraction of the total contaminant in each pool changes depending on sorption duration (Karickhoff 1980, DiToro *et al.* 1982, Karickhoff & Morris 1985a). To model this observation, a two-compartment kinetic model was deemed appropriate (Karickhoff & Morris 1985a); since then, additional kinetic models of partitioning have been employed to better describe the variety of complexities involved in partitioning, such as "solids concentration effects" and "multiple sorption sites on particles" (Elzerman & Coats 1987). Mechanistic models identify the sediment particle as a semisolid material and describe the sorption process by diffusion coefficients into the sediment matrix to account for the observed partitioning phenomena (Freeman & Cheung 1981, Wu & Gschwend 1986). These mechanistic models, which require a third phase to describe the overall partitioning, need more development before they can be very useful in bioavailability assessment.

Hysteresis effects exist in the time relations between adsorption and desorption of contaminants to sediment. Nonpolar organic compounds associate rapidly with sedimentary materials but desorb more slowly (Elzerman & Coats 1987, Witkowski *et al.* 1988). A recent quantitative model that examined the sorption and desorption of PCBs showed that the effects on partitioning, such as the solids concentration effect, result from employing equilibrium models instead of kinetic models (Witkowski *et al.* 1988). The importance of a third phase is not ruled out in kinetic models, but in some systems it may not be very important. From the above model, slow desorption relative to adsorption and the relatively slow uptake clearances from sediments (Landrum 1989) suggest small desorption rate constants for sediment sorbed contaminants. Thus, the time required to reach steady-state between the sediment and the interstitial water could be long, and the slow desorption rates will have a significant impact on the bioavailability of organic compounds from sediments.

Focusing on the kinetics, the only existing estimates of the desorption rate constants come from gas stripping experiments employing suspended sediments for some PCB congeners (Coats & Elzerman 1986), other chlorinated hydrocarbon compounds (Oliver 1985), and for certain PAH congeners (Karickhoff 1980, Karickhoff & Morris 1985a). This experimental design maximizes the rate of desorption by maintaining the maximum concentration gradient between the water and the sediment particles through gas purging of the water. These desorption data suggest that it takes months or even years, depending on sediment composition and compound characteristics, to achieve true equilibrium between interstitial water and sediment particles (Karickhoff 1980, Karickhoff & Morris 1985a, Oliver 1985, Coats & Elzerman 1986, Witkowski *et al.* 1988). The desorption rates reported for PAH (Karickhoff & Morris 1985a) were faster than those reported for the chlorinated compounds (Coats & Elzerman 1986), even when the relative K_{ow}s were considered. However, in gas stripping experiments employing humic materials as the sorptive phase, the desorption of PAH congeners was much slower than for chlorinated compounds of similar log K_{ow} (B. J. Eadie, Personal Communication, Great Lakes Environmental Research Laboratory, NOAA, Ann Arbor, MI). Further, the gas stripping of contaminants from humic materials was very dependent on the purge gas flow rate. Thus, comparisons between different results must be interpreted carefully. Based on experiments of desorption from humics with both chlorinated hydrocarbons and PAH under the same conditions, it seems likely that desorption of PAH will be slower and that equilibrium between interstitial water and sediment solids may require much longer contact times than for chlorinated compounds.

Based on a qualitative model (Landrum 1989), these small desorption rate constants help explain the slow rates of contaminant accumulation by benthos. Because of the apparent importance of the desorption constants, more effort to measure them is essential to development of quantitative models for contaminant accumulation from sediments. Further, because the currently measured desorption values are maximum desorption values (as a result of the experimental design), efforts are needed to determine the range of these constants under differing environmental conditions.

3.4 Role of Interstitial Water

From the literature, the dominant route for accumulation of sediment-associated contaminants is suggested to be via interstitial water (Muir *et al.* 1985, Oliver 1987, Shaw & Connell 1987, Knezovitch & Harrison 1988). However, two recent studies suggest that the interstitial water route may not dominate in all systems (Landrum 1989, Landrum *et al.* 1989). The measured partitioning between sediment interstitial water and sediment particles did not reflect the uptake clearances of contaminants belonging to different classes of compounds. The partition coefficient between the sediment interstitial water and sediment particles for benzo(a)pyrene (BaP) was approximately half that for tetrachlorobiphenyl (TCB). The uptake clearance for TCB was nearly ten times greater than for BaP (Landrum *et al.* 1989). The differences between the bioavailability of the two compounds can be better described by their apparent differences in the desorption rates from sediment than by their equilibrium partition coefficient. The significance of interstitial water as a source will be described in more detail below.

3.5 Biological Characteristics

The biological processes and characteristics that influence the transfer of contaminants from sediments include organism behavior, modes and rate of feeding, source of water — interstitial water versus overlying water — for respiration, and organism size (Neff 1984, Adams 1987, Knezovich *et al.* 1987, Reynoldson 1987, Rodgers *et al.* 1987). Critical parameters vary for the species involved; thus, each species must be considered separately.

Feeding behavior is an important feature in the accumulation of sediment-associated contaminants (Varanasi *et al.* 1985, Foster *et al.* 1987, McElroy & Means 1988, Schuyema *et al.* 1988), and in the case of the clam *Macoma nasuta*, the sediment-associated hexachlorobenzene comes almost completely from sediment particle ingestion (Boese *et al.* 1990). Because benthic organisms feed on fine sediment particles rather than bulk sediment and because the partitioning of contaminants is to the fine organic rich particles, the contaminant concentrations on the ingested particles will often be greater than measured bulk concentrations. Even low rates of ingestion or low assimilation efficiencies can result in significant amounts of accumulation via ingestion. This differential sorption between large and small particles must be accounted for before assimilation from ingestion can be determined (Lee *et al.* 1990). The importance of the fine material as a major contaminant source has been demonstrated for benthos in the Great Lakes (Eadie *et al.* 1985) and upper Mississippi River (Bailey & Rada 1984).

Behavioral factors influencing the transfer between sediment and organisms are well illustrated by the clam *Macoma nasuta*. *Macoma* essentially only uses overlying water for respiration (Windsor *et al.* 1990) but feeds on organic detrital material (Boese *et al.* 1990). Thus, its primary source of sediment-associated contaminants is dominated by the ingestion of particles and not by the ventilation of sediment interstitial water (Boese *et al.* 1990). Such discrimination between overlying and interstitial water may be especially important for tube dwellers, such as chironomids, hexagenia, and other organisms, that pump overlying water through their burrows. The extent to which organisms ventilate overlying water will govern the proportion of contaminant derived from interstitial fluids.

Reductions in the exposure of benthos to sediment-associated contaminants through behavioral modification can also occur for infaunal organisms such as oligochaetes. Oligochaetes can modify feeding behavior and even emerge from the sediments if the contaminants can be detected (Keilty *et al.* 1988a), thus modifying their exposure to both sediment interstitial water and to particle-associated contaminants by moving to zones of lesser contamination. In recent work, the oligochaete *Stylodrilius heringianus* was observed to burrow below a layer of fine particles, which presumably contained the greatest concentration of contaminants. This layer of fine sediments was created by allowing sediments to settle passively and therefore sort according to size. By contrast, in well-mixed sediments the introduced organisms burrowed but returned to the surface, presumably to avoid the toxic contaminants (White & Keilty 1988)

Exposure in natural systems is further modified by "conveyor-belt"-type bioturbation processes through particle-selective feeding, which tends to maintain contaminants in the bioactive layer of sediments (Robbins 1986). These "conveyor-belt" feeders consume particles deeper in the sediments and deposit them on the surface. In laboratory microcosms, oligochaetes have been observed to rework the sediment and enhance the concentration of contaminants in the upper layers with time, which demonstrates particle-selective transport (Karickhoff & Morris 1985b, Keilty *et al.* 1988b, 1988c). Organisms such as chironomids (Wood *et al.* 1987) have been shown to enhance the release of contaminants to overlying water. This release of contaminants from the sediments, which is likely to occur with many of the tube dwellers, would change the concentration to which organisms such as *Macoma* that ventilate overlying water are exposed. Thus, even though *Macoma* does not ventilate interstitial water, the action of the tube dwellers will tend to expose this and other organisms that live near or on the sediment surface to greater concentrations of desorbed contaminants by introducing some interstitial water into the overlying water.

4.0 MODELING BIOAVAILABILITY

In early modeling efforts, accumulation from sediments was treated as a first-order rate process with no definition of the source, sediment particles, or sediment interstitial water. Such models depended on empirical relationships between uptake clearance from sediments, contaminant characteristics, and sediment characteristics. These models provided reasonable initial estimates of the extent of the overall sources, overlying water or sediment-associated contaminants, to the bioaccumulation (Landrum *et al.* 1985b, Eadie *et al.* 1988). The empirical models suggested that features such as assimilation from ingestion and desorption from sediments would be important for mechanistic models. A conceptual framework including such processes has been developed (Landrum 1989).

Because sediment interstitial water has been strongly suggested as the primary source for sediment-associated contaminants (Pavlou & Weston 1984, Adams *et al.* 1985, Muir *et al.* 1985, Oliver 1987, Knezovitch & Harrison 1988), a mass balance box model (Figure 1) was developed that considered sediment as a box containing sediment particles, interstitial water, and organisms. This model, developed for *P. hoyi*, treats contaminant elimination as predominantly fecal, with compounds returning to the

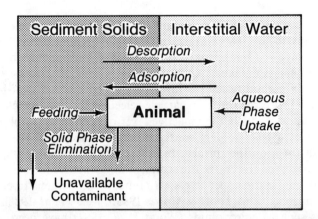

Figure 1.   Mass balance box model for the accumulation of organic contaminants for sediments.

sediment reservoir. This assumption was defensible because elimination of PAH is not measurable in the absence of sediment (Landrum 1982). It is not possible to tell, at this time, whether material eliminated via the fecal route is available to the organisms or is removed from the bioavailable pool because *P. hoyi* package their fecal pellets in a peritrophic membrane that essentially makes them large particles. We assumed for this model that the material was equally available after elimination or was such a small fraction of the total sediment that it was an unimportant source. The accumulation by *P. hoyi* comes both from interstitial water and from ingestion of sediment particles. The small pool of material associated with the DOM in the interstitial water can be assumed, for modeling purposes, to be part of the sediment particles. The uptake from interstitial water is considered to be limited by the size of the interstitial water pool to which the organisms are exposed and the desorption rate of contaminants from sediment. The indications of limitations for the interstitial pool size and the desorption rates come from studies of PAH uptake from sediments (Landrum 1989). The amount of compound accumulated from sediment particle ingestion is based on measured ingestion rates (M. A. Quigley, Personal Communication, Great Lakes Environmental Research Laboratory, NOAA, Ann Arbor, MI) and the measured fraction of fine grain material in the sediments. The amount accumulated from sediment ingestion is assumed to originate primarily from contaminants sorbed onto the fine grain sediments preferentially ingested by *P. hoyi*. The accumulation of contaminant from overlying water was ignored in the model because (1) previous studies demonstrated that contaminant dosed to sediments was accumulated with the same uptake clearance whether the accumulation was determined under static or flow-through conditions and (2) *P. hoyi* spends less than 10% of its time above the sediments (M. A. Quigley, Personal Communication, Great Lakes Environmental Research Laboratory, NOAA, Ann Arbor, MI). The quantity of contaminant in the water was allowed to vary through adsorption and desorption kinetics from sediments and the amount accumulated by *P. hoyi* (Equations 1 to 3).

$$dQa/dt = (K_a \cdot Qw) + (K_i \cdot Qs) - (K_e \cdot Qa) \qquad (1)$$

$$dQw/dt = (K_1 \cdot Qs) - (K_2 \cdot Qw) - (K_a \cdot Qw) \qquad (2)$$

$$dQs/dt = (K_2 \cdot Qw) - (K_1 \cdot Qs) + (K_e \cdot Qa) \qquad (3)$$

where Qa is the mass of contaminant in the organism on a mass per volume of sediment

Ka is the uptake rate constant from sediment for the organism (hr^{-1})

Qw is the mass of contaminant in the interstitial water on a mass per volume of sediment

K_i is the uptake rate constant from ingestion (hr^{-1})

Qs is the mass of contaminant per volume of sediment

K_e is the elimination rate constant (hr^{-1})

K_1 is the desorption rate constant from sediment particles (hr^{-1})

K_2 is the adsorption rate constant to sediment particles (hr^{-1})

t is time (hr)

The above equations represent a mass balance model and can be converted into equations with more normally measured concentration-based rate coefficients by the use of the following definitions. The density of organisms in the interstitial water (ρ) is defined as "organism wet weight per volume of interstitial water"; the fraction of interstitial water ($\emptyset$) is "the volume of interstitial water per volume of sediment". Thus, the volume of sediment particles is $\rho_s(1 - \emptyset)$, where ρ_s is the mean density of sediment particles, 2.5 ± 0.2 g mL^{-1} of sediment (Robbins 1980). The mass balance equations are then converted to concentration-based equations (4 to 7). When converting from the mass balance form of the equations to the concentration-based form of the equations, two of the rate constants (K_1 and K_e) do not change. Both constants represent the fractional change in contaminant concentration with time for their respective compartments and retain the units of hr^{-1}. To make their use in the concentration-based equations distinguishable from the mass-based equation and in the case of the elimination to use the more common nomenclature, K_1 is changed to K_{de} to reflect the rate constant for desorption and K_e is changed to K_d to reflect the depuration of the contaminant.

$$dCa/dt = (K_u \cdot Cw) + (K_f \cdot Cs) - (K_d \cdot Ca) \qquad (4)$$

where Ca is the concentration of contaminant in the organism (ng g^{-1}) and
equals $Qa/(\rho \cdot \emptyset)$

K_u is the uptake clearance constant (mL g^{-1} organism hr^{-1}) and equals K_a/ρ

K_f is the uptake clearance from ingestion (g dry sediment g^{-1} animal hr^{-1}) and equals the mass-specific feeding rate (g dry sediment g^{-1} organism hr^{-1}) times the assimilation efficiency, or $K_i \cdot \rho_s(1 - \emptyset)/(\rho \cdot \emptyset)$

Cs is the concentration of contaminant in the sediment (ng g^{-1}) and equals $Qs/[\rho_s \cdot (1 - \emptyset)]$

K_d is the elimination rate constant and equals K_e

t is time (hr)

The subsequent equation reflecting the change in interstitial water concentration is as follows:

$$dCw/dt = [K_{de} \cdot Cs \cdot \rho_s \cdot (1 - \emptyset)\,]/\emptyset - [K_p \cdot K_{de} \cdot Cw \cdot \rho_s \cdot (1 - \emptyset)]/\emptyset - (K_u \cdot \rho \cdot Cw) \tag{5}$$

where K_{de} is the desorption rate constant from sediment and is equal to K_1

K_p is the partition coefficient between the sediment and the interstitial water, and the remainder of the constants are as defined above

The substitution for K_2 is found by relating the adsorption and desorption rate at steady state to the partition coefficient (Equation 6).

$$Cs/Cw = K_p = K_2 \cdot \emptyset/[K_1\rho_s(1 - \emptyset)] \tag{6}$$

Finally, the equation for the change in concentration in the sediment with time is described by Equation 7.

$$dCs/dt = (K_d \cdot \rho \cdot \emptyset \cdot Ca)/[\rho_s \cdot (1 - \emptyset)] + K_p \cdot K_{de} \cdot Cw - \{[(K_f \cdot \rho \cdot \emptyset)/\rho_s \cdot (1 - \emptyset)] + K_{de}\} \cdot Cs \tag{7}$$

Because laboratory experiments have indicated that the measured contaminant concentrations in sediment do not change, in most cases, or change only slightly (Landrum 1989, Landrum et al. 1989) the sediments were assumed to be essentially an infinite source. Therefore, it is not necessary to include Equation 7 in the model. For the cases where the bioavailable concentration in the sediment is known to change, despite the constant or near constant chemically measured sediment concentrations (Landrum 1989), Cs was replaced by $Cs^o \cdot e^{-\lambda t}$, where Cs^o is the sediment concentration

Table 1. Parameter Estimates for Modeling the Accumulation of Sediments by *Pontoporeia hoyi*

Parameter	PHE[a]	PY[b]	TCB[c]	HCB[d]
Log K_{ow}	4.4[e]	5.2[e]	5.9[f]	6.7[f]
K_u (ml g^{-1} hr^{-1})	129[g]	199.2[g]	135[g]	53.5[h]
K_f^i (g sed. g^{-1} org. hr^{-1})	0.005	0.005	0.004	0.0029
K_d (hr^{-1})	0.0046[g]	0.0012[g]	0.0001[j]	0.0001[h]
K_{de} (hr^{-1})	0.18[k]	0.046[k]	0.005[l]	0.0024[l]
ρ_s^{m}	2.5	2.5	2.5	2.5
ϕ^n	0.698	0.698	0.698	0.698
K_p	273[n]	446[n]	1673[f]	1900[j]
λ^e (hr^{-1})	0.0055	0.003	0	0
K_s (g sed. g^{-1} org. hr^{-1})	0.041[e]	0.019[e]	0.018[f]	0.0057[e]
Assimilation efficiency	0.42[o]	0.42[o]	0.3355[o]	0.24[p]

[a]PHE = phenanthrene.
[b]PY = pyrene.
[c]TCB = tetrachlorobiphenyl.
[d]HCB = hexachlorobiphenyl.
[e]Landrum (1989).
[f]Landrum et al. (1989).
[g]Landrum (1988).
[h]Evans and Landrum (1989).
[i]K_f = feeding rate (0.012 g sed. g^{-1} org. hr^{-1}) • assimilation efficiency; feeding rate from Quigley, Personal Communication, Great Lakes Environmental Research Laboratory, NOAA, Ann Arbor, MI.
[j]Estimated.
[k]Karickhoff and Morris (1985a).
[l]Coats and Elzerman (1986).
[m]Robbins (1980).
[n]Landrum, unpublished data.
[o]Gobas et al. (1988).
[p]Klump et al. (1987).

at time equals zero and λ is the rate constant (hr^{-1}) for the apparent reduction in the bioavailable concentration with time.

The model was parameterized with values from the literature (Table 1). The uptake and elimination coefficients, K_u and K_d, were mean values from previous kinetic measurements of uptake from water and elimination in the presence of sediment for the PAH congeners and TCB (Landrum 1988) and for hexachlorobiphenyl (Evans & Landrum 1989). The value for ϕ (Landrum, unpublished data) was calculated from previous experiments designed to measure the uptake from sediments (Landrum 1989). K_{de} estimates for PCB congeners (Coats & Elzerman 1986) and for PAH congeners (Karickhoff 1980) were taken from the literature. The desorption rate constants were for sediments with organic carbon concentrations that were similar (approximately 1%) to that of the sediment used for measuring uptake rates in *P. hoyi*.

Both the feeding rate and the assimilation efficiency must be known to establish estimates for K_f, which is the product of the two terms. The feeding rate for *P. hoyi* was 0.012 g dry sediment g^{-1} organism hr^{-1}, calculated from the gut turnover and the characteristics of the sediment (Landrum 1989). This feeding rate is consistent with feeding rates found for Lake Ontario *P. hoyi* of 0.01 ± 0.002 g dry sediment g^{-1} organism hr^{-1} (Dermott & Corning 1988). In addition to the sediment throughput, the assimilation efficiency must be known to determine the flux of contaminant into the organism from this source.

Two approaches have been used to measure assimilation efficiency for benthic organisms. The first employs a dual radio tracer, ^{51}Cr and a ^{14}C-labeled contaminant. The ^{51}Cr associated with sediment is not well absorbed, while the contaminant will be partially absorbed. With this technique, the assimilation efficiency for 2,4,5,2',4',5'-hexachlorobiphenyl in oligochaetes ranged from 15 to 36% and was inversely proportional to the defecation rate (Klump *et al.* 1987). The second method measured the relative concentration of a radiolabeled contaminant in the food and the fecal pellets normalized to the organic carbon content. The assimilation of natural organic carbon by the organism must be known or measured and the concentration in the fecal matter must be adjusted for this loss. Using this technique, the assimilation efficiency for *Macoma nasuta* ranged from 39 to 57% for hexachlorobenzene (Lee *et al.* 1990).

These assimilation efficiencies are similar to those for other chlorinated compounds in fish, where the contaminants are associated with food (Gobas *et al.* 1988). The regression equation between contaminant log K_{ow} and assimilation efficiency (Gobas *et al.* 1988) was employed as a first estimate for *P. hoyi* assimilation efficiency from sediments when no other estimate was available.

The partition coefficient for the model was taken from measured partition coefficients between sediment interstitial water and sediment particles for TCB (Landrum *et al.* 1989), phenanthrene, and pyrene (Landrum, unpublished data). Because the partition coefficient had not been measured for hexachlorobiphenyl, it was estimated by multiplying the measured TCB partition coefficient by the ratio K_{ow} for hexachlorobiphenyl divided by K_{ow} for TCB.

With the above independent values to parameterize the model, an estimate of the optimum value of ρ, as determined by a least-squares fit to experimental data, was sought by model simulations based on data from some of the sediment uptake experiments used for kinetics determinations (Figure 2) (Landrum 1989, Landrum *et al.* 1989). In theory, as long as the density of organisms is below some critical value, the value where loading becomes limiting in the experimental design, ρ should be compound independent. For the present, determining ρ through the model will set a lower

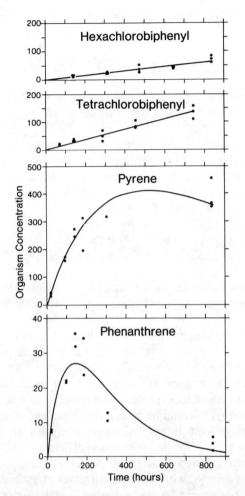

Figure 2. The fit of the model prediction to the experimental data for the using the optimized value of ρ for hexachlorobiphenyl, tetrachlorobiphenyl, pyrene, and phenanthrene.

bound on the amount of interstitial water to which the organism must be exposed to exhibit the measured rate of accumulation. Future efforts should attempt to measure ρ so that the variation in other parameters can be more fully examined.

It is important to begin such optimization with a good estimate; thus, an exact solution for ρ was determined for the steady-state condition by setting both Equations 4 and 5 to zero. The solution includes a required estimate for the organism contaminant concentration at steady state. This estimate can be obtained from a two-compartment model of accumulation from sediment (Landrum 1989, Landrum *et al.* 1989; Equation 7):

$$dCa/dt = (K_s \cdot C_s) - (K_d \cdot Ca) \tag{7}$$

where K_s is the measured uptake clearance from sediment (g dry sediment g^{-1} organism hr^{-1}) and the other terms are as previously defined.

Thus, the steady-state concentration of contaminant in organisms would be $(K_s \cdot Cs)/K_d$ and could be substituted into the equation to determine p. The K_s value used was the average measured sediment uptake clearance (Landrum 1989). With these substitutions, p has the following definition in terms of measured variables (Equation 8).

$$\rho = [K_{de} \cdot \rho_s \cdot (1 - \emptyset)] / \emptyset \cdot \{[K_u + (K_p \cdot K_f) - (K_p \cdot K_s)]/K_u \cdot (K_s - K_f)\} \tag{8}$$

While all of the values in the definition of ρ are important, the least well known are the assimilation efficiencies and desorption rate constants. Because ρ is directly proportional to K_{de}, as K_{de} decreases (all other factors remaining constant) the volume of interstitial water required for exposure will increase for a given uptake rate. As the accumulation from feeding becomes greater, the fraction resulting from interstitial water exposure becomes less and the volume of interstitial water required for exposure to maintain a fixed accumulation rate will decrease. For BaP, the model indicates that for *P. hoyi* ρ would be ∞ and the organism would not be obtaining any of its dose from interstitial water, assuming an ingestion rate of 0.012 g sediment g^{-1} hr^{-1} and an assimilation efficiency of 24% — the same as hexachlorobiphenyl, based on the two compounds having similar K_{ow} values. The above scenario, with our best estimates of assimilation efficiency, suggests that our estimates for the assimilation efficiency are inappropriate for BaP.

The optimized values for ρ (Table 2) range over an order of magnitude and the value for each compound depends on some estimates of various parameters for each contaminant, aside from the literature values available, because all of the necessary parameters have not been measured. These optimized values of ρ allowed the model to fit the accumulation data for the various sediment-associated contaminants (Figure 2). The estimated values for each of the contaminants are as follows: For hexachlorobiphenyl, the partitioning between sediments and interstitial water had not been measured but was estimated as described above. For TCB, K_d could not be measured (Landrum 1988) and was assumed to be one order of magnitude smaller than the smallest value previously measured, 0.0001 hr^{-1}. For all of the

Table 2. Estimated Densities of _Pontoporeia hoyi_ in Sediment Interstitial Water Based on Rates of Accumulation from Sediment

Compound	ρ (g wet organism mL^{-1} porewater)	Interstitial Water[a] Volume/Animal
2,4,5,2',4',5'-Hexachlorobiphenyl	0.736	8.15 μl
2,4,2',4'-Tetrachlorobiphenyl	0.26	23.1 μl
Pyrene	3.42	1.75 μl
Phenanthrene	4.91	1.22 μl

[a]This estimate assumes a mean organism weight of 6 mg wet weight.

compounds except hexachlorobiphenyl, the assimilation efficiency was estimated from the model of Gobas _et al._ (1988) proposed for fish.

An interesting feature of the ρ values is the implied small volume of interstitial water to which the organisms have to be exposed, based on current estimates of desorption rates, to demonstrate the measured rate of accumulation. At the estimated feeding rate, the organisms only "see" about three times their body mass as a maximum volume of interstitial water. If feeding were turned off, the volume would only be about four times their body mass.

To estimate the relative roles of uptake through interstitial water from direct desorption from particles, the model is run under with the optimum ρ value and all other parameters are held the same, except the K_f term is set at zero. This prevents any accumulation through the ingestion route. Projected organism concentrations obtained under conditions with no uptake from ingestion can be compared with projections where both uptake from ingestion and interstitial water occurred. The ratio of the concentration under the no-ingestion scenario compared to that with both ingestion and interstitial water times 100 gives the percentage of the accumulation attributable to desorption and uptake from interstitial water (Table 3). By the difference from 100%, the percentage of the uptake attributable to ingestion can be calculated from the percentage attributable to interstitial water (Table 3). For BaP, the expected percentage from ingestion was determined by comparing the amount of uptake that would be attributed to ingestion

Table 3. Estimated Contaminant Contribution from Interstitial Water and Sediment Ingestion for _Pontoporeia hoyi_

Compound	Source	
	Interstitial Water	Particle Ingestion
Benzo(a)pyrene	0%	100%
Hexachlorobiphenyl	52%	48%
Tetrachlorobiphenyl	80%	20%
Pyrene	74%	26%
Phenanthrene	88%	12%

with the total uptake observed. The percentage uptake from interstitial water was estimated by the difference from 100% (Table 3). The reason for the different approach for the calculation for BaP compared to the other compounds listed is that no estimate of K_{de} is available for the model.

It should be recognized that the relative percentages of accumulation reflect a balance between the flux due to ingestion and the flux due to uptake from interstitial water. If the desorption rates were greater than those used for parameterizing the model, as suggested by equilibrium partitioning advocates, the organisms would only need to contact a negligible amount of interstitial water to maintain the measured rate of accumulation and the balance of the two terms for the uptake would favor the interstitial water. Because desorption rates are determined by gas stripping, they represent the maximum desorption rates. Under natural conditions it is unlikely that the maximum thermodynamic gradient would be maintained, as occurs in these laboratory experimental conditions; therefore, the rate of desorption actually encountered by the organisms is likely smaller than would be measured under gas stripping. Further, in the sediment matrix, diffusion within the interstitial water may limit transfer of desorbed compounds to the organism. Thus, the volume of interstitial water contacted by the organism is likely larger than estimated above. Alternatively, if the desorption rate constant is infinitely slow, there was no diffusion within interstitial water and the organism could receive its interstitial water dose only by moving to fresh interstitial water, a maximum volume of interstitial water that the organism would need to contact to reach steady-state could be determined. For hexachlorobiphenyl, the volume would need to be about 560 ml, assuming no accumulation from food — admittedly an unlikely scenario. Thus, the bounds on the volume of interstitial water to which *P. hoyi* must be exposed to maintain the observed hexachlorobiphenyl uptake range from 8.2 μl to 560 ml, depending on the role of feeding and the desorption rate constant.

The model is extremely sensitive to both K_{de} and K_f. While these terms are very uncertain, particularly considering the limited data sets available, it is unlikely that K_{de} will be considerably larger than the values measured by gas stripping experiments. Therefore, accumulation through interstitial water cannot account for the rate of accumulation observed, particularly for the chlorinated biphenyls. To ascertain an accurate measure of the balance between the two routes of accumulation, improved data sets are required.

One approach to test the true limits for ρ could be accomplished through experiments designed to measure uptake clearance as a function of the organism density per gram of sediment. Once ρ was known, estimates for K_{de} under uptake conditions could be calculated. These could be compared with the maximum rate of desorption by gas stripping and could perhaps permit development of relationships between the gas stripping measurements and environmentally relevant desorption rates. Of course such estimates will still depend on gaining a better understanding of the feeding rate and assimilation efficiency.

5.0 SUMMARY

Our model demonstrates that the accumulation of contaminants from sediments is limited by two rate processes: ingestion and desorption from sediment particles. These processes dominate the flux of compound into the organism, depending on their individual magnitudes. For *P. hoyi* the accumulation from ingestion is most important for compounds that sorb most strongly to the sediments and that can account for more than 90% of the accumulation based on our model results. Ingestion of contaminated particles accounts for only approximately 12% of the accumulated phenanthrene. The model demonstrates that instantaneous equilibrium is not appropriate for estimating the amount of contaminant that is bioavailable or for determining the extent of accumulation of organic contaminants in organisms. More study is required to determine which process, desorption or ingestion, is most rate-limiting and to determine whether these processes are so limiting that the possible steady-state concentrations in organisms are kinetically limited.

ACKNOWLEDGMENTS

We wish to thank Lynn Herche for his advice on statistics and his review of a previous draft of the manuscript. We also wish to thank Wayne Gardner and Thomas Fontaine for their review of a previous draft of the manuscript.

REFERENCES

Adams, W. J., 1987. Bioavailability of neutral lipophilic organic chemicals contained on sediments: A review. In K. L. Dickson, A. W. Maki & W. A. Brungs (Eds.), *Fate and Effects of Sediment-Bound Chemicals in Aquatic Systems.* SETAC Special Publication Series. Pergamon Press, New York: 219–244.

Adams, W. J., R. A. Kimerle & R. G. Mosher, 1985. Aquatic safety assessmentof chemicals sorbed to sediments. In R. D. Cardwell, R. Purdy & R. C. Bahner (Eds.), *Aquatic Toxicology and Hazard Assessment: Seventh Symposium.* ASTM STP 854, American Society for Testing and Materials, Philadelphia, PA: 429–453.

Bailey, P. A. & R. G. Rada, 1984. Distribution of trace metals (Cd, Cr, Cu, Ni, Pb, Zn) in bottom sediments of navigation pools 4 (Lake Pepin), 5, and 9 of the upper Mississippi River. In J. G. Wiener, R. V. Anderson & D. R. McConville

(Eds.), *Contaminants in the Upper Mississippi River*. Butterworths, Boston: 119–138.

Black, M. C. & J. F. McCarthy, 1988. Dissolved organic macromolecules reduce the uptake of hydrophobic organic contaminants by the gills of rainbow trout (*Salmo gairdneri*). *Environ. Toxicol. Chem.* 7: 593–600.

Boese, B. L., H. Lee, D. T. Specht, R. C. Randall & M. Windsor. 1990. Comparison of aqueous and solid phase uptake for hexachlorobenzene in the tellinid clam, *Macoma nasuta* (Conrad): A mass balance approach. *Environ. Toxicol. Chem.* 9: 221–231.

Coats, J. T. & A. W. Elzerman, 1986. Desorption kinetics for selected PCB congeners from river sediments. *J. Contam. Hydrol.* 1: 191–210.

Couch, J. A. & J. C. Harshbarger, 1985. Effects of carcinogenic agents on aquatic animals: An environmental and experimental overview. *Environ. Carcinogenesis Rev.* 3: 63–105.

Dermott, R. M. & K. Corning, 1988. Seasonal ingestion rates of *Pontoporeia hoyi* (Amphipoda) in Lake Ontario. *Can. J. Fish. Aquat. Sci.* 45: 1886–1895.

DiToro, D. M., 1985. A particle interaction model of reversible chemical sorption. *Chemosphere* 14: 1503–1538.

DiToro, D. M., L. M. Horzempa, M. M. Casey & W. Richardson, 1982. Reversible and resistant components of PCB adsorption-desorption: Adsorption concentration effects. *J. Great Lakes Res.* 8: 336–349.

Eadie, B. J., W. R. Faust, P. F. Landrum & N. R. Morehead, 1985. Factors affecting bioconcentration of PAH by the dominant benthic organisms of the Great Lakes. In M. W. Cooke & A. J. Dennis (Eds.), *Polynuclear Aromatic Hydrocarbons: Eighth International Symposium on Mechanisms, Methods and Metabolism.* Battelle Press, Columbus, OH: 363–377.

Eadie, B. J., P. F. Landrum & W. R. Faust. 1988. Existence of a seasonal cycle of PAH concentration in the amphipod *Pontoporeia hoyi*. In M. W. Cooke & A. J. Dennis (Eds.), *Polynuclear Aromatic Hydrocarbons: A Decade of Progress.* Battelle Press, Columbus, OH: 195–209.

Elzerman, A. W. & J. T. Coats, 1987. Hydrophobic organic compounds on sediments: Equilibria and kinetics of sorption. In *Sources and Fates of Aquatic Pollutants*. Advances in Chemistry Series 216, American Chemical Society, Washington, D.C.: 263–318.

Evans, M. S. & P. F. Landrum, 1989. Comparative toxicokinetics of *Pontoporeia hoyi* and *Mysis relicta* for DDE, benzo(a)pyrene and 2,4,5,2',4',5'-hexachlorobiphenyl. *J. Great Lakes Res.* 15: 589–600.

Foster, G. D., S. M. Baksi & J. C. Means, 1987. Bioaccumulation of trace organic contaminants from sediment by baltic clams (*Macoma balthica*) and soft-shell clams (*Mya arenaria*). *Environ. Toxicol. Chem.* 6: 969–976.

Foster, G. D. & D. A. Wright, 1988. Unsubstituted polynuclear aromatic hydrocarbons in sediments, clams, and clam worms for Chesapeake Bay. *Mar. Pollut. Bull.* 19: 459–465.

Freeman, D. H. & L. S. Cheung, 1981. A gel partition model for organic desorption from a pond sediment. *Science* 214: 790–792.

Gardner, W. S., W. A. Frez, E. A. Cichocki & C. C. Parish, 1985. Micromethods for lipids in aquatic invertebrates. *Limnol. Oceanogr.* 30: 1100–1105.

Gobas, F. A. P. C., D. C. G. Muir & D. Mackay, 1988. Dynamics of dietary bioaccumulation and fecal elimination of hydrophobic chemicals in fish. *Chemosphere* 17: 943–962.

Gobas, F. A. P. C., D. C. Bedard, J. J. H. Ciborowski & G. D. Hassner, 1989. Bioaccumulation of chlorinated hydrocarbons by the mayfly *Hexagenia limbata* in Lake St. Clair. *J. Great Lakes Res.* 15: 581–588.

Gschwend, P. M. & S. Wu, 1985. On the constancy of sediment-water partition coefficients of hydrophobic organic pollutants. *Environ. Sci. Technol.* 19: 90–96.

IJC, 1987. *Guidance on Characterization of Toxic Substances Problems in Areas of Concern in the Great Lakes Basin.* Report to the Great Lakes Water Quality Board, International Joint Commission, Windsor, Ontario, Canada: 179 pp.

IJC, 1988. *Proceedures for the Assessment of Contaminated Sediment Problems in the Great Lakes.* Report to the Great Lakes Water Quality Board, International Joint Commission, Windsor, Ontario, Canada: 140 pp.

Karickhoff, S. W. 1980. Sorption kinetics of hydrophobic pollutants in natural sediments. In R. A. Baker (Ed.), *Contaminants and Sediments*, Vol. 2. Ann Arbor Science, Ann Arbor, MI: 193–206.

Karickhoff, S. W., D. S. Brown & T. A. Scott, 1979. Sorption of hydrophobic pollutants on natural sediments. *Water Res.* 13: 421–428.

Karickhoff, S. W. & K. R. Morris, 1985a. Sorption dynamics of hydrophobic pollutants in sediment suspensions. *Environ. Toxicol. Chem.* 4: 469–479.

Karickhoff, S. W. & K. R. Morris, 1985b. Impact of tubificid oligochaetes on pollutant transport in bottom sediment. *Environ. Sci. Technol.* 19: 51–56.

Keilty, T. J., D. S. White & P. F. Landrum, 1988a. Short-term lethality and sediment avoidance assays with endrin-contaminated sediment and two oligochaetes form Lake Michigan. *Arch. Environ. Contam. Toxicol.* 17: 95–101.

Keilty, T. J., D. S. White & P. F. Landrum, 1988b. Sublethal responses to endrin in sediment by *Stylodrilius heringianus* (Lumbriculidae) as measured by a [137]cesium marker layer technique. *Aquat. Toxicol.* 13: 251–270.

Keilty, T. J., D. S. White & P. F. Landrum, 1988c. Sublethal responses to endrin in sediment by *Limnodrilius hoffmeisteri* (Tubificidae), and in mixed culture with *Stylodrilius heringianus* (Lumbriculidae). *Aquat. Toxicol.* 13: 227–250.

Klump, J. V., J. R. Krezoski, M. E. Smith & J. L. Kaster, 1987. Dual tracer studies of the assimilation of an organic contaminant from sediments by depositing feeding oligochaetes. *Can. J. Fish. Aquat. Sci.* 44: 1574–1583.

Knezovitch, J. P, F. L. Harrison & R. G. Wilhelm, 1987. The bioavailability of sediment-sorbed organic chemicals: A review. *Water Air Soil Pollut.* 32: 233–245.

Knezovitch, J. P. & F. L. Harrison, 1988. The bioavailability of sediment- sorbed chlorobenzenes to larvae of the midge, *Chironomus decorus. Ecotoxicol. Environ. Safety* 15: 226–241.

Lake, J. L., N. Rubinstein & S. Pavignano, 1987. Predicting bioaccumulation: Development of a simple partitioning model for use as a screening tool for regulating ocean disposal wastes. In K. L. Dickson, A. W. Maki & W. A. Brungs (Eds.), *Fate and Effects of Sediment-Bound Chemicals in Aquatic Systems.* Pergamon Press, New York: 151–166.

Lamberson, J. O. & R. C. Swartz, 1988. Use of bioassays in determining the toxicity of sediment to benthic organisms. In M. S. Evans (Ed.), *Toxic Contaminants and Ecosystem Health: A Great Lakes Focus.* John Wiley and Sons, New York: 257–280.

Landrum, P. F., 1982. Uptake, depuration and biotransformation of anthracene by the scud, *Pontoporeia hoyi. Chemosphere* 11: 1049–1057.

Landrum, P. F., 1988. Toxicokinetics of organic xenobiotics in the amphipod, *Pontoporeia hoyi*: Role of physiological and environmental variables. *Aquat. Toxicol.* 12: 245–271.

Landrum, P. F., 1989. Bioavailability and toxicokinetics of polycyclic aromatic hydrocarbons sorbed to sediments for the amphipod, *Pontoporeia hoyi. Environ. Sci. Technol.* 23: 588–595.

Landrum, P. F., M. D. Reinhold, S. R. Nihart & B. J. Eadie, 1985a. Predicting the bioavailability of organic xenobiotics to *Pontoporeia hoyi* in the presence of humic and fulvic materials and natural dissolved organic matter. *Environ. Toxicol. Chem.* 4: 459–467.

Landrum, P. F., B. J. Eadie, W. R. Faust, N. R. Morehead & M. J. McCormick, 1985b. Role of sediment in the bioaccumulation of benzo(a)pyrene by the amphipod, *Pontoporeia hoyi.* In M. W. Cooke & A. J. Dennis (Eds.), *Polynuclear Aromatic Hydrocarbons: Eighth International Symposium on Mechanisms, Methods and Metabolism.* Battelle Press, Columbus, OH: 799–812.

Landrum, P. F., S. R. Nihart, B. J. Eadie & L. R. Herche, 1987. Reduction in bioavailability of organic contaminants to the amphipod *Pontoporeia hoyi* by dissolved organic matter of sediment interstitial waters. *Environ. Toxicol. Chem.* 6: 11–20.

Landrum, P. F. & R. Poore, 1988, Toxicokinetics of selected xenobiotics in *Hexagenia limbata. J. Great Lakes Res.* 14: 427–437.

Landrum, P. F., W. R. Faust & B. J. Eadie, 1989. Bioavailability and toxicity of a mixture of sediment-associated chlorinated hydrocarbons to the amphipod *Pontoporeia hoyi.* In U. M. Cowgill & L. R. Williams (Eds.), *Aquatic Toxicology and Hazard Assessment*, Vol. 12. ASTM STP 1027. American Society for Testing and Materials, Philadelphia: 315–329.

Lee, H., B. L. Boese, R. Randall & J. Pelletier, 1990. Method to determine the gut uptake efficiencies for hydrophobic pollutants in a deposit-feeding clam. *Environ. Toxicol. Chem.* 9: 215–219.

McCarthy, J. F. & B. D. Jimenez, 1985. Reduction in bioavailability to bluegills of polycyclic aromatic hydrocarbons bound to dissolved humic material. *Environ. Toxicol. Chem.* 4: 511–521.

McCarthy, J. F., B. D. Jimenez & T. Barbee, 1985. Effect of dissolved humic material on accumulation of polycyclic aromatic hydrocarbons: Structure activity relationships. *Aquatic Toxicol.* 18: 187–192.

McElroy, A. E. & J. C. Means, 1988. Factors affecting the bioavailability of hexachlorobiphenyls to benthic organisms. In W. J. Adams, G. A. Chapman & W. G.

Landis (Eds.), *Aquatic Toxicology and Hazard Assessment*, Vol. 10. ASTM STP 971. American Society for Testing and Materials, Philadelphia: 149–158.

McFarland, V. A. 1984. Activity-based evaluation of potential bioaccumulation for sediments. In R. L. Montgomery & J. W. Leach (Eds.), *Dredging and Dredged Material Disposal*, Vol. 1. American Society of Civil Engineers, New York: 461–467.

Muir, D. C. G., G. P. Rawn, B. E. Townsend, W. L. Lockhart & R. Greenhalgh, 1985. Bioconcentration of cypermethrin, deltamethrin, fenvalerate and permethrin by *Chironomus tentans* larvae in sediment and water. *Environ. Toxicol. Chem.* 4: 51–61.

Nalepa, T. F. & P. F. Landrum, 1988. Benthic invertebrates and contaminant levels in the Great Lakes: Effects, fates and role in cycling. In M. S. Evans (Ed.), *Toxic Contaminants and Ecosystem Health: A Great Lakes Focus.* John Wiley and Sons, New York: 77–102.

Neff, J., 1984. Bioaccumulation of organic micropollutants from sediments and suspended particulates by aquatic animals. *Fresenius Z. Anal. Chem.* 319: 132–136.

NOAA, 1987. *A Summary of Selected Data on Chemical Contaminants in Tissues Collected During 1984, 1985, and 1986.* National Status and Trends Program for Marine Environmental Quality, Progress Report. Technical Memorandum NOS OMA 38, National Oceanic and Atmospheric Administration, Rockville, MD: 104 pp.

NOAA, 1988. *A Summary of Selected Data on Chemical Contaminants in Sediments Collected During 1984, 1985, 1986 and 1987.* National Status and Trends Program for Marine Environmental Quality, Progress Report, NOAA Technical Memorandum NOS OMA 44, National Oceanic and Atmospheric Administration, Rockville, MD: 88 pp.

O'Connor, D. J. & J. P. Connolly, 1980. The effect of concentration of adsorbing solids on the partition coefficient. *Water Res.* 14:1517–1523.

Oliver, B. G., 1984. Uptake of chlorinated hydrocarbons from anthropogenically contaminated sediments by oligochaete worms. *Can. J. Fish. Aquat. Sci.* 41: 878–883.

Oliver, B. G., 1985. Desorption of chlorinated hydrocarbons from spiked and anthropogenically contaminated sediments. *Chemosphere* 14: 1087–1106.

Oliver, B. G., 1987. Biouptake of chlorinated hydrocarbons from laboratory- spiked and field sediments by oligochaete worms. *Environ. Sci. Technol.* 21: 785–790.

Pavlou, S. P. & D. P. Weston, 1984. *Initial Evaluation of Alternatives for Development of Sediment Related Criteria for Toxic Contaminants in Marine Water (Puget Sound). Phase II: Development and Testing the Sediment-Water Equilibrium Partitioning Approach.* Prepared by JRB Associates for the U. S. Environmental Protection Agency under EPA Contract No. 68-01-6388: 89 pp.

Pereira, W. E., C. E. Rostak, C. T. Chiou, T. I. Brinton, L. B. Barber, D. K. Demcheck & C. R. Demas, 1988. Contamination of estuarine water, biota, and sediment by halogenated organic compounds: A field study. *Environ. Sci. Technol.* 22: 772–778.

Reynoldson, T. B., 1987. Interactions between sediment contaminants and benthic organisms. *Hydrobiologia* 149: 53–66.

Robbins, J. A., 1980. *Sediments of Southern Lake Huron: Elemental Composition and Accumulation Rates.* EPA-600/3-80-080. U.S. Environmental Protection Agency, Environmental Research Laboratory, Duluth, MN: 48.

Robbins, J. A., 1986. A model for particle-selective transport of tracers with conveyor belt deposit feeders. *J. Geophys. Res.* 91: 8542–8558.

Rodgers, J. H., K. L. Dickson, F. Y. Saleh & C. A. Staples, 1987. Bioavailability of sediment-bound chemicals to aquatic organisms—Some theory, evidence and research needs. In K. L. Dickson, A. W. Maki & W. A. Brungs (Eds.), *Fate and Effects of Sediment-Bound Chemicals in Aquatic Systems.* SETAC Special Publication Series. Pergamon Press, New York: 219–244.

Schuyema, G. S., D. F. Krawczyk, W. L. Griffis, A. V. Nebeker, M. L. Robideaux, B. J. Brownawell & J. C. Westall, 1988. Comparative uptake of hexachlorobenzene by fathead minnows, amphipods, and oligochaete worms form water and sediment. *Environ. Toxicol. Chem.* 7: 1035–1045.

Shaw, G. R. & D. W. Connell, 1987. Comparative kinetics for bioaccumulation of polychlorinated biphenyls by the polychaete (*Capitella capitata*) and fish (*Mugil cephalus*). *Ecotoxicol. Environ. Safety* 13: 84–91.

Swartz, R. C., 1987. Toxicological methods for determining the effects of contaminated sediment on marine organisms. In K. L. Dickson, A. W. Maki & W. A. Brungs (Eds.), *Fate and Effects of Sediment-Bound Chemicals in Aquatic Systems.* Pergamon Press, New York: 183–197.

Swindoll, C. M. & F. M. Applehans, 1987. Factors influencing the accumulation of sediment-sorbed hexachlorobiphenyl by midge larvae. *Bull. Environ. Contam. Toxicol.* 39: 1055–1062.

Tatem, H. E., 1986. Bioaccumulation of polychlorinated biphenyls and metals from contaminated sediment by freshwater prawns, *Macrobracium rosenbergii* and clams, *Corbicula fluminea. Arch. Environ. Contam. Toxicol.* 15: 171–183.

Varanasi, U., W. L. Reichert, J. E. Stein, D. W. Brown & H. R. Sanborn, 1985. Bioavailability and biotransformation of aromatic hydrocarbons in benthic organisms exposed to sediment from an urban estuary. *Environ. Sci. Technol.* 19: 836–841.

Voice, T. C. & W. J. Weber, 1985. Sorbent concentration effects in liquid/solid partitioning. *Environ. Sci. Technol.* 19: 789–796.

White, D. S. & T. J. Keilty, 1988. Burrowing avoidance assays of contaminated Detroit River Sediments, using the freshwater oligochaete *Stylodrilius heringianus* (Lumbriculidae). *Arch. Environ. Contam. Toxicol.* 17: 673–681.

Windsor, M. H., B. L. Boese, H. Lee, R. C. Randall & D. T. Spect, 1990. Determination of the ventilation rates of interstitial and overlying water by the clam *Macoma nasuta. Environ. Toxicol. Chem.* 9: 209–213.

Witkowski, P. J., P. R. Jaffe & R. A. Ferrara, 1988. Sorption and desorption dynamics of Aroclor 1242 to natural sediment. *J. Contaminant Hydrol.* 2: 249–269.

Wood, L. W., G.-Y. Rhee, B. Bush & E. Barnard, 1987. Sediment desorption of PCB congeners and their bio-uptake by dipteran larvae. *Water Res.* 21: 875–884.

Wu, S. & P. M. Gschwend, 1986. Sorption of hydrophobic organic compounds to natural sediments and soils. *Environ. Sci. Technol.* 20: 717–725.

CHAPTER 9

Freshwater Sediment Quality Criteria: Toxicity Bioassessment

John P. Giesy and Robert A. Hoke

1.0 INTRODUCTION

In an address to the 1988 workshop *Toxic Substances: Approaches to Management*, which was held by the Office of Policy Analysis of the U.S. Environmental Protection Agency (Anon. 1988b), the director of that office, R. D. Morgenstern, stated that "contaminated sediments have the potential to become a significant regulatory issue with important science implications." He went on to state that ". . .in-place toxics pose a high risk to the environment on both a local and regional scale. We view toxic sediments as a potentially serious and costly environmental problem whose management may require participation and coordination of a wide range of players."

The extent of the contaminated sediment problem is unknown because comprehensive assessments have not been completed. We do know that there has been an accumulation of contaminants from agricultural, municipal, and industrial sources in the sediments of rivers, lakes, estuaries, and oceans. In the United States, the U.S. Environmental Protection Agency has identified 134 toxic hot spots where inplace pollutants are a serious problem (Anon. 1988b). Other surveys have identified 41 "areas of concern" in the North American Great Lakes (Anon. 1988c), 50 coastal sites from all areas of the United States, and 85 wildlife refuges where contaminated sediments are a problem (Anon. 1988b).

There are a number of toxic chemicals, potentially associated with sediments, which are not routinely monitored for and for which there is little or no toxicological information. Therefore, it is difficult to set chemically based sediment quality criteria for these compounds. It is also difficult to conduct chemical analyses if one is unsure of the types of chemicals for which one is monitoring. It has been estimated that there are approximately

63,000 chemicals in common use (Hunter *et al.* 1987) as of April 1989, and over 10 million chemicals have been documented in the American Chemical Society's Chemical Abstract Service (CAS) as of April 1990. In addition, other chemicals may form as side reactions or due to chemical reactions in the environment. Therefore, many chemicals potentially associated with sediments are not included in routine monitoring protocols.

The toxicity toward aquatic organisms of only a small number of chemicals other than pesticides has been determined. It has been estimated that only 5 to 10% of the known chemicals have been tested for their toxic properties, and less than 1% of the 50,000 or so compounds manufactured in the United States have been tested for their toxicity to aquatic organisms (Martell *et al.* 1988). The situation is complicated by the fact that organisms are exposed, simultaneously, to a number of contaminants in sediments (Giesy *et al.* 1990), and bioassays can often classify sediments as toxic even when the concentrations of standard contaminants are small (Ankley *et al.* 1990).

Aquatic sediments tend to become contaminated with both inorganic and organic chemicals, which are sorbed to particulate matter or in solution in sediment porewater (Salomons *et al.* 1987, Tessier & Campbell 1987). These contaminants can be accumulated by and directly affect benthic organisms (Ciborowski & Corkum 1988, Giesy *et al.* 1988a, 1990) or affect other aquatic organisms by becoming bioavailable through resuspension or leaching (Jones & Lee 1978, Malueg *et al.* 1983).

Because toxic discharges to surface waters have been more obvious and the potential for adverse effects due to these sources of contaminants seemed more acute, the effects of contaminants in the water column have received more attention than the effects of contaminated sediments (Dickson & Rodgers 1985). Investigation of the toxicity of sediments has been limited by the complexity of sediment-water column and sediment-biota interactions (Dickson *et al.* 1987, White 1988). A lack of knowledge concerning the precision of sediment toxicity tests has also prevented their more frequent use in sediment contaminant assessments (McIntyre 1984, Giesy & Allred 1985).

Here we will discuss methods for setting sediment quality criteria, with particular emphasis on site-specific, biological effects-based criteria, and present our rationale for the use of a battery of assays for assessing the toxicity of in-place pollutants. We will restrict our comments to tests of the effects of contaminants on bacterial metabolism or survival, or survival, growth, and reproduction of whole metazoans. Many assays which use cells of higher organisms or biochemical systems to investigate the toxicity, mutagenicity, and teratogenicity of environmental matrices have been developed but are not discussed here.

2.0 SEDIMENT TOXICITY CRITERIA

Ambient water quality criteria have been established in the United States for a relatively small number of contaminants, but no such criteria have yet been developed for sediments (Shea 1988). The United States Environmental Protection Agency (U.S. EPA) as well as advocacy groups have been pressing for the development of sediment quality criteria to facilitate decision-making about remediation, handling, and disposal of contaminated sediments. Agencies could use these criteria to classify the level of potential adverse effects of sediments relative to defined threshold values. This would eliminate much of the polemics of what constitutes a contaminated sediment. Presently the U.S. EPA is primarily interested in establishing nationally applicable numerical sediment quality criteria (Shea 1988).

The historical beginning of sediment quality criteria was in response to the need to manage dredged materials (Anon. 1977a,b). In general, these criteria relied on chemical analyses or on simple bioassays of acute toxicity of elutriates or sequential extracts. However, these types of simple criteria provide little information on the effects of contaminated whole sediments on biota (Hoke & Prater 1980).

The primary reason for the development of global, chemically based, numerical sediment quality criteria is to have uniform standards with a demonstrated scientific rationale for application at all locations. While it is understood that national or regional criteria, based on concentrations of chemicals, have some utility and are driven by the political, litigative, and engineering needs for some type of criteria, we feel that the protection of benthic communities from the subtle effects of complex mixtures of contaminants in sediments requires site-specific, biologically based criteria. Because of the limitations in the criteria development process discussed in this chapter, it is felt that site-specific criteria, using the triad approach, are more useful than global sediment quality criteria. Global criteria may be easier to apply and for managers and regulators to understand, but they will not provide as much protection and understanding of individual locations.

In some locations, the use of chemical analyses to determine potential adverse effects and map the distribution of the contamination can be done because there is primarily one dominant contaminant, such as PCBs in Waukegan Harbor, WI, U.S.A. or the Hudson River, NY, U.S.A. However, the presence of specific contaminants or exceedence of established criteria for concentrations of chemicals in sediments is not always the primary concern relative to sediment contaminants. In Puget Sound of Washington State, U.S.A., observed biological effects are a greater concern than are concentrations of typical sediment contaminants (Anon. 1988a). An alternative approach would be to assign numerical criteria to biological response variables. These could be agreed upon and applied in a uniform manner, as could numerical chemically based criteria, but would account for the incomplete chemical information available for most locations and integrate interactions among contaminants. This approach could also include assays to measure bioconcentration and mutagenic potential as well as lethality.

A number of procedures, based on both chemical and biological analyses, have been proposed as the basis for the development of sediment quality criteria. These include:

1. the background concentration approach
2. the water quality criteria approach
3. the equilibrium partitioning approach
   a. empirical
   b. sediment-water
   c. sediment-biota
   d. sediment-water-biota
4. the field bioassay approach
5. the screening level concentration approach
6. the apparent effects threshold approach
7. the spiked bioassay approach

A comprehensive history of the development of these approaches and rationale for their use has been compiled as part of the background development of sediment quality objectives (Anon. 1983, 1985, 1986a, b, c, d, 1987a, b, c, d, e, f, g, 1988a, b, c, d, e, f, g, Neff *et al.* 1985, Persaud *et al.* 1989). Here we briefly describe each major approach and what we consider to be its advantages and disadvantages.

The U.S. EPA is presently considering only three approaches for the development of numerical sediment quality criteria. These include: 1) the water quality criteria approach, 2) an equilibrium partitioning approach, and 3) site-specific bioassays (Shea 1988). The first two approaches are chemically based methods and the third is a biologically based method. We feel that the first two techniques have little utility for setting local sediment quality criteria and that it would be dangerous to establish sediment quality criteria without considering biological responses. The chemically based methods may be useful for setting global target guidelines but should be supplemented with biologically based local or regional criteria.

2.1 Sediment Background Concentration Approach

In the sediment background (BC) approach, concentrations of contaminants in a sediment of interest are compared to the concentrations of that chemical or element in sediments which have been deemed, based on the presence of indicator organisms, to be of suitable quality (Mudroch *et al.* 1986). This can be done several ways. One method is to adopt the concentration in sediment at a remote reference site (Persaud *et al.* 1989). For metals, however, natural concentrations vary greatly among geographic regions, and the most appropriate background concentrations would be historical concentrations in a particular region, as determined from sediment cores.

The BC approach is useful only because it requires little information and can serve as a benchmark against which to evaluate concentrations of heavy metals. This method is useless for synthetic organic compounds but may have some utility for naturally occurring organic compounds such as polycyclic aromatic hydrocarbons (PAH). However, it is difficult to establish an acceptable reference sediment, and the background approach does not have a strong biological basis. For these reasons we do not advocate the use of the BC approach.

2.2 Apparent Effects Threshold Approach

The apparent effects threshold (AET) approach determines the concentration of a particular toxicant above which a statistically significant effect would be observed (Anon. 1986a). It is assumed that when the apparent effects threshold concentration of a toxicant is exceeded under field conditions any observed effects are attributable to the contaminant of interest. Below the threshold concentration, effects would be attributed to other toxicants. The apparent effects concentrations are determined empirically from an examination of synoptic observations of sediment chemistry and biological effects on a defined species and endpoint at a specified level of statistical significance. The AET method assumes that a toxicological Leibig's law of the minimum can be invoked and a single toxicant identified as the cause of all the observed adverse effects. For this to be a reasonable assumption, no sediments could contain contaminants with the same or similar joint action modes of toxicity. This assumption would seldom be met in sediments containing several metals or neutral halogenated hydrocarbons, which are known to have similar modes of action, respectively. The AET approach also assumes the most toxic chemical or element in a mixture can be identified without error and its concentration accurately quantified.

The AET approach has both advantages and disadvantages. The greatest advantage of the technique is the development of a criterion from field conditions using both chemical and biological information. However, the AET method is limited by the fact that a rather large data set is required, preferably for the same species and endpoint and collected in the same manner. An additional limitation of the AET approach for establishing sediment quality criteria is that the criteria are developed on a chemical by chemical basis, and the method does not account for the effects of multiple toxicants. For this reason the AET approach does not have much predictive power or potential to adequately protect local populations.

Normalization to remove the effects of among-location variation of co-variates is also difficult. If the data set is large and there is sufficient information on these covariates, the data used in the AET criterion develop-ment can be corrected to remove this variability and determine a more accurate, global estimate of the apparent effects threshold. In addition to the effects of nontoxic covariates, the criteria developed with the AET approach can be obscured by the presence of noncovarying toxicants or covarying toxicants. In the first case, uncertainty would be added to the estimate of the apparent effects threshold. In the second case, if the data set was not sufficiently great the cause of the apparent effects would be misas-signed. Thus, the existence of multiple toxicants, all occurring at concentra-tions near their threshold for effects, could adversely affect both the accu-racy and precision of criteria developed by the AET approach.

Sediment toxicity can be estimated by comparing the concentrations of toxic substances associated with sediments with known dose-response rela-tionships. However, unknown factors which affect these comparisons include the efficiency of the extractions and analyses (Bellar *et al.* 1980, Hoke & Prater 1980, Samoiloff *et al.* 1983), the availability of the toxicants to biota (Babich & Stotsky 1977, Ward & Young 1984, Oliver 1985), and the potential interactions among toxicants. Therefore, measured contaminant concentrations may not accurately reflect the potential toxicity of sediments to organisms.

We do not feel that the AET approach is a particularly useful method for establishing sediment quality criteria. There are too many uncertainties. Criteria established by using this method would also need to be validated for each contaminant by the use of either the field bioassay or spiked bioassay approach.

2.3 Screening Level Concentration Approach

The screening level concentration (SLC) approach is similar to the appar-ent effects threshold technique in that it compares the distribution of ben-thic invertebrates with the concentrations of contaminants in the same sedi-ments (Neff *et al.* 1985). The calculation of sediment quality criteria with the SLC approach is a two-step process. The first step is to calculate the 95th percentile of the frequency distribution of chemical concentrations at a minimum of ten locations and for a minimum of ten different species. Because a minimum of 100 species-location pairs must be examined for each chemical, the SLC is a very data-intensive method. It is assumed that the data set includes locations containing the full range of contaminant concen-trations to which each species is tolerant so that a good estimate of the threshold concentration can be obtained. SLCs have been developed primarily for nonpolar organic compounds and estimate the greatest concentration of a particular toxicant which will be tolerated by 95% of the benthic invertebrate species included in the data set. However, the technique can, with appropriate normalization, be adopted for use with polar organic compounds or inorganic toxicants.

The SLC approach is similar to the AET approach and suffers from the same limitations. Both approaches also suffer from the disadvantage that unless all physicochemical parameters, other than toxicants, and sources of colonizing individuals are identified, it is difficult to demonstrate that differences among populations or in endpoint responses are due to toxic substances in the sediments. Short-term chemical and physical stressors such as chlorine, pH, or temperature can eradicate populations of benthic organisms without leaving toxic residues in the sediments. The absence of macroinvertebrates in sediments does not necessarily implicate sediment toxicity as the causative factor, and presence-absence data are not indicative of the potential toxicity of deeper sediments, which are generally not colonized by benthic invertebrates but may be uncovered or resuspended and come into contact with biota. Therefore, the presence or absence of benthic invertebrates alone may not provide much insight into the toxicity of sediments.

Since the SLC is a correlation method, it does not assume causal relationships between concentrations of particular contaminants and observed effects. Thus, there is little effect of edaphic parameters. The SLC approach is also less biased by the simultaneous exposure of organisms to multiple contaminants. However, because synergisms are not considered and organisms could be responding to unidentified or unquantified toxicants, the SLC always results in sediment quality criteria which are conservative. In fact, the derived criteria may be too protective.

As with all of the empirical approaches relating biological effects to chemical concentrations, in order to derive global sediment quality criteria expressed as concentrations of specific chemicals, normalization to correct for contaminant availability will decrease variability and allow for a more accurate estimate of the sediment quality criterion. Because the SLC approach was designed for use primarily with organic chemicals, the most appropriate normalization would be to total organic carbon (TOC) content of the sediment.

2.4 Equilibrium Partitioning Approach

There are several equilibrium partitioning (EP) approaches, which consider the equilibria between the concentrations of contaminants in the solid phase of the sediment and the interstitial water or with the concentration of contaminants in organisms. With this information one can calculate the apparent lethal concentration (LC), effective concentration (EC), lethal dose (LD), or effective dose (ED) values by comparing the observed concentrations to water quality criteria or to previously determined toxicity values. The EP approach is not limited to toxic effects but also can be used to predict the bioaccumulation potential so that estimates of other, more long-term effects can be made.

The theory on which the EP approach is based assumes contaminants associated with the solid phase of sediments will reach an equilibrium with the porewater and associated organisms which can be predicted from the physical and chemical properties of the sediment and chemical of interest. The intent of the EP approach is to correct for differences in dose of different chemicals in different sediments. This is done by correcting for the "speciation" of the contaminants so that better prediction of biological effects can be made from bulk sediment chemical analyses. The EP approach has been developed primarily for organic compounds but theoretically could be used for inorganic toxicants, such as metals.

The advantage of the EP method is that it corrects for differences in dose by using very simple chemical parameters of the sediment. For organic compounds this is done by use of the sediment-water partition coefficient and organic carbon content of the sediment. The EP method assumes that the primary vector of bioaccumulation is via the porewater and not the solid sediment phase. Studies of the accumulation of sediment-sorbed hexachlorobiphenyl by midge larvae support the hypothesis that accumulation can be accounted for primarily by the partitioning model of exposure; however, biological processes such as ingestion of particulates could account for some accumulation (Swindoll & Applehans 1987). The technique can be carried further by using the bioconcentration factor (BCF) to predict the accumulation of the compound into organisms after the concentration in interstitial water has been predicted from the carbon-normalized partitioning coefficient (K_{OC}).

The EP approach suffers from the same limitations of all of the chemically based global, numerical, sediment quality criteria in that it does not consider unmeasured chemicals or interactions among chemicals and cannot be used for chemicals for which there is no toxicological information. It is suggested that the EP approach has utility for establishing general criteria but that it will not substitute entirely for site-specific criteria.

2.5 Spiked Bioassay Approach

The spiked bioassay (SB) approach is analogous to bioassays to determine the effects of pure chemicals or mixtures of toxicants on organisms which have been used to set surface water quality criteria. In the SB approach, the effects of toxicants are determined when the toxicant of interest is added (spiked) at different concentrations to whole sediments (Cairns *et al.* 1984, Giesy *et al.* 1990).

The results of bioassays with benthic organisms can then be compared to the concentrations of individual toxicants in the sediment to determine where on the dose-response relationship the sediment is for each chemical. Attempts to use this technique to infer the potential for effect and the concentration to which sediments would need to be diluted to protect benthic organisms have not been successful because of the potentially large number of contaminants present at most locations and possible interactions among contaminants and with the sediment (Lee & Jones 1982, Chapman 1986, Chapman 1987, Chapman *et al.* 1987). Dose-response relationships for toxicants associated with sediments can be determined by spiking control or reference sediments of a defined character with known concentrations of single toxicants or mixtures. Results may then be used to predict the effects of these contaminants on benthic organisms (Cairns *et al.* 1984). While these techniques provide useful information and should be maintained as methods for sediment toxicity evaluation, they are limited and do not provide sufficient information to answer all of the pertinent questions about the in situ toxicity of contaminant mixtures in sediments (Chapman *et al.* 1987, Giesy *et al.* 1988a,b, Giesy *et al.* 1990).

The dose-response relationship characterizes the toxicity of a mixture of contaminants in a sediment and permits quantification of the adverse effects threshold concentration of a sediment for an organism (Giesy *et al.* 1988a, Giesy *et al.* 1990). This information helps answer several questions, including:

1. What is the volume of sediment which exceeds a threshold toxicity value?
2. Where are the most toxic sites for remedial action prioritization?
3. How much would a toxic sediment need to be diluted before it no longer threatens benthic invertebrates?
4. How toxic would the sediment be if it were resuspended into and diluted by the water column (as in dredging operations)?
5. What is the toxic potential of deep sediments not directly toxic to benthic organisms in the surface substratum, but which may be uncovered or resuspended (Chapman 1987)?
6. What time would be required, given a known rate of degradation, for the contaminant concentration to be reduced below the adverse effect threshold value for an organism?
7. What effect would the addition of more of the same or different toxicants have on the toxic potential of sediments?
8. What is the sediment toxicity trend associated with a given class of organisms?

In addition, the use of dose-response functions allows one to determine the probability of an effect and the potential proportion of the population affected and to place statistical confidence intervals on these values. It is the probability function which is needed for predictive modeling.

In order to determine the relative toxicity of sediments using dose-response assays, a number of questions need to be addressed (Adams & Darby 1980). These include:

1. What assay organism or organisms should be used?
2. What positive or negative control treatments should be used?
3. What endpoints should be monitored?
4. What are the relationships among different assays and endpoints?
5. What is the minimum number of assays to be included in a battery of tests to provide maximum nonredundant information?
6. What are the relative sensitivities of the possible bioassays?
7. What are the appropriate test conditions, and how do the test conditions affect the results and interpretation of the assays?

To determine the dose-response relationship for a sediment or porewater (interstitial water), one must perform dilutions. Sediments can be classified as exhibiting acute or chronic toxicity to an organism without performing dilutions, but such an assessment will not answer the questions posed earlier. The preparation of sediment dilutions is not as simple as the preparation of effluent or porewater dilutions. Sediments should be diluted with a reference sediment which is not as contaminated as the sediment to be tested (Chapman 1987). Questions arise as to what sediment should be used for a reference, how samples should be mixed, and how long they should be allowed to equilibrate before use in a bioassay. Recently, Giesy *et al.* (1990) demonstrated that assay results differed depending on the dilution technique used, but conclusions concerning classification of a sediment as either very toxic or relatively nontoxic were similar, regardless of the dilution technique.

Due to the complications of making dilutions of whole sediment, the use of porewater extracts and elutriates has been proposed (Anon. 1977a, Bahnick *et al.* 1980, Nebeker *et al.* 1984b, Giesy *et al.* 1988a, 1990). Aqueous solutions of the contaminants associated with sediments can be made by 1) extracting the porewater (Batley & Giles 1980, Bellar *et al.* 1980), 2) making an elutriate (Anon. 1977a, Hoke & Prater 1980), or 3) making differential extractions to selectively remove particular classes of toxic substances (Jenne *et al.* 1980, Samoiloff *et al.* 1983).

Elutriates may give different results than tests with porewaters. We have observed that elutriates are often more toxic than porewaters (Hoke *et al.* 1990b). However, elutriates may be useful in determining the potential toxicity of resuspended sediments in the water column. The testing of elutriates to determine the toxicity of sediments is not universally accepted, and the utility of these tests has not been thoroughly explored (Hoke & Prater 1980, Seelye & Mac 1982, Chapman *et al.* 1987). Therefore, we feel that bioassays of porewaters are the most useful. The utility of porewater bioassays also has not been fully explored. However, recent investigations support the hypothesis that a significant proportion of benthic macroinvertebrates' exposure to sediment contaminants is via the sediment porewater.

The use of porewaters or elutriates as toxicant solutions has facilitated the testing of standard nonbenthic bioassay organisms, such as *Daphnia magna* (Malueg *et al.* 1984a,b, Schuytema *et al.* 1984, Giesy *et al.* 1988a), *Ceriodaphnia dubia*, *Pimephales promelas* (Hoke *et al.* 1990b), and the bacterium *Photobacterium phosphoreum* (Giesy *et al.* 1988a,b). The use of these organisms has facilitated the comparison of porewater or elutriate assay results with established surface water criteria. In addition, these techniques preclude many of the problems associated with sediment dilutions and allow for use of an appropriate control treatment in a defined medium (Nebeker *et al.* 1984b). The relationship between the results of porewater or elutriate tests and those conducted with whole sediments is unclear (Bahnick *et al.* 1980, Cairns *et al.* 1984, Nebeker *et al.* 1984b) because the partitioning of toxicants from the solid phase into the porewater or elutriate is dependent on many chemical and physical processes (Sinex *et al.* 1980, Knezovich *et al.* 1987) which are not well understood.

In a conceptual sediment toxicity exposure model (Figure 1), benthic organisms can be exposed to toxicants in either of two ways: directly from sediment-bound contaminants or via the porewater. The simplification of sediment toxicity testing by porewater exposure, rather than whole sediment exposure, is based on the assumption that organisms receive most of their exposure through contact with the porewater (Nebeker *et al.* 1984b). The concentrations of toxic materials in the porewater are in a dynamic equilibrium with the solid phase and its associated contaminants. Most contami-

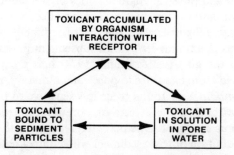

Figure 1. Schematic representation of the dynamic equilibrium between the sediment solid phase, porewater, and organisms (Giesy *et al.* 1989).

nants associated with sediments have limited solubility in water, and it is possible to have a considerable excess of contaminant associated with the solid phase of a sediment such that observed toxicity may not be well correlated with the concentrations of toxic materials in the bulk sediment (Patrick *et al.* 1977, Shaner & Knight 1985, DiToro 1989). Thus, porewaters and concentrations of their associated contaminants have been suggested as a more proximate measure of sediment toxicity.

In a study of the toxicity of cadmium in sediment to *D. magna*, the free or dissolved, uncomplexed cadmium was found to be the toxic form of the metal (Schuytema *et al.* 1984). The authors concluded that "cadmium adsorbed to the sediment had negligible toxicity" and that the toxic effects of metals associated with sediments could best be predicted from the concentrations of unbound metals in the porewater. In a study of the toxicity of copper in sediments to several organisms, including *D. magna* and *Chironomus tentans*, it was found that the toxicity of copper was attributable to soluble copper and not to sediment-bound copper (Cairns *et al.* 1984). The toxicity of cadmium, mercury, and the insecticide dieldrin to *D. magna* and *C. tentans* was more highly correlated with the concentration in porewater than that in bulk sediments (van de Guchte & Mass-Diepeveen 1988). Similarly, in a study of accumulation of chlorobenzenes from sediments, it was reported that bioaccumulation by larvae of the midge *Chironomus decorus* was dependent on the concentration of the chemicals in the interstitial water (Knezovich & Harrison 1988). Knezovich *et al.* (1987) recently reviewed the results of a number of studies, conducted with organic or inorganic toxicants and a number of bioassay species, to determine the relative importance of the sediment and interstitial water phases as pathways of exposure. This review concluded that the relative importance of interstitial water and whole sediments as sources of contaminants varies greatly, depending on organism and toxicant type, but both can serve as significant sources of contaminants under different conditions.

The validity of a simple porewater exposure model is likely to be dependent on chemical and physical characteristics of the sediment, type of contaminants present, and the test species. For instance, burrowing mayflies of the genus *Hexagenia* digest large quantities of sediment (Zimmerman *et al.* 1975) and detritus (Cummins 1973) with subsequent potential for release of contaminants in the gut (Lee & Plumb 1974) and greater exposure than would be observed from exposure to porewater alone. In a study by Giesy *et al.* (1990), *H. limbata* were found to be less sensitive to the toxic effects of whole sediments than to the effects of porewater extracted from a subsample of the same sediment. One potential reason for the observed effects may be that the dilution of toxic sediment with control sediment reduced the amount of toxicant available to *H. limbata* by adding organic material. In such an exposure model, there would be sufficient contaminant associated with the bulk sediment to result in similar concentrations in the porewater extracts of diluted sediments, while the amount directly available from the sediment would be diminished. Due to their habitat requirements for a suitable sediment type in which to burrow, exposure to porewater in the absence of sediment may have been stressful in and of itself. However, in the study by Giesy *et al.* (1990) artificial burrows were provided for the *Hexagenia* (Fremling & Mauck 1980). Thus, it is doubtful that the organisms were unduly stressed by the absence of sediment.

The isolation of interstitial water is not a trivial matter, and a number of techniques have been proposed to minimize artifacts in toxicity assays as a result of sample extraction and isolation methods (Knezovich *et al.* 1987). Specifically, it has been suggested that centrifugation, the technique which we endorse, is not suitable for analysis of hydrophobic chemicals because they may rapidly partition back onto the sediment (Karickhoff 1980). If the concentration of toxicant associated with the solid and liquid phases is in equilibrum, this should not be a problem.

The major utility of the SB approach is that it provides a dose-response relationship, which is necessary for simulation models to extrapolate to other conditions and to predict the proportion of response in a population. The method allows one to assess the probability that observed effects are caused by a particular toxicant. Also, the probability of effect by a toxicant in a sediment can be evaluated by adding an additional amount of the suspected toxicant to sediment. In the first case, if the suspected sediment is actually causing toxicity, the physicochemical characteristics of the two sediments are similar, and the suspected toxicant is the only cause of observed effects, one should be able to reproduce the observed effects in the reference sediment. If the effects cannot be duplicated there may be additional causative agents in the sediment of interest. In the second method, where an additional quantity of the toxicant of interest is added (as in Figure 2) to the sediment of interest, if that is the only toxicant causing effects or if that toxicant has the same mode of toxic action and relative potency as other toxicants present, the dose-response relationship should be a linear extrapolation of the response of the original sediment (Figure 2). If a response other than a directly linear response is observed (lines B or C), the suspected toxicant is probably not solely responsible for the effects observed in the sediment of interest (SI in Figure 2). The response given on the abscissa could be any parameter, either continuous or discontinuous, which is expressed by the bioassay organism. This could be either a population- or individual-level response, such as mortality rate or inhibition of weight gain, respectively. The exposure units must be appropriate to relate the concentrations of spiked and suspected toxicant in the original sediment. That is, if the effects of the sediment of interest are plotted against the total concentration of a particular metal, the concentration in

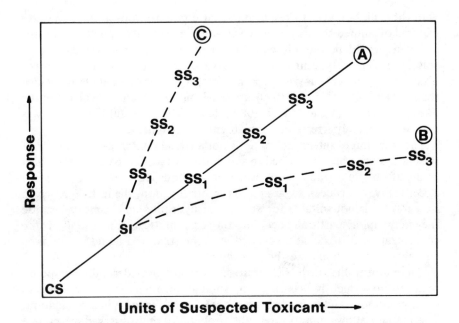

Figure 2.   Schematic representation of the response of a population of organisms to the addition of a known toxicant to a contaminated sediment. CS is the response to an uncontaminated control sediment and SI is the response to the sediment of interest. Lines A, B, and C represent theoretical responses of the population to three additions (SS_1, SS_2, and SS_3) of the suspected toxicant.

the extrapolated portion of the curve must be the sum of the added and original metal or total concentration of metals corrected for relative potencies (Giesy *et al.* 1990).

The interpretation of the SB approach is complicated by differences in bioavailability of toxicant which are possible in the sediment of interest. Specifically, one must be certain that the toxicant spike has had sufficient time to equilibrate with toxicant present in the sediment originally. Also, it is possible that the dose-response relationship will be nonlinear due to saturation of binding sites. This should not be a problem if the range of additions is small. Spiking of contaminated sediments would not be relevant if the toxicity of the contaminated sediment was so great that no additional effects could be elicited.

Spiked sediment bioassays are very effort-intensive and, thus, will not be useful as sediment toxicity screening tests. However, we do feel that the SB approach can be useful in identifying or confirming the effects of suspected toxicants in sediments. The spiked sediment approach is useful for determining the toxicity of compounds potentially associated with sediments under laboratory conditions prior to their release to the aquatic environment. This type of assay will also be useful for determining the toxicity of a few ubiquitous toxicants or representative classes of toxicants to a few cosmopolitan benthic invertebrates species and in calibrating the responses of several species.

2.6 Water Quality Criteria Approach

In the water quality (WQ) approach, concentrations of individual contaminants in porewater (interstitial water) are compared to existing water quality criteria. This method has the advantage that it relies on an extensive body of toxicological information for many chemicals and species. Even though final water quality criteria have been developed for only approximately 50 chemicals, that is more than the number of sediment quality criteria that have been developed. Also, information exists with which to evaluate many more chemicals.

There are several disadvantages to the WQ approach. The numerical criteria used in this approach were developed for sediment-free systems and, thus, do not consider the amelioration of biological effects due to binding of contaminants, which reduces their bioavailability. This means that sediment quality criteria produced in this way will be conservative and overprotective. Water quality criteria are also based on the responses of nectonic and not benthic organisms. This is not a serious limitation of the technique, and if no other information is available an estimate of the hazard of sediments can be made using the WQ approach.

3.0 SUMMARY — CHEMICAL, CONCENTRATION-BASED SEDIMENT QUALITY CRITERIA

The greatest limitation of chemical, concentration-based, global sediment quality criteria is the fact that so few criteria have been derived. The first sediment quality guidelines were developed in 1973 by the Federal Water Quality Administration. These were then adopted by the U.S. EPA and have been referred to as the "Jensen Criteria". This first set of sediment quality criteria only considered seven contaminants (Table 1). The criteria were applied such that if any one numerical value was exceeded the sediment was classified as polluted and had to be treated as such.

In 1973 the U.S. EPA published criteria and regulations for managing marine dredged sediments (Anon. 1973a,b). Other early sediment quality criteria were developed by the U.S. EPA and U.S. Army Corps of Engineers (ACOE) (Anon. 1977a,b). The criteria were developed to help assess the impact of freshwater and marine dredged materials and, when coupled with site-specific sediment bioassays, have been the standard reference used for

Table 1. Federal Water Quality Administration Criteria for Maximum Allowable Contaminants in Dredged Material

Parameter	Criterion[a]
Volatile solids	6.0
Chemical oxygen demand	5.0
Total Kjeldahl nitrogen	0.10
Oil and grease	0.15
Mercury	0.0001
Lead	0.005
Zinc	0.005

Source: Anon. (1973).
[a]All values as % dry weight.

regulating contaminated sediments for 10 years (Table 2). These regulations and principles are still in effect.

Recently several agencies have developed additional global, numerical sediment quality criteria. The Wisconsin Department of Natural Resources has developed criteria for some metals, PCBs, and a few pesticides (Table 3, Sullivan et al. 1985). The Ontario Ministry of Environment (OME) also has published sediment management guidelines (Persaud et al. 1989). In their development of global criteria, the OME compared the results of several techniques and reported three guidelines: 1) the no effect level, 2) the lowest effect level, and 3) a limit of tolerance level (Table 4). The OME sediment

Table 2. U.S. EPA, Region V, Guidelines for Classifying Sediments of Great Lakes Harbors[a]

	Nonpolluted	Moderately Polluted	Heavily Polluted
Volatile solids	<5%	5%–8%	>8%
COD	<40,000	40,000–80,000	>80,000
TKN	<1,000	1,000–2,000	>2,000
Oil and grease (hexane solubles)	<1,000	1,000–2,000	>2,000
Lead	<40	40–60	>60
Zinc	<90	90–200	>200
Mercury	<1.0	N.A.	>1.0
Ammonia	<75	75–200	>200
Cyanide	<0.10	0.10–0.25	>0.25
Phosphorus	<420	420–650	>650
Iron	<17,000	17,000–25,000	>25,000
Nickel	<20	20–50	>50
Manganese	<300	300–500	>500
Arsenic	<3	3–8	>8
Cadmium	–	–	>6
Chromium	<25	25–75	>75
Barium	<20	20–60	>60
Copper	<25	25–50	>50

Source: Anon. (1977a).
[a]All concentrations as μg/kg, dry weight.

Table 3. Wisconsin Department of Natural Resources Sediment Quality Criteria

Contaminant	Guideline[a]
Arsenic	10
Cadmium	1
Chromium	100
Copper	100
Lead	50
Mercury	0.1
Nickel	100
Zinc	100
Heptachlor	0.05
Endrin	0.05
Aldrin	0.01
Chlordane	0.01
PCBs	0.05
Dieldrin	0.01
δ–BHC	0.05

Source: Sullivan et al. (1985).
[a] $\mu g/g$, dry weight.

quality criteria are based on overt toxicity to benthic invertebrates and do not consider other potential effects such as bioaccumulation and subsequent effects on longer-lived species. The Ontario Ministry of Environment also derived a list of criteria for the classification of sediments into categories for subsequent disposal (Table 5). Similarly, Hart et al. (1988) proposed a set of sediment quality guidelines (Table 6).

A number of criteria and rationales have been presented for the development of sediment quality criteria (Persaud et al. 1989, Table 7). It is useful to determine potential sediment quality criteria with as many approaches as possible, ascertain if there is some consensus of appropriate criteria among the approaches, and then choose an appropriate criterion (Hart et al. 1988).

A useful method for determining if the observed toxicity of sediments is due to primarily one contaminant or a class of contaminants with a similar mode of action is to plot the toxic units for the whole sediment as a function of concentrations of possible contaminants (Figure 3). The toxic units can be based on either lethality (LC_{50}) or another nonlethal response (EC_{50}). When such a plot is made, if all of the observed toxicity at each location is due to the suspected toxicant one would observe a linear relationship. Locations above the 1:1 correspondence line indicate that there is toxicity not accounted for by the suspected toxicant. Points which fall below the 1:1 correspondence line indicate that there is an antagonistic effect. This is generally due to bioavailability differences among locations, and the effect can often be removed by accounting for the activity (speciation) of the toxicants of interest. This is particularly true of metals, which can be bound and precipitated, or ammonia, the toxicity of which is very pH-dependent.

Table 4. Sediment Quality Criteria (μg/g Dry Weight for Metals and Nutrients) Proposed by the Ontario Ministry of Environment

Metals	No Effect Level	Lowest Effect Level	Limit of Tolerance Level
As	4.0	5.5	33.0
Cd	0.6	1.0	10.0
Cr	22.0	31.0	111.0
Cu	15.0	25.0	114.0
Fe%	2.0	3.0	4.0
Pb	23.0	31.0	250.0
Mn	400.0	457.0	1110.0
Hg	0.1	0.12	2.0
Ni	15.0	31.0	90.0
Zn	65.0	110.0	800.0

Organics[a]			μg Contaminants/g Carbon (TOC)
δ-Chlordane	0.001	0.005	6.6
Heptachlor	0.001	0.002	0.5
Endrin	0.002	0.003	33.1
Aldrin	0.001	0.007	128.4
Mirex	0.001	0.002	9.1
Chlordane	0.001	0.008	6.2
p,p-DDT	0.005	0.009	13.6
p,p-DDD	0.002	0.008	9.0
p,p-DDE	0.003	0.005	21.3
o,p-DDT	0.001	0.006	11.3
PCB 1254	–	0.058	34.4
PCB 1248	–	0.034	150.5
PCB 1016	–	0.007	53.3
PCB (Total)	0.020	0.041	69.8
Dieldrin	0.006	0.019	59.0
BHC	–	0.003	11.8
δ-BHC	0.002	0.003	25.0
β-BHC	0.001	0.005	21.0
α-BHC	0.001	0.006	10.4
HCB	0.001	0.020	47.6
Heptachlor epoxide	0.001	0.005	5.5

Nutrients			
TOC %		1.0	10.0
TKN		545.0	4800.0
T.P.		600.0	2050.0
Solvent extractables (oils and greases)		2400.0	–

Source: Persaud et al. (1989).
[a]Values in this column are multiplied by the actual TOC content of the sediments, e.g., @ TOC of 5%, the total PCB value is 70 × 0.05 or 3.5 ppm.

3.1 AMMONIA EXAMPLE

The results of an analogous approach to identifying the compounds responsible for the toxicity of porewater are given in Figures 4 and 5. Ankley et al. (1990) observed a statistically significant relationship between

Table 5. Dredged Material Disposal Classification Criteria[a] Used by the Ontario Ministry of Environment

Parameter	Open Water Disposal	Unrestricted Land Use	Restricted Land Use
Cadmium	1.0	1.6	4.0
Lead	50.0	60.0	500.0
Mercury	0.3	0.5	0.5
PCBs	0.05	<2.0	>2.0
Loss on ignition	6.0		
Oil and grease	1,500.0		
Total phosphorus	1,000.0		
Total Kjeldahl nitrogen	2,000.0		
Ammonia	100.0		
Grain size			
Visual description			
Arsenic	8.0	14.0	20.0
Copper	25.0	100.0	100.0
Zinc	100.0	220.0	500.0
Chromium	25.0	120.0	120.0
Iron	10,000.0	350,000.0	350,000.0
Nickel	25.0	32.0	60.0
Cobalt	50.0	20.0	25.0
Silver	0.5		
Cyanide	0.1		
Molybdenum		4.0	4.0
Selenium		1.6	2.0

Source: Anon. (1988g).
[a] $\mu g/kg$.

total ammonia concentrations in the porewaters from 13 locations on the lower Fox River and Green Bay, U.S.A. and the responses in toxicity assays (Figure 4). Toxicity was also observed to be completely removed from the porewaters by passing them over a zeolite column and retesting using the same assay. This suggested that ammonia was responsible for observed toxicity, although other cations could also potentially be removed by the zeolite resin.

In a subsequent study (Hoke et al. 1990c) using subsamples of the sediments originally used by Ankley et al. (1990), un-ionized ammonia did not account for all observed toxicity (Figure 5). In fact, when a toxic units approach was applied, the combined concentrations of un-ionized ammonia, copper, and zinc did not account for the toxicity observed in the Microtox® assay. This approach assumed that all measured metals were present in the toxic forms.

Ammonia is a particularly pertinent example compound because it appears to be present at toxic concentrations in a variety of sediments (Ankley et al. 1990). It undoubtedly exists in aquatic ecosystems as a result of both natural processes and human activities. Because ammonia appears

Table 6. Sediment Quality Guidelines Proposed by Beak Consultants Ltd.

Contaminant	Proposed Guideline (μg/g)[a]	Method of Derivation[b]
Arsenic	17	SLC
Cadmium	2.5	Background
Chromium	100	Bioassay
Copper	85	Bioassay
Iron (%)	5.9	Background
Lead	55	Background
Manganese (%)	0.12	Background
Mercury	0.6	Background
Nickel	92	SLC
Zinc	143	Background
Heptachlor	0.008	SLC
Endrin	0.012	SLC
Mirex	0.028	SLC
Aldrin	0.008	SLC
Chlordane	0.03	SLC
DDT (total)	0.02	Partitioning
p,p-DDT	0.036	SLC
p,p-DDD	0.032	SLC
p,p-DDE	0.02	SLC
p,p-DDT	0.024	SLC
PCBs	0.16	SLC
PCB 1254	0.23	SLC
PCB 1248	0.14	SLC
PCB 1016	0.03	SLC
Dieldrin	0.076	SLC
δ-BHC	0.0113	SLC
β-BHC	0.012	SLC
α-BHC	0.02	sLC
HCB	0.08	SLC
Heptachlor epoxide	0.018	SLC
δ-Chlordane	0.002	SLC

Source: Hart et al. (1988).
[a]Except as noted.
[b]Proposed guidelines based on SLC and partitioning were derived in μg/g C and adjusted to bulk sediment basis assuming an average 4% total organic carbon.

to be a common contaminant in sediments, its effects must be accounted for before other causes of toxicity can be identified.

4.0 SITE-SPECIFIC SEDIMENT QUALITY ASSESSMENT

Before effective remediation of sediments can be accomplished, all continuing sources of contamination need to be eliminated. If this cannot be done, sediments will simply be recontaminated. The initial step in evaluating an area is to make use of all available historical information (Figure 6). This should include not only monitoring information but information on potential sources of contaminants.

Table 7. Criteria for Selecting a Sediment Quality Guideline Method

Criterion 1	The method should consider a range of contaminant levels which is wide enough to determine the level at which ecotoxic effects become noticeable.
Criterion 2	The method should be based on cause-effect relationships between a specific contaminant and benthic organisms in a multicontaminant medium.
Criterion 3	The SQG should be derived from chronic effects since acute levels do not offer adequate protection.
Criterion 4	The method should be capable of incorporating a wide range of environmental factors that could have a bearing on the presence or absence of organisms in a given area.
Criterion 5	The method must be scientifically sound and understandable.
Criterion 6	The method must produce guidelines that relate to conditions that prevail in the natural environment.
Criterion 7	The method must be capable of deriving sediment guidelines from existing information bases.

Source: Persaud *et al.* (1989).

4.1 Field Bioassay Approach

Both chemical and biological information are required to understand the causes of sediment toxicity (Figure 7). Site-specific bioassays should be an

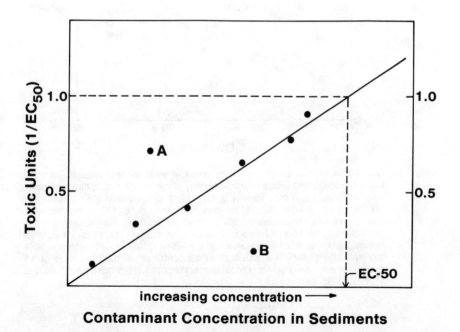

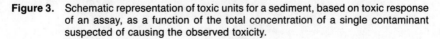

Figure 3. Schematic representation of toxic units for a sediment, based on toxic response of an assay, as a function of the total concentration of a single contaminant suspected of causing the observed toxicity.

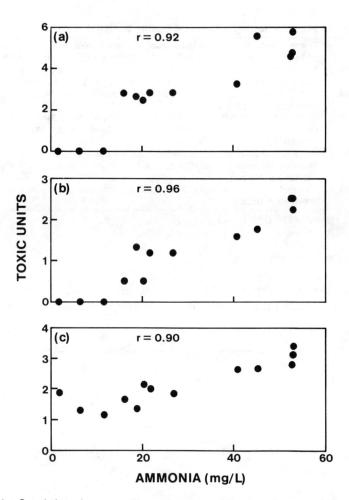

Figure 4. Correlation of concentrations of ammonia in sediment porewaters from the lower Fox River and Green Bay with toxicity of the samples to fathead minnows and *Ceriodaphnia dubia*. Toxicity is expressed as toxic units (i.e., $100/LC_{50}$ or $100/EC_{50}$) for (a) 96-hr fathead minnow mortality, (b) 48-hr *C. dubia* mortality, and (c) 168-hr *C. dubia* reproduction. When no mortality was observed in the 96-hr fathead minnow or the 48-hr *C. dubia* exposure, a value of zero toxic units was assigned. In the two instances in the 48-hr *C. dubia* tests where <50% mortality was observed at a porewater concentration of 100%, a value of 0.5 toxic units was assigned for computational purposes. (Reprinted from Ankley *et al.* 1990. With permission.)

CUMULATIVE TOXIC UNITS

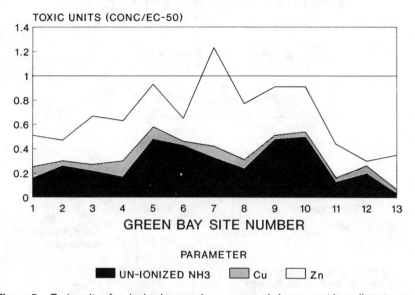

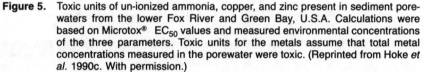

Figure 5. Toxic units of un-ionized ammonia, copper, and zinc present in sediment pore-waters from the lower Fox River and Green Bay, U.S.A. Calculations were based on Microtox® EC_{50} values and measured environmental concentrations of the three parameters. Toxic units for the metals assume that total metal concentrations measured in the porewater were toxic. (Reprinted from Hoke *et al.* 1990c. With permission.)

integral part of a comprehensive site evaluation plan. However, the first stage should be to make measurements of simple chemical and physical properties of sediments to determine if they are suitable for colonization by benthic organisms (Figure 8, Table 8). Certainly, greater understanding could be gained by more detailed analyses. However, the suitability of the sediments to support benthic invertebrate communities should be determined before conducting detailed chemical and biological assays to ascertain the concentrations and toxicity of contaminants. Benthic invertebrates may often be restricted from colonizing sediments due to contaminants such as sulfides or ammonia (Ankley *et al.* 1990) which may also mask the effects of other contaminants. After the general suitability of sediments has been assessed, a battery of simple bioassays should be used to determine the toxic potency of the sediments.

This bioassay approach is often referred to as the field bioassay (FB) approach because bioassays are conducted on sediments collected from the field rather than on laboratory-spiked sediments. These types of assays can be used in two ways, to simply determine the toxicity of the sediments or to

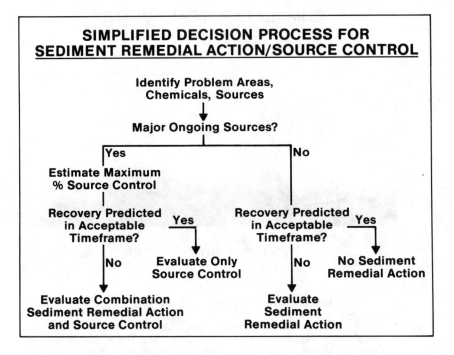

Figure 6.   Decision process for sediment remedial action developed by U.S. EPA (Anon. 1988b).

relate the observed effects to measured concentrations of contaminants in the sediments. The second approach is similar to the apparent effects threshold and screening level approaches for the development of global sediment quality criteria. However, instead of enumerating and identifying naturally occurring individuals in the sediment, surrogate species are tested in the laboratory. The responses of test organisms are then compared to the responses of individuals of the same species tested in reference sediments. The global sediment quality values are established by correlating concentrations of toxicants with effects that are statistically significantly different from those observed in the controls.

It is difficult to use endemic species in toxicity bioassays. This is due, in part, to the lack of consistency of species among geographic regions, lack of background toxicological information, and lack of methods for culturing and dosing of benthic organisms in sediments. Therefore, biomonitoring has been advocated to develop site-specific sediment quality criteria (Anon. 1985). However, the FB approach has the advantage that by using a single surrogate species a great deal of variability and uncertainty in the endpoints

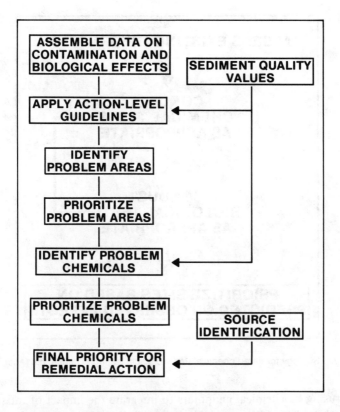

Figure 7. Generalized strategy for assessing sediment remedial action priorities developed by U.S. EPA (Anon. 1988b).

can be removed because the test organisms are of more similar genetic stock and nutritional and toxicant exposure history.

Even though the FB approach can be used to establish global sediment quality criteria, since the responses of the bioassay organisms are to the total mixture of toxicants, the FB approach suffers from the same limitations as the AET and SLC approaches and should not be used alone to establish global sediment quality criteria. The greatest utility of the FB technique is in its ability to identify sediments of toxicological concern. Bioassays alone can direct cleanup while chemical analyses alone are insufficient to do so (Athey *et al.* 1989).

The simple assays used can be of three types. These include: 1) assays to determine the direct toxicity of the sediments to populations of endemic or surrogate bacteria, algae, invertebrates, or vertebrates; 2) assays to determine the potential for bioaccumulation of contaminants in the sediments or long-term, adverse effects such as mutagenesis or teratogenesis; and 3)

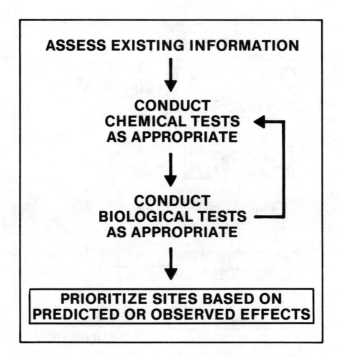

Figure 8.   Strategy for tiered chemical and biological testing of sediments (Anon. 1988b).

assays using biochemical endpoints to integrate the impact of mixtures of chemicals with the same mode of toxic action.

In addition to their use in screening sediments for toxicity, simple toxicity assays can be used to direct the allocation of resources for sediment chemistry analyses. The differential responses of assays in a battery of tests can give insight into the causes of sediment toxicity based on the differential sensitivities of the test species. Simple bioassays also can be combined with chemical and physical fractionation techniques to determine the most probable causes of observed toxicity of sediments. Later in this chapter we will describe a scheme of toxicity identification and evaluation (TIE) which has been adapted to help make the identification of the causes of sediment toxicity more efficient.

Table 8. Simple Chemical and Physical Parameters to be Measured Prior to Conducting Bioassays

Chemical	Physical
Dissolved oxygen	Particle size
pH	
Total ammonia	
Total sulfides	

5.0 CHARACTERISTICS OF AN IDEAL SEDIMENT BIOASSAY

Ideally, sediment toxicity bioassays and assay organisms should have the following characteristics:

1. Test organisms should be easy to culture and maintain in the laboratory and be available for tests at any time.
2. The responses of control organisms should be predictable and constant.
3. The assay organisms should respond similarly to many classes of toxicants, or organisms with sufficiently different responses should be included in a battery of tests (Chapman *et al.* 1987).
4. The results of the assays should be related to ecologically relevant processes under field conditions.
5. The results of bioassays should be related to sediment or water quality standards and criteria.
6. The bioassays should be applicable to a number of sediment types and environments.
7. The assays and dilution methods should be chosen to provide information which is correlated with observed adverse effects on organisms under field conditions (Anon. 1977a, Malueg *et al.* 1984a,b, Nebeker *et al.* 1984b).
8. The assays should be rapid, replicable, inexpensive, and easily implemented so that large areas of concern can be surveyed rapidly with good resolution.
9. The assays should be standardized to facilitate widespread use.
10. They should be sensitive enough to identify potential problem sediments yet discriminatory enough to permit ranking of the relative toxicity of many samples.

The advantages of standardized protocols for effluent and pure compound toxicity tests as presented by Davis (1977) are as follows: 1) use of uniform test protocols among laboratories, 2) increased data accuracy, 3) facilitation of test replication, and 4) increased comparative value of test data. These concepts also apply to sediment toxicity tests. The use of standard protocols has greater regulatory and legal impact by virtue of the verification process to which such protocols are subjected and the publication of the protocols themselves in both the open and "gray" literature. Standardization does, however, increase the risk that creative new approaches to problems may be ignored, particularly if these new approaches are at odds with existing regulatory agency policy (Davis 1977). Typical standard methods also may not address such concerns as delayed toxicity (Buikema & Benfield 1979) or potentiation of contaminant effects as a result of the stress of a standard test environment (Buikema *et al.* 1982). The potential also exists for inadequate characterization of contaminant effects in a changing natural environment as a result of the optimization of all conditions in a standard test (Geckler *et al.* 1976). By increasing compa-

Table 9. Characteristics of Ideal Sediment Bioassays

- Rapid
- Simple
- Replicable
- Inexpensive
- Standardized
- Sensitive
- Discriminatory
- Ecologically relevant
- Relatable to field effects
- Useful in developing, and relatable to, regulatory standards

rability and reliability of test data, standard methods will facilitate investigations of the relative toxicity of different sediments and the sensitivity of different assays.

Sediment assays, particularly screening assays, must be rapid, simple, and inexpensive (Table 9). Toxicity assays provide information on the effects of contaminants on the test species and are envisioned as a tool for delimiting the extent of further investigations. Rapid, simple, and inexpensive sediment screening assays facilitate the comparative evaluation of a large number of samples in a timely manner, without the expense and lack of biological effect data inherent in chemical analyses. Prioritization of further investigations can then be based on the results of the screening assays and chemical analyses utilized to determine probable contaminants and their concentrations responsible for the observed biological effects (Table 8). The discriminatory ability of a particular assay is also an important consideration in assay selection. Some assays exhibit quantal "all or none" type responses, resulting in the classification of a particular sediment simply as either toxic or nontoxic. For comparative evaluation of effects, it is useful to be able to rank toxic sediment samples relative to one another. The discriminatory power of an assay is the principal factor involved in the ranking of "toxic" responses and is dependent on the measured response variable. The use of a continuous response variable such as weight gain, as in the *Chironomus tentans* 10-day growth reduction assay (Giesy *et al.* 1988a), rather than a quantal response variable such as lethality increases the discriminatory power of an assay.

Ecological relevance and correlation to field effects and regulatory standards are characteristics less likely to be found in screening assays but are important for definitive assays or assays designed to address site-specific questions using indigenous organisms. Definitive assays generally are used to develop dose-response relationships and estimate the proportion of a population affected by a given contaminant level. Although mortality of an organism would prohibit further reproductive contributions to the population, gross effects such as mortality may not be sensitive enough to deter-

mine subtle ecological effects. Growth and reproduction effects are generally more sensitive test endpoints and may be indicative of ecological
changes which result in altered species distribution and community structure in field situations. Assays using surrogate species, such as *D. magna* or
Ceriodaphnia dubia, and assays using important indigenous organisms such
as *Chironomus* sp. are valuable tools in sediment toxicity investigations.
The use of surrogate species facilitates assessment of reproductive impairment using sediment extracts which eliminate the physical problems encountered in the development of dose-response relationships using whole sediments (Giesy *et al*. 1990). The use of sediment extracts also enables
comparisons to be made between contaminant concentrations in the
extracts and existing water quality criteria.

6.0 APPROPRIATE SEDIMENT TOXICITY ASSAYS AND ASSAY ORGANISMS

Many assay procedures have been developed in an attempt to standardize
results so that sediment toxicity bioassays are more reproducible, have
greater applicability among classes of chemicals (APHA 1985, Chapman *et
al*. 1987, Depinto *et al*. 1987, Giesy *et al*. 1988a), and are more simple and
more readily interpreted (Nebeker *et al*. 1984b). While a number of organisms have been used in bioassays of sediment toxicity, a few, such as the
midge, *Chironomus tentans*, and the water flea, *D. magna*, have been
demonstrated to be particularly useful (Nebeker *et al*. 1984b, Giesy *et al*.
1988a, 1990). The response of one test organism to single compounds or
mixtures is often correlated with that of other species, but seldom is the
correlation perfect (Giesy *et al*. 1988a). Therefore, no single bioassay can be
expected to be adequate for the detection of potential adverse effects of
complex mixtures of contaminants (Chapman 1987, Giesy *et al*. 1990), due
to the relative sensitivities and natural history characteristics of different
bioassay organisms. We have selected assays which, based on previous work
by our laboratory (Giesy *et al*. 1988a,b, 1990) and others (Nebeker *et al*.
1984a,b), meet many of the above requirements. Below we discuss the
disadvantages and advantages of the assays and organisms which we propose for sediment toxicity evaluations.

6.1 Bacteria

A number of assays have been developed using the responses of bacteria
to pollutants (Goatcher *et al*. 1984, Johnson & Romanenko 1984, Liu &
Dutka 1984, Coleman & Qureshi 1985, Burton & Lanza 1987, Burton *et al*.
1987) because of the importance of degradation processes in the geochemi-

cal carbon and nutrient cycles. In addition to their inherent importance, bacteria have been used as surrogate organisms to measure biotic responses in rapid screening tests.

The total activity of endemic microbial populations has been suggested as a rapid, effective tool for assessing the viability of populations of bacteria in sediments (Burton & Stemmer 1988). In this method, a known concentration of a specific substrate is added to a sediment sample or sediment extract and the activities of specific bacterial enzymes, such as glucosidase and galactosidase, are measured. We do not advocate the use of these assay systems because they can be influenced by many parameters in addition to the toxicity of chemicals in sediments.

The use of bacteria as surrogate assay organisms is predicated on the assumption that some biochemical and physiological systems are evolutionarily conservative, and toxicants elicit observed effects due to interactions with biomolecules which are similar in many different organisms. However, because of the differences in modes of action of toxicants and physiologies and biochemistries of organisms, one would not expect all organisms to respond similarly to a range of toxic chemicals. For this reason, bacteria, algae, and animals may exhibit similar sensitivities to some chemicals but differential sensitivities to others. When the toxicity of 156 pollutants on unicellular organisms was examined, 23 exhibited a pronounced selective toxic action on bacteria, while 47 were more toxic to algae and 43 had the greatest effect on protozoans (Bringmann & Kuhn 1980).

Bacteria, in general, are equally or less sensitive to metals than are plant or animal cells (Babich & Stotzky 1985). *Photobacterium phosphoreum* is much less sensitive to both mercury and cadmium than is *D. magna* (DeZwart & Slooff 1983); however, marine bacteria such as *P. phosphoreum* are particularly sensitive to the toxic effects of metals such as copper (Gillespie & Vaccaro 1978, Sunda & Gillespie 1979). The low sensitivities to metals reported for *P. phosphoreum* may have been due, in part, to the use of an inappropriate ionic strength adjustor in the Microtox® assay (Hinwood & McCormick 1987).

Bacteria have generally been thought to be tolerant of pollution by petroleum hydrocarbons (Adams 1985). However, bacteria can be inhibited by exposure to crude oils (Hodson *et al.* 1977, Baker & Griffiths 1984). Bacteria are known to be very tolerant of some organic compounds which are extremely toxic to crustaceans or fish. The LC_{50} for malathion to *D. magna* is 1.8 g/L, while a solution of 10 g/L actually promotes growth of some bacteria (Jones *et al.* 1984). The insecticide Lindane is approximately 300 times more toxic to guppies than to *P. phosphoreum* (Hermanns *et al.* 1985). Thus, for compounds like Lindane, which have specific modes of action for vertebrate animals, some inhibition would be observed in the *P. phosphoreum* assay, but this response would underestimate the effects on higher organisms such as fish. The commonly used herbicides Simazine and Endothall also had no effects on numbers or function of aquatic bacteria (Beckmann *et al.* 1984) but do affect algae and aquatic angiosperms. Bacteria are also not very sensitive to some other chlorinated organic compounds, such as solvents, PCBs, and insecticides (Vitkus *et al.* 1985). Alternatively, bacteria are known to be much more sensitive to organic compounds such as antibiotics (DeZwart & Slooff 1983).

Because of this type of variation in sensitivities among compounds and deviation from the responses of higher organisms, the use of microbial bioassays for rapid assessment of sediment has not been widely accepted. Rather, it has been suggested that sediment microbial activity should be used as part of a battery of assays for assessing the toxicity of sediments (Archibald 1982, Chapman *et al.* 1982a, Munawar *et al.* 1984, Obst 1985, Bedford *et al.* 1987).

6.1.1 Microtox® Bacterial Assay

The Microtox® assay is a bacterial luminescence bioassay developed by Beckman, Inc. in 1977 (Bulich 1984) as a rapid screening alternative to standard acute toxicity testing with fish or invertebrates. This test is based on the reduction in bioluminescence of the marine bacterium *Photobacterium phosphoreum* (NRRL B-11177) by toxicants. The test is simple, replicable, can be completed in a short period of time, and results in a dose-response relationship (Figure 9). The assay has been studied extensively and the results compared to acute bioassays with both fish and invertebrates for a large number of pure compounds (Figure 10) and aqueous complex mixtures (Figure 11) (Bulich *et al.* 1981, Lebsack *et al.* 1981, Curtis *et al.* 1982, Qureshi *et al.* 1982, 1984, Indorato *et al.* 1984, Schiewe *et al.* 1985, Hermanns *et al.* 1985, Nacci *et al.* 1986, Tarkpea *et al.* 1986). These comparisons have demonstrated a general agreement between toxicity values determined by fathead minnow and *D. magna* acute assays and the Microtox® assay, both within and among laboratories (Greene *et al.* 1985). However a recent investigation by Ankley *et al.* (1990) has demonstrated that Microtox® is less sensitive to the effects of ammonia in sediment porewater than *Ceriodaphnia dubia* or *Pimephales promelas*.

The inhibitory effects of porewater from Detroit River sediments in the Microtox® test were compared to the lethality of *Daphnia magna* and the effects of the whole sediment on larval growth of *Chironomus tentans* (Giesy *et al.* 1988a, 1990). While the results of the three assays were not completely congruent (Figures 12–16), it was established that they were highly intercorrelated and that the most toxic and nontoxic locations were well discriminated by the Microtox® assay (Table 10, Figures 17–19). Extensive surveys of the chemical constituents of the sediments which caused the

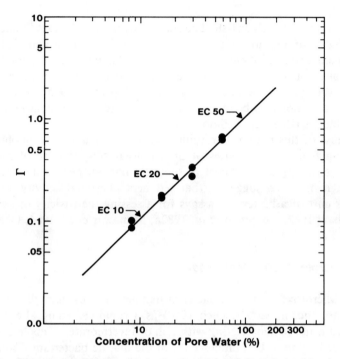

Figure 9.   Dose-response relationship for the decrease in light emission from the *P. phos-phoreum* Microtox® assay as a function of the concentration of porewater (interstitial water) from sediments.

greatest toxic effects in the assays have been conducted by the U.S. Environmental Protection Agency (Pranckevicius 1986). These studies revealed potentially toxic concentrations of metals, PCBs, other industrial organic chemicals, and PAH. Therefore, any number of toxic chemicals could be responsible for the observed effects.

Sediments which have smaller grain sizes and a higher organic carbon content tend to become more greatly contaminated by both metals and organic compounds (Knezovich *et al.* 1987). For this reason, it seems reasonable that the sedimentation zones of the Detroit River would be contaminated with a mixture of toxic contaminants from multiple sources as opposed to single contaminants. Therefore, the Microtox® bacterial bioassay was useful in rapidly identifying greatly contaminated and relatively clean areas (Giesy *et al.* 1988a,b). An EC_{10} value of 25% porewater in the Microtox® assay caused approximately a 30% reduction in the growth of *Chironomus tentans* in assays which used the corresponding whole sediment. An EC_{10} value of 25% porewater was also the toxicity of sediment which corresponded to that at locations where no benthic insects were

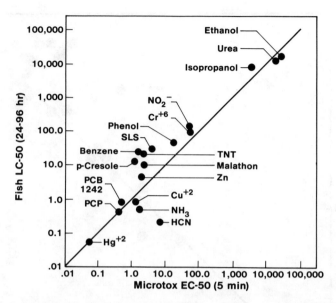

Figure 10. Comparison of the response of the *Photobacterium phosphoreum* (Microtox®) assay with the acute toxicity to fish of 18 reference toxicants. The contaminants represent a range of modes of toxic action. (After Bulich *et al.* 1981.)

observed to be living in the Detroit River (Giesy *et al.* 1988a, Rosiu *et al.* 1989). Therefore, all 85 locations in the lower Detroit River, which were classified as very or moderately toxic based on the Microtox® assay, were too toxic to support benthic insects, snails, or clams.

We are aware of only a limited number of studies which have used the Microtox® assay to investigate the toxicity of sediments (Atkinson *et al.* 1985, Schiewe *et al.* 1985, Williams *et al.* 1986, Dutka *et al.* 1988, Dutka & Kwan 1988, Giesy *et al.* 1988a,b, Ankley *et al.* 1990). Although it is not sensitive to all compounds (Hermanns *et al.* 1985, Ankley *et al.* 1990), the Microtox® assay has a relative sensitivity to many compounds which is similar to most species of insects, crustaceans, protozoans, molluscs, and fish and has been recommended as the first stage in a tiered testing scheme (Slooff 1985). Other microbial assays exist (*Spirillium*, Polytox®, activated sludge respiration inhibition) and could potentially be used for sediment assessment. In pure compound assays, however, the Microtox® assay has generally been demonstrated to be both the most sensitive and the most replicable of the microbial assays (Greene *et al.* 1985, Elnabarawy *et al.* 1988). Therefore, we feel that the Microtox® assay should be included in a battery of assays for the screening evaluation of sediment toxicity.

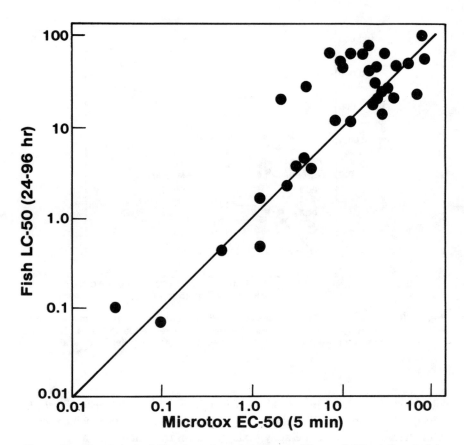

Figure 11. Comparison of *Photobacterium phosphoreum* (Microtox®) EC_{50} values to fish LC_{50} values for complex effluents. Toxicity is expressed as percent effluent on a volume/volume basis. (After Bulich *et al.* 1981.)

6.2 Algae

The selection of planktonic algae for toxicity testing is based on their importance in aquatic ecosystems (Munawar & Munawar 1987). Phytoplankton are intimately coupled to energy, nutrient, and geochemical cycling in aquatic ecosystems (Hutchinson 1957). Phytoplankton represent the base of the aquatic food chain and convert solar energy into carbon biomass, which is available for utilization by higher trophic levels. Phytoplankton also have a significant influence on the cycling and availability of nitrogen, phosphorus, heavy metals, and other trace elements (Andreae 1978, Effler & Driscoll 1985, Fisher *et al.* 1987), and they can influence processes, such as photolysis, which affect the fate of compounds in water

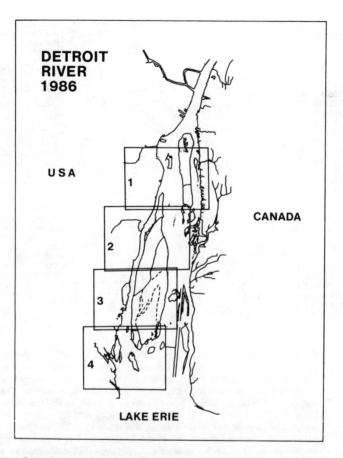

Figure 12. Sampling locations for sediment toxicity investigations (Figures 13–16) in the lower Detroit River, Michigan, U.S.A.

(Zepp & Schlotzhauer 1983). Plant and algal cells are sensitive to a number of compounds, such as herbicides, to which many other organisms are relatively insensitive (Beckmann *et al.* 1984). It is for these reasons that we feel that an algal assay should be included in a protocol which includes a battery of bioassays for sediment assessment.

6.2.1 Selenastrum capricornutum Assay

When comparing sediment toxicity to algae from a number of widely different sources, it is advantageous to use the same species of algae for the comparison and reporting of relative effects (Anon. 1978). *Selenastrum capricornutum* Printz is a green alga (Chlorophyceae) of the order Chlorococcales and is the standard algal species used for determining the toxicity

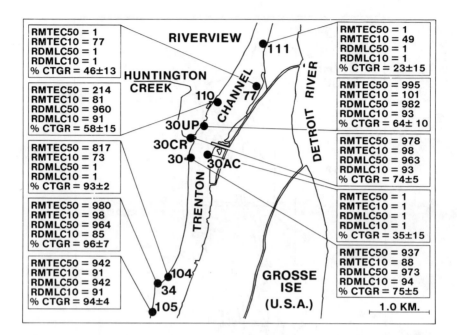

Figure 13. Comparison of responses of three bioassays to toxicity of sediments in the lower Detroit River, U.S.A. and Canada in 1986. The response of the midge *Chironomus tentans* is given as the mean reduction in weight relative to weight gain in sediment from station #83, the reference location. The relative response of *Photobacterium phosphoreum* is given as RMTEC50 = 1,001 − EC_{50} and RMTEC10 = 101 − EC_{10}, with EC_{50} and EC_{10} given as % pore-water. The greater the relative value, the more toxic the sediment. Absolute values are indicated at the top of the histogram. The response of *Daphnia magna* is reported as the relative LC_{50} (RDMLC-50) calculated in the same manner as for the Microtox® assay.

of waters and effluents (Anon. 1978, ASTM 1986). *Selenastrum capricornutum* is an excellent bioassay organism because it is easy to culture, has a short generation time, and toxicity tests can be performed with small sample volumes and without costly equipment (Van Coillie *et al.* 1983). Phytoplankton typically are more sensitive to certain toxicants, such as metals and some organic compounds, than higher trophic level organisms (Blanck 1984, Thomas *et al.* 1986). *Selenastrum capricornutum* was found to be more sensitive than either *Daphnia magna* or Microtox® to soil elutriate samples from a wide range of hazardous waste sites contaminated with heavy metals, pesticides, PAHs, and other toxic compounds (Thomas *et al.* 1986). In addition, algae are extremely sensitive to compounds, such as herbicides, which have a mode of toxic action specific to plant cells. Several phytoplankton assays have been adapted to assess the toxicity of sediments

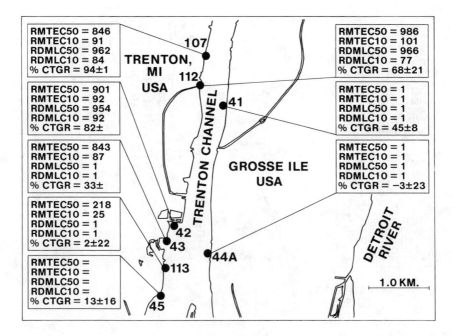

Figure 14. Panel 2 of responses of three assays to Detroit River sediments. Description of values is given in legend of Figure 13.

to algal cells (Munawar 1982, Munawar *et al.* 1983, 1984, 1985, Blaise *et al.* 1986); however, these assays all violate one or more of our criteria for sediment toxicity assays (Table 9). For this reason, we suggest that a rapid, easily interpreted assay be developed and then calibrated to the standard flask assay.

We feel that three assay techniques have potential as rapid phytoplankton assays. These include: 1) microplate techniques, which measure ATP concentration or $^{14}CO_2$ assimilation (Blaise *et al.* 1986); 2) flow cytometry, which can measure size of cells, or biochemical integrity (Olson *et al.* 1983, Berglund & Eversman 1988, Berglund *et al.* 1988); and 3) whole fixed cell, ETS response. In the latter technique, the effects of pollutants are assessed by using a redox meter to measure the response of the electron transport system of algal cells fixed on a membrane (Rawson *et al.* 1987).

We feel all of these assays merit further development and assessment for possible inclusion in a battery of sediment toxicity bioassay tests. Each of the proposed assay systems is rapid, sensitive, requires much less elutriate or porewater than other assays, and can be conducted in a short enough time to avoid changes in the chemical characteristics of porewater or elutriates.

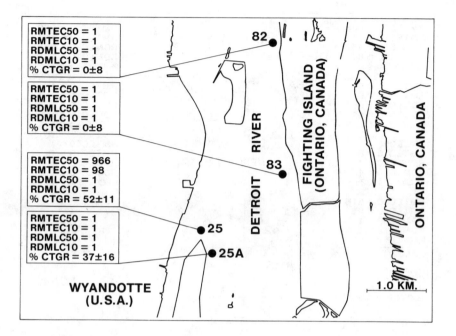

Figure 15.   Panel 3 of responses of three assays to Detroit River sediments. Description of values is given in legend of Figure 13.

6.3 Cladocerans

6.3.1 Daphnia sp. Assay

Cladoceran species have numerous advantages for toxicity testing, including the fact that they are among the most sensitive species to a wide variety of environmental contaminants (Hall *et al.* 1986). Cladocerans also are important in food chains, both as consumers of bacteria and phytoplankton and as prey for many species of fish. Various species of the genus *Daphnia*, particularly *D. magna* and *D. pulex*, have been used in aquatic toxicity tests, and a large data base exists for the effects of pure compounds on *D. magna* (Buikema *et al.* 1980). There are, however, several disadvantages to the use of these species. The absence of *D. magna* as a significant part of zooplankton communities at many locations, difficulties conducting life-cycle chronic tests due to high control mortality, and the existence of culturing difficulties have all been discussed as negative aspects in the use of this species (Mount & Norberg 1984).

While *D. magna* is not a benthic organism, it has been selected as an appropriate organism to assay the acute toxicity of porewater extracts

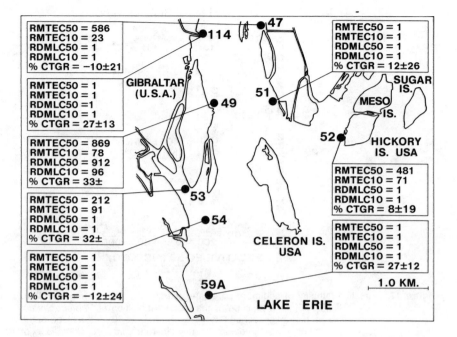

Figure 16. Panel 4 of the response of three assays to Detroit River sediments. Descriptions of values are given in legend of Figure 13.

because it is a standard bioassay organism for determining the toxicity of effluents and standard rearing and assay methods have been established (Buikema *et al.* 1980, Goulden *et al.* 1982). *Daphnia* have been used as an indicator of aquatic pollution, primarily in acute tests, because of their advantages for this type of testing (Anderson 1944, Leonard 1979). Acute

Table 10. Pearson Product-Moment Correlations (r) for Among-Assay Comparisons for Sediments and Sediment Porewaters from the Lower Detroit River (N = 30)[a]

	%CTGR[b]	MTEC50[c]	MTEC10[d]	DMLC50[e]
DMLC10[f]	−0.668	0.507	0.544	0.996
DMLC50	−0.687	0.514	0.560	
MTEC10	−0.713	0.561		
MTEC50	−0.509			

Source: Giesy *et al.* (1988a).

[a] $p > F$, < 0.0001 for all regressions.
[b] % CTGR = % *C. tentans* growth reduction, relative to control location.
[c] MTEC50 = Microtox®, % porewater to reduce light emission by 50%.
[d] MTEC10 = Microtox®, % porewater to reduce light emission by 10%.
[e] DMLC50 = *D. magna*, % porewater to kill 50%.
[f] DMLC10 = *D. magna*, % porewater to kill 10%.

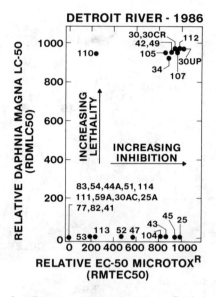

Figure 17.   Relative lethality of sediment porewater extracts to *Daphnia magna* (48-hr LC_{50}, RDMLC50) as a function of the relative inhibition of *Photobacterium phosphoreum* bioluminescence (15-min EC_{50}, RMTEC50). If the slope of the log-probit relationship was not significantly different from zero, or the LC_{50} or EC_{50} greater than 100%, the RDMLC50 and RMTEC50 were set to 1.0. (Reprinted from Giesy *et al.* 1988a. With permission.)

toxicity tests with *D. magna* are reproducible (Lewis & Weber 1985, Gersich *et al.* 1986). The toxicities of many compounds, elements, and mixtures have been measured under a great variety of conditions using *D. magna* (Johnson & Finley 1980, LeBlanc 1980, Hermanns *et al.* 1984, Cowgill 1987) and the relative sensitivities between *D. magna* and other organisms are known (Lewis & Perry 1981, Gersich 1984, Kangarot & Ray 1987). Therefore, a large body of information exists with which to compare results from other species. Acute lethality tests with *D. magna* seem appropriate for use in sediment toxicity bioassays because this organism is used to establish surface water criteria (Leeuivangh 1978, Anon. 1980, Nebeker *et al.* 1983, Schuytema *et al.* 1984, Van Leeuwen *et al.* 1985) and the results could be compared to existing toxicity criteria (Malueg *et al.* 1983, 1984a,b, Nebeker *et al.* 1984b, 1986).

Individuals of the genus *Daphnia* are known to be very sensitive to the toxic effects of metals (Canton & Adema 1978, Baudouin & Scoppa 1974, Bertram & Hart 1979, Braginsky & Scherban 1978). For instance, the LC_{50} values for cadmium and zinc have been found to be as small as 5 and 68 μg/ L, respectively (Attar & Maly 1982). When the acute toxicities of cadmium

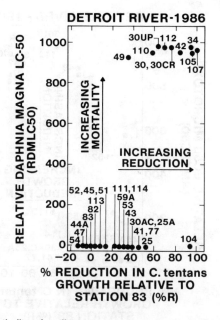

Figure 18. Relative lethality of sediment porewater extracts to *Daphnia magna* (48-hr RMDLC50) as a function of the percent reduction in *Chironomus tentans* growth (10 days) relative to that at station 83 (% CTGR). (Reprinted from Giesy *et al.* 1988a. With permission.)

and zinc toward *Daphnia* sp. were compared to that toward other invertebrates and toward vertebrates it was concluded that *Daphnia* sp. were as sensitive or more sensitive to these metals than were other organisms (Attar & Maly 1982). Furthermore, it was reported that "*Daphnia magna* is far more sensitive to these metals than any other daphnids" (Attar & Maly 1982).

When the acute toxicity of sediment porewaters on *D. magna* was compared to the effects on *Chironomus tentans* weight gain in a chronic laboratory test as well as to the *in situ* distribution of benthic invertebrates, it was found that the concentration of porewater causing 100% lethality within 48 hr in the *D. magna* assay corresponded to the degree of toxicity which caused a 30% inhibition of *C. tentans* weight gain (Giesy *et al.* 1988a). More importantly, this was the threshold degree of toxicity required to exclude other species of macrozoobenthos such as insects, clams, snails, and amphipods from sediments under field conditions. Thus, it can be concluded that the *D. magna* acute lethality test is useful for screening sediments for toxicity and should be included in a battery of assays for screening evaluations.

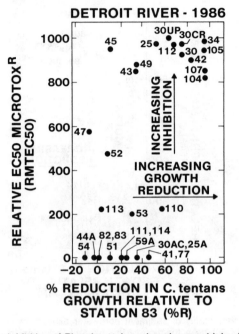

Figure 19.   Relative inhibition of *Photobacterium phosphoreum* bioluminescence (15 min, RMTEC50) as a function of the percent reduction in *Chironomus tentans* growth (10 days) relative to that at station 83 (% CTGR). (Reprinted from Giesy *et al.* 1988a. With permission.)

6.3.2 Ceriodaphnia sp. Assay

Ceriodaphnia dubia has been suggested as a potential cladoceran test species which obviates the chronic testing disadvantages reported for *D. magna* and which also has several additional advantages as a test species when compared with other cladocerans (Mount & Norberg 1984, Burton *et al.* 1987). Because of its small size, *C. dubia* can be tested in smaller volumes of test solution, large broods are produced, ephippia are relatively infrequent in crowded cultures, testing at 25°C is possible, and the species has a wide edaphic distribution in comparison to *D. magna*. Mount & Norberg (1984) published culturing procedures and methods for 7-day survival and reproduction assays. In subsequent years, the *C. dubia* assay has been used in the U.S. EPA effluent monitoring program and is currently a required monitoring assay in some municipal and industrial NPDES permitting protocols. *C. dubia* is generally more sensitive to the toxic effects of chemicals than *D. magna* or *D. pulex* (Winner 1988, Takahashi *et al.* 1987, Elnabarawy *et al.* 1986, Cowgill *et al.* 1984). Although *Ceriodaphnia* sp. are useful assay organisms, the purpose of the investigation should determine whether or not they are included in a battery of screening evaluation assays. If information on the potential reproductive effects of contaminants is desired, this assay should be included in the battery of screening assays. However, if the purpose of the investigation is to map toxicity for prioritization of further biological and chemical analyses, the 48-hr *Daphnia* sp. acute assay should be used.

6.4 Benthic Insects

6.4.1 Chironomus tentans Assay

Chironomus tentans (Diptera:Chironomidae) is a representative of a group of insects known as the midges, which are widely distributed in freshwater sediments during their larval stage of development. Chironomids often comprise a significant proportion of the benthic biomass and are important in the cycling of residues into and from the sediments due to bioturbation (Gerould *et al.* 1983). *C. tentans* spends most of its life cycle in a tunnel in the upper few centimeters of sediment (Adams & Heidolph 1985). *C. tentans* can be satisfactorily reared in the laboratory and has been used as a bioassay organism (Wentsel *et al.* 1977, 1978, Batac-Catalan & White 1982, Mosher & Adams 1982, Mosher *et al.* 1982, Sasa & Yasuno 1982, Cairns *et al.* 1984, Nebecker *et al.* 1984a,b, Adams *et al.* 1985, Ziegenfuss & Adams 1985, 1986, Khan *et al.* 1986).

Chironomus sp. larvae have been used in partial life cycle (Kosalwat & Knight 1987b) and whole life cycle (Hatakeyama 1987) bioassays, in which the effect of toxic sediments on survival and growth of adults, as well as on the fecundity and survival of the F1 generation, were determined. In these studies, the most common practice is to collect and enumerate all of the individuals which hatch from the eggs produced by exposed adults.

Whole-life toxicity tests are difficult to conduct and time-consuming, and many possibilities exist for the introduction of errors. It is difficult to begin experiments with eggs or small larvae because they are difficult to handle and enumerate. For this reason, egg masses and groups of individual instars must be added to assay chambers containing a fixed volume and surface area of sediment, which can lead to density-dependent effects in the assay. Density-dependent effects on growth and survival can be further exacerbated by mortality in assays, such that surviving individuals grow larger, produce more eggs, and reduce the sensitivity of the test. Collecting and mating emerging adults and enumerating eggs produced are also difficult. Adults can become trapped in the surface film and die or escape while being captured for mating. Some may lay eggs in the exposure chamber before they can be captured and mated under controlled conditions. All of these logistical problems may contribute to a great amount of variability in the *C. tentans* whole-life assay.

Based on acute toxicity studies (48-hr LC_{50}) of several dipteran species, midges have been found to be between 50% and 75% as sensitive as inhibition of bioluminescence of *Photobacterium phosphoreum* (15-min EC_{50}) to a range of toxicants. Chironomids have been found to be sensitive to the effects of metals, such as copper (Kosalwat & Knight 1987a,b, Anderson 1980). In studies conducted in our laboratory, we observed little lethality of *C. tentans*. However, when the chronic effects on weight gain of *C. tentans* were considered, it exhibited a sensitivity which was similar to the Microtox® 15-min EC_{50} (Giesy *et al.* 1988a). In this same study, we observed that a growth reduction of approximately 30–40% in the *C. tentans* whole sediment toxicity bioassay corresponds to the lack of viable benthic invertebrate communities during faunal surveys of the same sediments.

The most sensitive life stage of chironomids is the larva (Powlesland & George 1986). Furthermore, when larvae of *C. riparius* were exposed to nickel it was found that growth of larvae was the most sensitive endpoint and that the MATC was significantly lower than the LC_{50} (Powlesland & George 1986). Larval growth of *C. decorus* was found to be the most sensitive indicator of chronic exposure to metals (Kosalwat & Knight 1987a,b). These authors suggested that eggs were protected from the toxic effects of copper by their shell. When the midge *Tanytarsus dissimilis* was exposed to copper, cadmium, and lead there was no effect on growth at concentrations below the LC_{50} (Anderson *et al.* 1980). In the case of *C. riparius* the first instar was found to be the most sensitive to the effects of nickel, but second instar larvae were almost as sensitive (Powlesland & George 1986). The second instar larvae of *C. tentans* (13 days posthatching) has been found to be the most sensitive life stage (Nebeker *et al.* 1984a). Similarly, the second instar of *C. riparius* is the most sensitive life stage to the effects of the metals, mercury, cadmium, and the pesticide dieldrin (van de Guchte & Mass-diepeveen 1988).

Weight gain seems to be a sensitive and discriminatory endpoint. When the results of the *C. tentans* assay were compared to those of the *P. phosphoreum* and *D. magna* assays, *C. tentans* growth was an equally sensitive endpoint and more discriminatory than lethality of *D. magna* (Giesy *et al.* 1988a, 1990). When the midge *Polypedium nubifer* (Chironomidae) was chronically exposed to cadmium, decreased growth of the first and second instars was observed at concentrations which were less than those causing no decrease in the percent emergence, number of eggs layed, or survival of F_1 individuals (Hatakeyama 1987). This indicates that, at least for cadmium, measurement of growth of the second instar should be sufficient to protect midges from the adverse effects of cadmium.

Therefore, after considering relative sensitivity of life stages, ease of handling, probability of obtaining a valid test, and ability to measure relevant, sensitive endpoints accurately, we recommend an assay that measures weight gain and survival of *C. tentans* beginning with second instar larvae (12 days posthatching) and continuing for 10 days. We feel this is a simple, sensitive, reproducible test with a low probability of error or artifact and recommend that it be used in the screening evaluation of sediment toxicity.

6.5 Fish

6.5.1 Pimephales promelas Assay

Pimephales promelas, the fathead minnow, has been used extensively in early life stage testing because it has a cosmopolitan distribution and is an important part of most aquatic food chains (Macek & Sleight 1977, McKim 1977, Iwan & Cella 1981). It can be readily cultured in the laboratory, and eggs and larvae are available to start tests continuously (Norberg & Mount 1985). In addition, other life stages of the species have been used in acute and partial life cycle assays, and a large data base exists for the effects of pure compounds on these life stages (Brooks *et al.* 1984, Geiger *et al.* 1985, Mayer & Ellersieck 1986, Johnson & Finley 1980). Norberg & Mount (1985) first proposed the use of the *P. promelas* 7-day chronic survival and growth assay as a rapid method of assessment of the sublethal effects of effluents and complex mixtures. Other investigators adopted the assay for the examination of the effects of sediment extracts (Dawson *et al.* 1988), and our laboratory has recently completed a study in which the assay was used to assess the effect of sediment elutriates. The *Pimephales promelas* assay is suggested for use in characterizing the chronic effects of sediment porewaters on the survival and growth of a representative nektonic vertebrate but not as part of a battery of assays for the screening evaluation of sediment toxicity.

Not all bioassays are equally efficient or useful for sediment toxicity screening. In Table 11 we present relative rankings of some commonly used assays for 10 different attributes as well as overall scores of utility for screening the toxicity of sediments.

7.0 MAXIMIZATION OF INFORMATION GAIN

Appropriate test designs are dependent on the purpose of an investigation, and methods of data analysis are coupled to the test design. Experimental design such as replication of experimental units, randomization procedures, and sampling of test organisms is an integral, but often neglected, portion of any toxicity testing protocol. Ideally, the investigator decides

Table 11. Relative Ranking[a] of the Characteristics of Currently Used and Proposed Sediment Toxicity Screening Bioassays

Assay	Reference	Rapid	Simple	Replicable	Inexpensive	Standardized	Sensitive	Descriminatory	Ecologically Relevant	Relatable to Field Effects	Relatable to Regulatory Standards	Total Score[b]
Polytox®	Elnabarawy et al. 1988	4	4	2	4	4	2	4	1	1	1	27
Microtox®	Bulich 1984 Giesy et al. 1988b	4	4	4	4	4	3	4	1	1	1	30
Protozoan assay	Cairns 1979	3	2	2	2	1	3	3	4	2	1	23
Algal flask assay	Anon. 1978	3	3	3	3	4	3	3	2	2	2	28
Algal microplate assay	Blaise et al. 1986	4	3	3	3	2	3	2	2	2	2	26
Algal flow cytometry	Berglund et al. 1988	4	2	3	1	2	3	3	2	2	2	24
Algal ETS assay	Rawson et al. 1987	4	4	2	4	2	3	2	2	2	2	27
Oligochaete assay	Horning 1980 Keilty et al. 1988a Chapman & Brinkhurst 1984	3	3	2	3	2	2	2	4	4	2	27
Mollusc assay	Daube et al. 1985	1	2	2	2	2	2	2	4	4	2	23
Amphipod assay	Nebeker et al. 1986 Nebeker & Miller 1988	2	2	2	2	2	2	4	4	4	2	26
Daphnia magna assay	Schuytema et al. 1984 Nebeker et al. 1984b, 1986	3	3	4	3	4	3	2	3	3	4	32
Ceriodaphnia dubia assay	Hoke et al. 1990	3	2	3	3	3	4	2	3	3	3	29
Hexagenia assay	Prater & Anderson 1977 Malueg et al. 1983 Giesy et al. 1990	2	1	2	1	2	4	1	4	4	2	23
Chironomus tentans assay	Wentsel et al. 1977 Adams et al. 1985 Giesy et al. 1988a Hoke et al. 1990	2	2	3	2	3	2	4	4	4	2	28
Pimephales promelas assay	Dawson et al. 1988	2	2	3	2	4	3	2	3	3	4	28

Source: Giesy & Hoke (1989).
[a]1 = poor, 2 = fair, 3 = good, 4 = excellent.
[b]Maximum total score possible = 40.

whether the goal of a study is to develop comparative data, definitive dose-response data, or site- or species-specific data on the effects of sediment contaminants. Once this decision has been made, the issue of appropriate test species and endpoints can be evaluated. Screening assays used to develop comparative data will generally be acute assays because of time, economics, and number of samples to be examined. Definitive dose-response and site- or species-specific information may be developed using either acute, subchronic, or chronic assays. Lethality is the most common endpoint in acute assays, while subchronic and chronic assays typically examine growth or reproductive impairment, as well as lethality.

Methods of data analysis may include various types of hypothesis testing such as *t*-tests or analysis of variance, probit analysis, or various types of multivariate analyses. Determination of the comparative effects of contaminated sediments generally involves statistical evaluation of the null hypothesis that no difference exists between the mean of the response variable in sediments from different locations. Power analysis for the optimization of sample replication is an integral component of the experimental design. To maximize the utility of the results of bioassays, one should determine the statistical power of the bioassay. Specifically, the sources of variation during a test should be identified and unexplained within- and among-test variation minimized (Giesy & Allred 1985). The degree of resolution required for a particular test must also be defined. Once these estimates and decisions have been made, the power of the test can be optimized by selecting the required replication to demonstrate some defined difference between the responses in two or more tests (Giesy & Hoke 1989).

Analysis of variance followed by mean separation tests or the use of orthogonal contrasts (Steele & Torrie 1980, Sokal & Rohlf 1981) is a useful data analysis technique for comparative sediment toxicity data for which dose-response relationships are not developed (Hoke *et al.* 1990a). Williams (1987) has also presented a discussion of the use of analysis of variance for unbalanced designs which are often found in biological data. Various methods exist for the development of dose-response relationships, including both graphical and statistical techniques (Litchfield & Wilcoxon 1949, Stephan 1977, Finney 1971, 1978). It is suggested that whenever possible a dose-response relationship be generated from bioassay information.

7.1 Rationale for a Battery of Tests

As discussed above, the responses of a number of assay organisms to the effects of a mixture of toxicants in sediments generally result in some degree of correlation among assays (Table 10, Giesy *et al.* 1988a, Dawson *et al.* 1988). However, it has been observed that there is not perfect correlation between tests because of differences in the relative sensitivities of test organisms and the differences in relative concentrations of toxicants from location to location (Table 10, Giesy *et al.* 1988a). Thus, to maximize predictive power, it is suggested that a battery of screening tests should be performed. It is suggested to include *P. phosphoreum* (Microtox®) and *D. magna* as a minimum set of assays.

The NOEL (no observed effect level) for toxicity to *D. magna* has been found to be correlated with the NOEL for the fathead minnow for chemicals with a variety of structures (r = 0.79) and highly correlated when only chemicals of similar structure are considered (r = 0.98, Maki 1979). Therefore, it is probably unnecessary to include both of these species in a battery of tests. *D. magna* is easier to culture and use in tests, so it is recommended that *D. magna* be included in the battery of tests. Also, the results of the *D. magna* and *Ceriodaphnia dubia* assays may be similar, but these species are typically used in acute and chronic assays, respectively. After sites exhibiting toxicity have been identified using the the proposed screening battery of assays, the 7-day *Ceriodaphnia dubia* and *Pimephales promelas* assays could be used to assess the potential reproductive and growth effects caused by contaminants from the toxic sites.

When the responses of three assays, including the bacterium *Photobacterium phosphoreum*, embryos of the oyster, and the marine amphipod *Rhepoxynius abronius* exposed to contaminated sediments, were compared, there was "a high level of agreement among the three bioassays; however, individual correlations suggested considerable variation among the bioassays" (Williams *et al.* 1986). These authors concluded that a range of toxicity tests should be used to obtain the maximum range of sediment toxicities resulting from different relative and absolute concentrations of contaminants.

Giesy *et al.* (1988a) compared the responses of a battery of three bioassays, *Chironomus tentans, D. magna,* and *P. phosphoreum,* to sediments from the Detroit River, Michigan, U.S.A., which are known to be contaminated by a variety of chemicals, including metals, petroleum hydrocarbons, and synthetic organic chemicals (Giesy *et al.* 1990). In general, toxic sediments were identified as toxic in all three assays (Figures 17–19). For instance, sediment porewater from some locations caused maximum responses in both the *D. magna* and *P. phosphoreum* assays (Figure 17). However, while there was some correlation among the results of the three bioassays, there was also variation in the response among assays at different locations. For instance, the response of the *D. magna* lethality assay was quantal, which resulted in an all or nothing type of response. Sediments from some locations caused almost complete inhibition of bioluminescence in the *P. phosphoreum* assay, while causing no lethality to *D. magna.* Only one location, station no. 110, caused almost complete lethality in the *D. magna* assay but little effect on bioluminescence of *P. phosphoreum.* Thus, the *D. magna* acute lethality bioassay was less sensitive and less discriminatory than the *P. phosphoreum assay*.

Similarly, *Chironomus tentans* weight gain in whole sediment was more sensitive to and discriminatory among the effects of sediments from the Detroit River than was the *D. magna* assay (Figure 17). Sediment from which porewater caused no lethality of *D. magna* caused as much as a 50% reduction in the weight gain of *C. tentans*. Alternatively, when almost complete lethality of *D. magna* was observed, the reduction in weight gain exhibited more power to discriminate among the mixtures of toxicants present in the sediments.

The correlations among the responses of the three assays suggest that any of the three assays could have been used to identify the most toxic sediments. Even though the *D. magna* acute lethality assay was less discriminatory than the other two assays, the LC_{50} value was equivalent to the toxicity which restricted colonization of the sediments by chironomids under field conditions (Giesy *et al.* 1988a, Fallon & Horvath 1985). It was also determined that the degree of toxicity of Detroit River sediments required to restrict colonization by chironomids was equivalent to approximately a 25-35% reduction in weight gain of *C. tentans* (Giesy *et al.* 1988a). Similarly, a reduction of 25% in weight gain of *C. tentans* in laboratory bioassays, relative to that on reference sediments, was approximately the threshold above which colonization by chironomids of sediments in Toledo Harbor and western Lake Erie was also restricted (Giesy & Hoke 1988).

Even when a battery of tests is used, one cannot expect to have perfect predictability. When the results of acute and chronic toxicity tests for a number of different species were compared, it was found that 25-30% of the test chemicals caused effects in at least one species which were not expected based on a battery of tests containing a standard algal, daphnid, and fish species (Slooff 1985). However, when the data on a variety of chemicals were considered together, the difference among species of aquatic organisms was rather small. For this reason it has been suggested that a rather small number of test species would be necessary to screen for the effects of environmental contaminants.

When a battery of tests is conducted the results of each test will contain some unique and some redundant information. A group of two or more bioassays can be used to classify the relative toxicity and relative similarities of the toxicity of three or more sediments. The goal of using multiple assays is to produce a canonical classification variable which maximizes the information about sediment toxicity while minimizing the probability of misclassifying the toxicity of a sediment. Multivariate techniques are available which facilitate the use of data from several assays to develop empirical descriptors that maximize explained variation and minimize redundant information (Cooley & Lohnes 1971, Devillers *et al.* 1988). Using this

Table 12. Results of Principal Components Analysis to Predict the Toxicity of Detroit River Sediment

	% CTGR[a]	RMTEC50[b]	RDMLC50
Simple statistics			
Mean	40.32	459.58	314.98
Standard deviation	30.74	433.68	449.77
Covariances			
% CTGR	944.97	9,349.62	10,730.82
RMTEC50		188,077.10	138,109.80
RDMLC50			202,292.90
Total variance = 391,315			

		Variance explained	
	Eigenvalue	Proportion	Cumulative
Principal components			
PC1	334,0894	0.85370	0.85370
PC2	56,899	0.14540	0.99915
PC3	331	0.00085	1.00000

	PC1	PC2	PC3
Eigenvectors			
% CTGR	0.042644	−0.010784	0.999032
RMTEC50	0.688001	0.725390	−0.021538
RDMLC50	0.724456	−0.688253	−0.038353

Source: Giesy *et al.* (1988a).
[a]% CTGR, percent reduction in *Chironomus tentans* growth relative to that at a reference station.
[b]RMTEC50, relative Microtox® EC_{50}.
[c]RDMLC50, relative *Daphnia magna* LC_{50}.

approach, Giesy *et al.* (1988a) developed dose-response relationships via probit analysis for three assays (Microtox®, *Daphnia magna*, and *Chironomus tentans*) of sediment toxicity in the Detroit River. Dose-response data from each assay were then combined in a multivariate principal components analysis. Microtox® and *D. magna* assays produced quantal, "all or none" type responses to sediment contaminants, while the response of *C. tentans* growth was a continuous variable. Sensitivity among the assays was approximately the same, but the *C. tentans* growth assay was more discriminatory. In the principal components analysis, each assay considered separately accounted for approximately 60–70% of the explained variance, while any two assays in combination accounted for greater than 95% of the explained variance (Table 12). Therefore, one of the assays could have been deleted from the investigation with no appreciable loss of information and with a concurrent reduction in total study expense. An algal assay could be included in the battery if one had reason to suspect that herbicides or chemicals which are algistatic would be present in the sediment. It has been found that for sediments contaminated with conventional toxicants the responses of algal assays are highly correlated with the responses of the

Microtox® assay, and both assays are not needed (Giesy *et al.* unpublished data).

A battery of tests need not include an exhaustive list of assays. It is suggested that two assays, the *D. magna* and *Photobacterium phosphoreum*, be used, because in combination they provide great predictive power and also meet the criteria for effective assays, which were proposed earlier. If one wants to use a chronic, whole-sediment assay, the *C. tentans* assay is a sensitive, discriminatory assay with good predictive power. The assays which have been found useful and are proposed here are not necessarily unique; many other batteries of tests can be defined which give good predictive and discriminatory power and meet the criteria of useful and effective assays.

There are several multivariate statistical methods which can be used to combine the nonredundant information provided by a battery of tests, while removing redundant information (Morrison 1976, Cooley & Lohnes 1971). Giesy *et al.* (1988a) used principal components analysis to investigate the "among location" variance in toxicity with a battery of tests. Briefly, principal components analysis is a multivariate statistical procedure that transforms the possible indices of toxicity into new orthogonal parameters. When the results of three assays are included in the battery there would be three principal components, identified as PC1, PC2, and PC3, which summarize the information contained in the original intercorrelated data set.

The relative importance or weight of the principal components can be examined by comparing the magnitude of the eigenvectors of the three principal components. In the case of the Detroit River sediment toxicity assessment, one principal component explained approximately 85% of the total variation (Table 12). The addition of a second principal component explained more than 99.9% of the observed variation. Thus, it was concluded that while no single bioassay explained all of the variation in toxicity among locations, only two assays were required to explain essentially all of the variation.

PC1, which explained the greatest proportion of the variation (Table 12), was almost equally weighted for contributions by the relative Microtox® EC_{50} (RMTEC50) and relative *D. magna* LC_{50} (RDMLC50), which indicates that these two parameters explained similar variance components. PC2, which was orthogonal (uncorrelated with PC1), explained the differences in these two parameters. These two parameters were "loaded to" the first two principal components because both the *P. phosphoreum* and *D. magna* assays tend to result in "all or nothing" responses.

PC3 contained information contributed almost exclusively by the percent reduction in *C. tentans* growth (Table 12), which indicates that this parameter is the most discriminatory of the three assays. However, because so many of the locations were either toxic or nontoxic, based on the other two assays, this third principal component contributed little to the model for classifying sediments based on their toxicity. The principal components analysis of the sediment toxicity information from the Detroit River suggests that a model containing the results of the *D. magna* and *P. phosphoreum* assays would explain more than 99.9% of the variation in toxicity among locations and that it would be needless to conduct a larger battery of tests.

Principal components analysis represents a normalization of the original data. Hence, any models produced by the regression of principal components are not appropriate for predicting sediment toxicities. Rather, multiple regression techniques should be used to predict some dependent variables from independent variables. For instance, biomass production of a benthic invertebrate species could be predicted from the results of the acute toxicity assays once such a model has been calibrated. A technique which can be used to classify sediments as either toxic or nontoxic is canonical correlation drawn from a model, such as the one presented here, based on a classification data set. Then, subsequent measurements in two or more bioassays of sediments could be used to place sediments into two or more classifications. This uses the information embodied in the responses of several different assays to predict the "overall inherent toxicity" of sediments. Using several assay organisms increases the probability of correctly classifying sediments that would be expected to be toxic to a community of benthic organisms under field conditions. For instance, the possibility of classifying a location as nontoxic based on the *P. phosphoreum* assay alone, when in fact it was found to be very toxic by the *D. magna* assay, would be much diminished. Also, the classifications could be calibrated to correspond to other thresholds, such as the toxicity potential required to restrict a particular group of benthic invertebrates such as mayflies, snails, clams, or midges from a particular sediment. Because a dose-response relationship was developed for each assay, probabilities for exceeding a threshold could be assigned. The principal components analysis indicated that such a model would be more than 99.9% accurate when the results of both the *D. magna* and *P. phosphoreum* assays were included, but the results of either assay alone would be insufficient to accurately classify the toxicity of the sediments.

A centroid hierarchical cluster analysis could be used to determine which locations were most similar, based on the variation explained using the nonredundant information provided by several toxicity assays (Figure 20). As with the principal components analysis, most of the variation among locations can be explained by including the results of the *D. magna* and *P. phosphoreum* assays of the Detroit River sediment alone (Table 13).

The type and number of assays conducted depend on the type and number of contaminants associated with the sediment as well as sediment type.

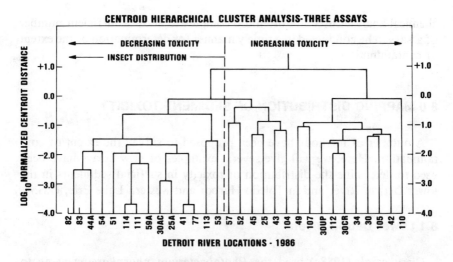

CENTROID HIERARCHICAL CLUSTER ANALYSIS-THREE ASSAYS

DETROIT RIVER LOCATIONS · 1986

Figure 20. Tree diagram of centroid cluster analysis of the toxicities of 30 Detroit River, U.S.A. and Canada sediments, based on the *Daphnia magna* 48-hr lethality test, the *Chironomus tentans* 10-day growth reduction test, and the *Photobacterium phosphoreum* 15-min bioluminescence reduction test. (Reprinted from Giesy *et al.* 1988a. With permission.)

The work of Giesy *et al.* (1988a) indicates that a battery of at least two acute toxicity assays was needed to accurately classify Detroit River sediments. While all of the assays were not necessary to classify sediments from the Detroit River as to toxicity, other locations contaminated with a different set of toxicants may require a different set of assays.

In our opinion, the assessment of many contaminated sites is excessive and interferes with the implementation of remedial action. The assays and

Table 13. Centroid Hierarchical Cluster Analysis of the Toxicity of 30 Detroit River Sediments, Based on Three Assays: *Daphia magna* 48-hr Lethality Test; *Chironomus tentans* 10-day Growth Reduction Test, and *Photobacterium phosphoreum* 15-min Bioluminescence Reduction Assay

Assays Included	Eigenvalue	Cumulative Variance Explained
D. magna	336,909	0.83438
D. magna *P. phosphoreum*	66,390	0.99880
D. magna *P. phosphoreum* *C. tentans*	484	1.00000

Source: Giesy *et al.* (1988a).

chemical analyses used are often too complex to allow a sufficient number of assays to be conducted in a timely manner for determination of the extent of contamination.

8.0 MAPPING DISTRIBUTION OF SEDIMENT TOXICITY

A number of methods have been proposed to survey the extent of contamination. Here, we will give several examples of how assays have been used to determine the distribution of toxicity in surficial sediments in the lower Detroit River and in Toledo Harbor and western Lake Erie, U.S.A.

8.1 Lower Detroit River

Giesy *et al.* (1988a) used the *Photobacterium phosphoreum* assay to assess the toxicity of sediments in the lower Detroit River. Because this assay is simple and rapid, they were able to assess the toxicity of porewater from surficial sediments at 136 locations (Figures 21–22).

Toxicity of sediments ranged from undetectable to very great. Of the 136 sediments tested, 25 were found to have EC_{10} (effective concentration to reduce bacterial bioluminescence by 10%) values which were less than 10% porewater extract. These sediments were classified as being very toxic. These sediments were located primarily in the Rouge River and along the western shore of the Trenton Channel (Figures 21–22). Additional sediments found to be very toxic were from locations in the main channel of the Detroit River, north of Grosse Ile, and in several isolated locations, such as station nos. 53, 183, 42, and 121. Sediments from 60 locations were classified as moderately toxic because they had EC_{10} values of greater than 10% and less than 40% porewater. Moderately toxic sediments were observed throughout the lower Detroit River system on both sides of Grosse Ile (Figure 22). Ten locations yielded sediment porewater which exhibited EC_{10} values of greater than 40% and less than 80% porewater. These samples were classified as slightly toxic. Of the 136 sediments sampled, 41 were found to be nontoxic. These sediments exhibited EC_{10} values of greater than 80% porewater extract or had nonsignificant slopes in the log-probit regression. The area along the western shore of Fighting Island, including station nos. 82, 83, 130, and 136, represented a relatively large area of nontoxic sediment. The sediments from the eastern shore of the Trenton Channel were generally less toxic than those from the western shore. An area of nontoxic sediment was also located in the Rouge River and, in general, sediments from the eastern side of Grosse Ile were less toxic than those on the western side.

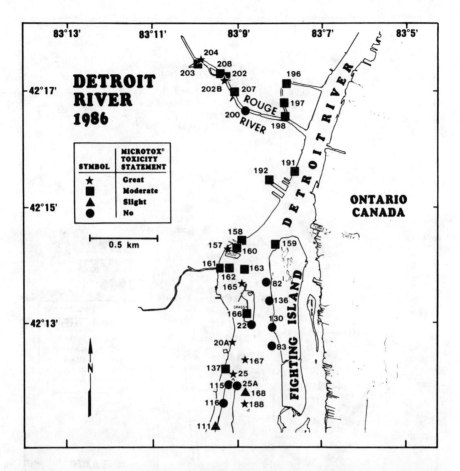

Figure 21. Distribution of toxicity of porewater extracts, isolated from sediments collected from 36 locations in the Rouge and Detroit Rivers, U.S.A. and Canada, 1986. The toxicity, as determined by *Photobacterium phosphoreum* bioluminescence, was classified into four categories based on concentrations of porewater extract to elicit a 10% reduction in bioluminescence; great—$EC_{10} < 10\%$, moderate—$10\% < EC_{10} < 40\%$, slight—$40\% < EC_{10} < 80\%$, and nontoxic $EC_{10} > 80\%$. (Reprinted from Giesy *et al.* 1988b. With permission.)

8.2 Western Lake Erie and Toledo Harbor

Giesy & Hoke (1988) used the *P. phosphoreum* assay to evaluate the distribution of toxicity in surficial sediments from western Lake Erie and the Maumee River, Ohio, U.S.A. The assay was used to determine the potential toxicity of sediments to be dredged in a channel maintenance project.

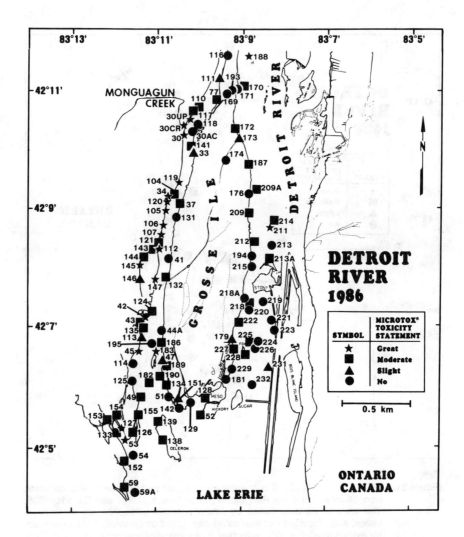

Figure 22.   Panel 2, distribution of toxicity of porewater extracts isolated from sediments collected from the Detroit River, U.S.A. and Canada, 1986. See legend to Figure 21 for description of symbols. (Reprinted from Giesy *et al.* 1988b. With permission.)

Sediment elutriates from some point source and navigation channel locations in the Maumee River caused significant reductions in bioluminescence of *P. phosphoreum*, while elutriates from sediments in the navigation channel or other locations in the western basin of Lake Erie caused no significant reductions in bioluminescence of *P. phosphoreum* (Figure 23). Giesy &

Hoke (1988) calibrated the *P. phosphoreum* assay to the effects of metals, which were thought to be the dominant toxicants, and to the results of other assays to effectively determine the locations of toxic sediments on a real-time basis so that dredging operations could be guided. This would not have been possible if the operator has been forced to wait for complex and lengthy chemical analyses or sediment toxicity tests.

9.0 REMEDIAL ACTION PLANNING

The utility of sediment assays in remedial action planning can be divided into three components: 1) mapping sediment toxicity, both horizontally and vertically; 2) prioritizing sites for further analyses or potential remediation; and 3) assessment of the effectiveness of remedial action. Sediment toxicity assays of several designs conducted with various species have been used to assess the areal extent of sediment contamination in recent investigations (Prater & Anderson 1977, Bahnick *et al.* 1980, Hoke & Prater 1980, Chapman & Fink 1984, Malueg *et al.* 1984a,b, Chapman *et al.* 1987, Giesy *et al.* 1988a). The *Chironomus tentans* assay has also been used to assess the extent of vertical contamination in Detroit River sediments (Rosiu *et al.* 1989). In the Detroit River study using *C. tentans*, assay results were used to prioritize the need for dredging to remove "hot spots" of sediment contamination. Based on estimates of the areal and vertical extent of sediment toxicity, preliminary estimates were calculated of volumes of sediment for potential dredging along with the associated costs of dredging.

A similar approach has been advocated for remedial action planning at

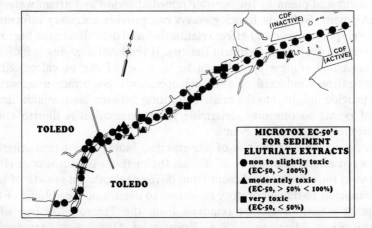

Figure 23.   Distribution of toxicity of porewater extracts isolated from sediments collected from the Maumee River and western Lake Erie, U.S.A., 1987. (Reprinted from Giesy & Hoke 1988. With permission.)

hazardous waste sites (Athey *et al.* 1987). Acute toxicity bioassays were found to be useful, cost-effective tools in designing hazardous waste site remediation programs to determine the distribution and toxicity of mixtures of contaminants. Information on potential causes of the observed effects may also be gained, depending on the classes of contaminants present and the assay and test species used in the assessment. The only question for which assays provide no information is what quantities of contaminants are present at a given location. The two most attractive features of bioassays in remedial action plans are that they provide direct measures of biological effect on the test organism and they are relatively inexpensive and rapid when compared to faunal assessments or chemical analyses. This facilitates the screening of a relatively large number of samples and the judicious use of expensive chemical analyses. Sediment toxicity assays can also be used to assess the effectiveness of remediation in terms of toxicity removal. To the best of our knowledge, however, no such use of these assays has been made to date.

Surveys of surficial sediments are not sufficient for determining the toxicity of deeper sediments and calculating the volume of sediment to be removed during remediation. In addition, survey data cannot identify the toxicity in deeper sediments where organisms do not occur naturally. This must be done with bioassays, which are structured to determine the toxic potency of sediments that were uncovered and available for colonization with benthic invertebrates under aerobic conditions. Because buried sediments have the potential to be uncovered and resuspended as a result of natural and human activities, characterizing the three-dimensional distribution of toxic sediments is necessary for the effective management of in-place pollutants and planning for possible remedial action and urban waterfront development. Sediment toxicity assays can provide necessary information by developing a cause and effect relationship and quantifying the horizontal and vertical profiles of sediment toxicity. If remedial dredging is indicated, the mass and volume of sediment to be removed can be calculated and prioritization conducted of dredging operations. Such measurements can also provide insight into the relative duration between contaminant depositional events so one may determine if contamination is due to current occurrences or historical events.

As an example of the use of site-specific, biologically based criteria to direct remedial actions, we will discuss the methods of Rosiu *et al.*(1989), who used the effects of sediment from different depths on growth of larval *C. tentans* in 10-day laboratory exposures to determine the volume of toxic sediment which should be removed from the Trenton Channel of the Detroit River, Michigan, U.S.A. Rosiu *et al.* (1989) also compared the results of *C. tentans* bioassays with the distribution of benthic macroinvertebrates in the Trenton Channel. The horizontal and vertical distributions of toxic sediments were mapped and estimates made of the mass of toxic sediments that would need to be removed to allow rehabilitation of benthic macroinvertebrate communities. Estimates were also provided on the cost of several levels of remedial action based on current prices for handling toxic sediments.

Sediment cores from 12 locations within the Trenton Channel and Trenton Channel delta of the Detroit River, Michigan, U.S.A (Figure 24) were collected in the summer of 1987 using 2-in.-diameter stainless steel tubes with a Wildco® Hand Core Sampler (Wildlife Supply Co., Saginaw, Michigan). Surficial sediment from an additional site supporting a diverse benthic invertebrate community, including individuals of the genus *Chironomus*, was used as a reference (control) sediment. Major sedimentation zones had been defined by previous studies (Fallon & Horvath 1985), and minor sedi-

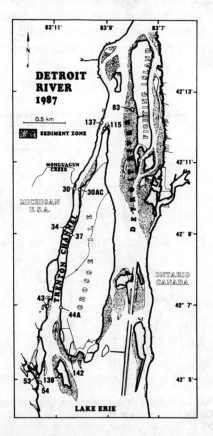

Figure 24. Sampling locations and locations of sedimentation zones (stippled areas) in the Trenton Channel of the lower Detroit River, U.S.A., 1987. (Reprinted from Rosiu *et al.* 1989. With permission.)

mentation zones were identified using a National Ocean Service 1987 bathymetric map of the lower Detroit River (Figure 24). Cores extended to a depth of 25 cm or until a hard, compacted clay layer was reached; cores ranged from 10 to 25 cm in actual length. Five composite cores were collected at each location and extruded from the core sampler into a clear, graduated Lexan® plastic tube where they were sectioned into 5-cm intervals.

Statistically significant (ANOVA, $p < 0.05$) reductions in weight gain, relative to control, were observed among *C. tentans* exposed to sediments from one or more depths at 8 of the 12 stations. Weight gains in sediment from other locations were not significantly different ($p > 0.05$) from control and were therefore determined to be nontoxic (Figure 25).

A uniform toxicity profile was observed at all depths for stations 30, 34, and 137 (Figure 25), which suggests a continuing source of contamination or recent mixing of sediments at these locations. In this case, the management of continuing point-source pollutants should be the first priority in remediation. A toxicity profile with a steadily increasing pattern of toxicity

DETROIT RIVER 1987
Chironomus tentans BIOASSAY

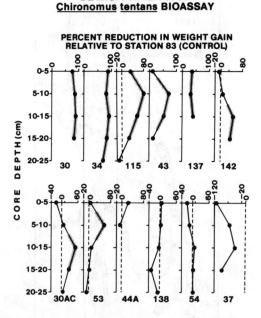

Figure 25. Vertical toxicity profiles of sediment core depth vs percent reduction in weight gain of *Chironomus tentans* relative to control. Highlighted lines profile growth reduction of 25% or greater, which was observed to be the threshold for restriction of colonization of surficial sediments. Negative values represent greater weight gain than that on control sediments. (Reprinted from Rosiu *et al.* 1989. With permission.)

with increasing depth could be due to chemical processes or confirm the success of management strategies for the reduction of contaminant inputs.

At stations 115, 43, 142, 30AC, and 53, the toxicity in profiles peaked at the mid-core intervals (5–10 cm or 10–15 cm) and then diminished with increasing core depth (10–25 cm) to levels near or below that of the control (Figure 25). The surficial sediment from stations 115 or 43 did inhibit growth of *C. tentans* relative to control; however, the sediments immediately below the surficial zone (5–10 cm) inhibited growth *C. tentans* more than the surficial sediments (Figure 25). Without toxicity data from sediment cores, the distribution of benthic macroinvertebrates in surficial sediments at these locations would have led one to underestimate the toxic potential of the underlying sediments. The composition of benthic species at station 43 suggested a questionable habitat quality, yet still better quality than was suggested by the toxicity data for *C. tentans* in the 5–10 cm depth from the sediment core (Figure 25). At station 142 the effect was even more dramatic. A maximum *C. tentans* weight gain reduction of 46.7% was observed at the 10–15 cm depth; however, the overlying surficial sediment was nontoxic at the 0–5 cm (–6.9%) and 5–10 cm (8.9%) depths and supported a diverse community of macroinvertebrates. Therefore, at stations 142 and 30AC, sediments that were determined to be toxic by the *C. tentans* bioassay were observed buried under as much as 10 cm of nontoxic sediments. At 9 of the 12 locations, a pattern of diminishing toxicity with increasing depth suggested that a 25-cm core length was sufficient to describe the entire range of toxicity observed at those locations in the Trenton Channel.

Due to diagenic and bioturbation processes, it is often very difficult to interpret the observed vertical distribution of toxicity or contaminant concentrations in sediments (Keilty *et al.* 1988b). However, we feel that coupled with sediment chemistry, bioassays can be used to successfully determine the extent of contaminated sediments for remedial action planning. With sufficient sampling, sediment core profiles of toxicity can provide data for calculating estimates of the spatial extent and volume of contaminated sediment in a region. These figures will facilitate subsequent calculation of the required costs of dredging and volume of appropriate containment facilities.

9.1 Remediation of Sediment Toxicity in the Trenton Channel

Since the presence of contaminated sediments in the Trenton Channel was confirmed to have adverse effects in the chironomid growth assay, Rosiu *et al.* (1989) calculated the volume of toxic sediment which would have to be removed from those locations. Outside the major sedimentation zones, the rate of sedimentation in the Trenton Channel is not great due to the channel configuration and the velocity of the flow (Kreis 1988). Suspended particles which enter the Trenton Channel have an average travel time of 8 hr before they are deposited in the delta and the western basin of Lake Erie. Contaminants originating on the western channel shoreline remain isolated near that shore and appear downstream as a toxic gradient (Giesy *et al.* 1988a,b). The total number of sedimentation zones in the channel is small because of the high flow velocity. In waterways with larger, more extensive, or complex sedimentation patterns, more sampling may be necessary to accurately ascertain the three-dimensional distribution of toxicity.

Rosiu *et al.* (1989) used maps of horizontal and vertical sediment toxicity to calculate the maximum and minimum sediment volumes which would have to be removed from the Trenton Channel to improve the quality of the benthic habitat to a specified level. The maximum volume of sediment to be removed from the Trenton Channel was reported to be approximately 231,000 m^3. This would remove all sediments causing a statistically significant inhibition of growth in *C. tentans*, relative to reference (control) sediment. This maximum volume included sedimentation zones surrounding stations 115, 43, 53, 44A, and 142 to a depth of 0.5 m, a ribbon of sedimentation averaging 20 m wide and 0.5 m deep for the remaining eastern and western lengths of the Trenton Channel, and the submerged delta at Gibraltar Bay to a depth of 1 m (Figure 24). Dredging is limited in its ability to effectively remove less than some minimum depth of material. The 0.5-m dredge depths used in the calculations are approximately that minimum operational limit and constitute a slight overestimate of the actual volumes of toxic sediment requiring removal. The minimum volume of sediment needing removal was estimated to be approximately 4,246 m^3 (Table 14). This calculation includes those sediments which were found to exceed the toxic threshold for the *C. tentans* bioassay, plus nontoxic sediments overlying toxic sediments at stations 30AC, 142, and 53 (Figure 25). This minimum volume is the total of the volumes for sedimentation zones represented by stations 137, 115, 30, 30AC, 34, 142, and 53 to a sediment depth of 0.5 m (Table 14).

The combined 1988 cost of dredging and disposal was estimated to be $357.61/m^3 for contaminated sediment ($327/yard^3, personal communication with John Adams, U.S. Army Corps of Engineers, Buffalo, NY). The minimum and maximum costs of remediation would be approximately $1.5 and $82.6 million, respectively. These amounts are approximately 100 times greater than the costs for dredging and disposal of uncontaminated sediments, due to costs for the contained disposal facility (CDF) required for storage of contaminated dredge spoils. The toxic nature of these sediments may also require specialized dredging operations (i.e. nonoverflow hydraulic removal) which would reduce the resuspension of sediment but also

Table 14. Minimum Area, Volume, and Cost Estimates for Dredging Toxic Sedimentation Zones in the Trenton Channel Containing Sediments Exhibiting a Greater than 25% Reduction in Weight Gain Relative to Control in the *C. tentans* Bioassay

Station	Area (m^2)	Dredge Volume (m^3)	Cost[a] ($U.S.)
137	588	294	$ 105,137.
115	780	390	$ 139,468.
30	540	270	$ 96,555.
30AC	687	344	$ 123,018.
34	519	260	$ 92,979.
43	1268	634	$ 226,725.
142	2855	1428	$ 510,667.
53	1251	626	$ 223,864.
Sediment removed from toxic sedimentation zones		Totals: 4246	$1,518,413.

Source: Rosiu *et al.* (1989).
[a]Estimates based on a cost of $357.61/$m^3$ (327/yard^3) for toxic sediment removal and confined disposal.

increase the costs of dredging. As a result, cost-effective remediation is best achieved by the removal of sediment that has been identified as contaminated material.

10.0 TOXICITY IDENTIFICATION AND EVALUATION

After the toxicity of sediments has been identified, the role of the aquatic toxicologist is to determine the cause of the toxicity so that 1) the source can be identified and eliminated and 2) the appropriate responsible parties can be identified and appropriate remedial action technologies can be employed. The current practices of exhaustive chemical analyses are very inefficient, and we feel that toxicity identification and evaluation (TIE) methods can be used to guide the chemical analyses of contaminated sediments. The use of these techniques minimizes the chance that a toxic chemical will be overlooked and streamlines and directs the chemical analyses.

Recently, the U.S. EPA issued guidance documents for the use of procedures to characterize and identify sources of acute toxicity in complex effluents (Mount & Anderson-Carnahan 1988a,b, Mount 1988). Although the procedures address matrix and multiple toxicant effects on toxicity of effluents to aquatic organisms, the major strength of the procedures is the development of direct causal relationships between chemical contaminants and observed effects on organisms.

The U.S. EPA toxicity identification and evaluation scheme is divided into three tiered phases. Specific compounds or classes of compounds are removed or rendered biologically unavailable prior to toxicity testing in

Phase I. Characterization of the physical and chemical properties of the toxicants potentially present in the sample via a standard series of chemical/physical manipulations and toxicity tests is the goal of Phase I (Figure 26). Adjustment of pH, aeration, filtration, reverse-phase chromatography, chelation of metals by EDTA, and oxidant reduction by additions of sodium thiosulfate are the primary techniques used for removal or inactivation of potential toxicants. The toxicity of altered sample fractions is then compared to the toxicity of the original, unaltered sample. Routine chemical measurements, including pH, hardness, conductivity, and dissolved oxygen, are also made in each step to facilitate design of sample manipulations and as an aid to interpretation of toxicity test data. Correct application of Phase I procedures permits tentative identification of compounds which are cationic metals, nonpolar organics, polar organics, volatile compounds, oxidant compounds, or compounds whose toxicity is pH dependent.

Procedures used during Phase II rely on Phase I results to guide subsequent fractionation and chemical identification of suspected toxicants (Mount & Anderson-Carnahan 1988b). Toxic subsamples from Phase I are chemically analyzed and lists compiled of all compounds identified in each subsample. Literature values or predictions from structure activity models are then used to develop LC_{50} values for identified compounds. Concentra-

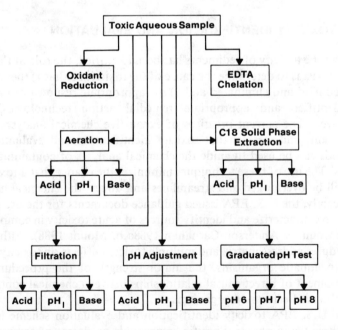

Figure 26. Proposed TIE scheme for complex effluents (Anon. l988a) also applicable to sediment porewaters and elutriates.

tions of the identified chemicals and their LC_{50} values are then compared and a list made of potential suspect toxicants. Refinement of the list is done by determining LC_{50} values for all suspect toxicants using the TIE test species.

Confirmation of the suspect toxicant(s) as the actual toxicant(s) is the goal of Phase III (Mount 1988). No one Phase III approach is adequate for confirmation of causal toxic compounds; therefore, multiple confirmation procedures are used. Confirmation techniques employed during Phase III include correlation of observed and expected toxicity, observations based on differential species sensitivity, observations of poisoning symptoms characteristic of a given compound or class of compounds, standard additions (spiking) of suspect toxicant(s), and mass balance techniques.

These TIE approaches have been applied successfully to a variety of acutely toxic aqueous samples including complex effluents, river water, and sediment porewater (Burkhard & Ankley 1989). We feel that the TIE procedures developed by the U.S. EPA for complex effluents can be adapted to identify the compounds responsible for toxicity observed in porewaters or elutriates from contaminated sediments. Previous studies have demonstrated that toxicity and/or bioaccumulation of sediment-associated contaminants such as cadmium, zinc, mercury, kepone, fluoranthrene, and various organochlorines, such as chlorobenzenes, by benthic macroinvertebrates is highly correlated with the concentrations of these chemicals in porewater (Adams et al. 1985, Swartz et al. 1985, 1988, Knezovich & Harrison 1988, Connell et al. 1988). Thus, porewater has been recommended as an appropriate fraction for sediment toxicity evaluations (DiToro 1988, Giesy & Hoke 1990). Recent studies utilizing toxic porewater from sediment samples from the lower Fox River have shown this fraction to be useful for TIE work. In addition to porewater, TIE techniques can be used with sediment elutriates which may be particularly representative of situations where sediment resuspension is of concern (Anon. 1977a). However, previous studies have shown marked differences in the toxicity of elutriates versus porewater from split sediment samples (Hoke et al. 1990b).

A number of procedures have been used to isolate porewater from sediment samples. These include compression (squeezing) techniques such as displacement of water from sediment via the use of inert gases, centrifugation of bulk sediment, direct sampling of porewater through the use of dialysis membranes, and microsyringe sampling (Knezovich & Harrison 1988, Knezovich et al. 1987, Sly 1988). The most representative porewater samples probably are obtained using the latter two procedures; however, sample volumes generated using these techniques are too small to be useful for toxicity tests and associated TIE work. To date, there has been no critical evaluation of the relative advantages/disadvantages of the former two porewater preparation procedures in terms of toxicity assessment, so to recommend the use of one over another would be premature. Elutriates can be obtained using standard procedures (Anon. 1977a). Briefly, elutriates can be prepared by suspending bulk sediment in laboratory dilution water at a ratio of 1:4 (v/v) followed by aeration for 30 min and centrifugation for 10 min at 6,000–7,000 × g. The resultant supernatant can be decanted and used for toxicity tests and TIE evaluations.

11.0 CONCLUSIONS

At this time, there is no consensus on the correct approach for the assessment of the toxicity of sediment-associated contaminants. Numerous assay protocols have been developed for the assessment of sediment toxicity, and a variety of species, representing the major classes of aquatic organisms, have been utilized as test organisms. Much useful information has been gained, but no comprehensive data set exists for a particular assay or test organism. Questions also remain concerning whether aquatic organisms are exposed to sediment contaminants primarily through direct exposure to contaminated sediments or via contaminants in porewater.

In recognition of these concerns, we feel that a battery of screening evaluation assays should be employed for the assessment of sediment toxicity. The proposed battery of assays includes Microtox®, an algal assay, the 10-day *Chironomus tentans* growth assay, and the 48-hr *Daphnia magna* acute lethality assay. This battery of assays contains tests which are simple, reproducible, inexpensive, ecologically relevant, and relatable to regulatory criteria. These assays also use organisms with different individual sensitivities to various classes of environmental contaminants. The use of benthic and surrogate species also facilitates the investigation of the effects of contaminants in whole sediments and their associated porewaters.

The use of a battery of assays with a range of sensitivities for rapid screening of sediments will enable larger numbers of sediments to be analyzed and decrease the probability of incorrectly identifying a toxic sediment as a nontoxic sediment. Rapid identification of toxic vs nontoxic sediments will also facilitate the prioritization of further investigations and permit the more judicious use of expensive reproduction and growth assays (*Ceriodaphnia dubia, Pimephales promelas*) and chemical analysis for determination of the potential causes and ecological effects of the observed toxicity. Finally, the proposed battery of assays will provide a rapid means of assessing the effectiveness of the remedial actions employed to deal with contaminated sediments. The proposed battery of assays is not meant to be definitive but simply represents the current state of our knowledge concerning the use of bioassays for the assessment of sediment toxicity. Several assays using test organisms such as protozoans, amphipods, oligochaetes, and mollusc embryos or larvae show promise as rapid screening tools for contaminated sediments. Further investigation of these techniques is required, however, before they can be used for the routine assessment of sediment toxicity.

The ultimate goal of any sediment toxicity assessment is to identify the causal agent(s) of the observed effects. TIE procedures developed for use on complex effluents show promise as a means of determining causes of toxicity observed in sediment porewater assays. Identification of the causal agents for observed toxicity coupled with determinations of the volume of contaminated sediment will facilitate the selection of the most cost-effective, efficient, and environmentally acceptable remedial action alternatives for contaminated sediments.

12.0 SUMMARY

Criteria for classifying the causes and intensity of the toxic effects of sediments are necessary to 1) regulate releases of chemicals to aquatic systems, 2) determine the causes of observed effects and identify sources of effects, 3) determine which sediments exhibit unacceptable effects on benthic and pelagic organisms, 4) set priorities for remedial action, and 5) determine the most appropriate remedial actions. Assessment of the effects of contaminants in sediment environments is complicated by the fact that organisms can be exposed, simultaneously, to multiple toxicants as well as naturally occurring extremes of pH, dissolved oxygen, and redox potential. A number of methods have been suggested to set sediment toxicity (quality) criteria. Each of the proposed methods has positive and negative attributes. Here we discuss the relative merits of each approach and present in detail the rationale and methods for the bioassay approach. We propose the use of a battery of rapid, simple, reproducible bioassays in conjunction with chemical and physical manipulations to determine the intensity and causes of toxicity of sediments. The assays recommended for inclusion in the screening battery for evaluation of sediment toxicity are: 1) the bioluminescence inhibition test with *Photobacterium phosphoreum* (Microtox®), 2) an algal assay using *Selenastrum capricornutum* Prinz, 3) a benthic invertebrate chronic growth test with *Chironomus tentans*, and 4) an acute lethality test with the cladoceran *Daphnia magna*. Finally, we propose a toxicity identification and evaluation (TIE) procedure which uses selective fractionations and chemical and physical adjustments of sediment porewaters to isolate and identify the primary toxic agents in sediments.

ACKNOWLEDGMENTS

This manuscript was prepared with support from a cooperative agreement between the U.S. Environmental Protection Agency (EPA) and Michigan State University (MSU) under agreement no. DL 85-002-06 and the Michigan Agricultural Experiment Station, from which it is contribution no. 13125. The patience of J. Thompson while typing both an earlier and this final version of the manuscript is appreciated.

REFERENCES

Adams, D.D. & D.A. Darby, 1980. A dilution mixing model for dredged sediments in freshwater ecosystems. In R.A. Baker (Ed.), *Contaminants and Sediments*, Vol. 1. Ann Arbor Science Publishers, Inc., Ann Arbor, MI: 373–392.

Adams, J.C., 1985. Non-toxicity of an oil shale process water to *Escherichia coli*. *Microbiologica* 8: 211–215.

Adams, W.J. & B.B. Heidolph, 1985. Short-cut chronic toxicity estimates using *Daphnia magna*. In R.D. Cardwell, R. Purdy & R.C. Bahner (Eds.), *Aquatic Toxicology and Hazard Assessment*. ASTM STP 854. American Society for Testing and Materials, Philadelphia, PA: 97–103.

Adams, W.J., R.A. Kimerle & R.G. Mosher, 1985. Aquatic safety assessment of chemicals sorbed to sediments. In R.D. Cardwell, R. Purdy & R.C. Bahner, (Eds.), *Aquatic Toxicology and Hazard Assessment*. ASTM STP 854. American Society for Testing and Materials, Philadelphia, PA: 429–453.

American Public Health Association/American Water Works Association/Water Pollution Control Federation (APHA), 1985. *Standard Methods for the Examination of Water and Wastewater*, 16th ed. APHA, Washington, D.C.: 615–714.

American Society for Testing and Materials (ASTM), 1986. Proposed new standard guide for conducting 96-h toxicity tests with microalgae, Draft 11. American Society for Testing and Materials, Philadelphia, PA: 33 pp.

Anderson, B.G., 1944. The toxicity thresholds of various substances found in industrial wastes as determined by the use of *Daphnia magna*. *Sewage Works J.* 16: 1156–1165.

Anderson, R.L., C.T. Walbridge & J.T. Fiandt, 1980. Survival and growth of *Tanytarsus dissimilis* (Chironomidae) exposed to copper, cadmium, zinc and lead. *Arch. Environ. Contam. Toxicol.* 9: 329–335.

Andreae, M.O., 1978. Distribution and speciation of arsenic in natural waters and some marine algae. *Deep-Sea Res.* 25: 391–402.

Ankley, G.T., A. Katko & J. Arthur, 1990. Identification of ammonia as an important sediment-associated toxicant in the lower Fox River and Green Bay, Wisconsin. *Environ. Toxicol. Chem.* 9: 312–322.

Anon., 1973. Ocean dumping: Final regulations and criteria. U.S. Federal Register. 38(198).

Anon., 1977a. *Ecological Evaluation of Proposed Discharge of Dredged or Fill Material into Navigable Water*. Interim Guidance for Implementation of Section

404(b) (1) of Public Law 92–500 (Federal Water Pollution Control Act Amendments of 1972). Misc. Paper D-76-17. U.S. Army Corps of Engineers, Waterways Experiment Station, Vicksburg, MS: 1-EZ.

Anon., 1977b. Implementation Manual for Section 103 of Public Law 92-532 (Marine Protection, Research and Sanctuaries Act of 1972, July 1977 [Second printing, April 1978]). U.S. Army Corps of Engineers, Environmental Effects Laboratory, Waterways Experiment Station, Vicksburg, MS.

Anon., 1978. *The Selanastrum capricornutum Printz Algal Assay Bottle Test*. EPA-600/9–78–018. U.S. Environmental Protection Agency, Corvallis, OR.

Anon., 1980. Water quality criteria documents availability. *Fed. Regist.* 45: 79318–79379.

Anon., 1983. *Initial Evaluation of Alternatives for Development of Sediment Related Criteria for Toxic Contaminants in Marine Waters (Puget Sound) Phase I: Development and Testing of the Sediment − Water Equilibrium Partitioning Approach*. U.S. Environmental Protection Agency, Office of Water Regulations and Standards, Criteria and Standards Division, Washington, D.C.

Anon., 1985. *Sediment Quality Criteria Development Workshop*. U.S. Environmental Protection Agency, Office of Water Regulations and Standards, Criteria and Standards Division, Washington, D.C.

Anon., 1986a. *Development of Sediment Quality Values for Puget Sound*, Vol. 1. DACW 67-85-0029 (TC 3090-02; Task 6). Tetra Tech., Inc., Bellevue, WA.

Anon., 1986b. *Sediment Quality Validation: Calculations of Screening Level Concentrations from Field Data*. U.S. Environmental Protection Agency, Office of Water Regulations and Standards, Criteria and Standards Division, Washington, D.C.

Anon., 1986c. *Elaboration of Sediment Normalization Theory for Non-Polar Organic Contaminants*. U.S. Environmental Protection Agency, Office of Water Regulations and Standards, Criteria and Standards Division, Washington, D.C.

Anon., 1986d. *Protocol for Sediment Toxicity Testing for Non-Polar Organic Compounds*. U.S. Environmental Protection Agency, Office of Water Regulations and Standards, Criteria and Standards Division, Washington, D.C.

Anon., 1987a. *Sediment Quality Criteria for Metals III. Review of Data on Complexation of Trace Metals by Particulate Organic Carbon*. U.S. Environmental Protection Agency, Office of Water Regulations and Standards, Criteria and Standards Division, Washington, D.C.

Anon., 1987b. *Sediment Quality Criteria for Metals IV. Surface Complexation and Acidity Constants for Modeling Cadmium and Zinc Adsorption Onto Iron Oxides*. U.S. Environmental Protection Agency, Office of Water Regulations and Standards, Criteria and Standards Division, Washington, D.C.

Anon., 1987c. *Sediment Quality Criteria for Metals II. Review of Methods for Quantitative Determination of Important Adsorbants and Sorbed Metals in Sediments*. U.S. Environmental Protection Agency, Office of Water Regulations and Standards Division, Washington, D.C.

Anon., 1987d. *Evaluation of the Equilibrium Partition Theory for Estimating the Toxicity of the Non-Polar Organic Compound DDT to the Sediment Dwelling Organism Rhepoxynius abronius*. U.S. Environmental Protection Agency,

Office of Water Regulations and Standards, Criteria and Standards Division, Washington, D.C.

Anon., 1987e. *Sediment Quality Criteria for Metals V. Optimization of Extraction Methods for Determining the Quantity of Sorbents and Adsorbed Metals in Sediments*. U.S. Environmental Protection Agency, Office of Water Regulations and Standards, Criteria and Standards Division, Washington, D.C.

Anon., 1987f. *Sediment Quality Criteria Methodology Validation: Uncertainty Analysis of Sediment Normalization Theory for Non-Polar Organic Contaminants*. U.S. Environmental Protection Agency, Office of Water Regulations and Standards, Criteria and Standards Division, Washington, D.C.

Anon., 1987g. *Regulatory Applications of Sediment Criteria*. U.S. Environmental Protection Agency, Office of Water Regulations and Standards, Criteria and Standards Division, Washington, D.C.

Anon., 1988a. *Sediment Quality Values Refinement: Volume I - Data Appendices - 1988 Update and Evaluation of Puget Sound AET*. Puget Sound Estuary Program and U.S. Environmental Protection Agency, Region 10—Office of Puget Sound, Seattle, WA.

Anon., 1988b. *Toxic Sediments: Approaches to Management*. EPA 68-01-7002. U.S. Environmental Protection Agency, Science, Policy Integration Branch, Washington, D.C.: 90 pp.

Anon., 1988c. *Options for the Remediation of Contaminated Sediments in the Great Lakes*. Great Lakes Water Quality Board, International Joint Commission, Sediment Subcommittee, Windsor, Ontario, Canada: 78 pp.

Anon., 1988d. *Interim Sediment Criteria Values for Non-Polar Hydrophobic Organic Contaminants*. U.S. Environmental Protection Agency, Office of Water Regulations and Standards, Criteria and Standards Division, Washington, D.C.

Anon., 1988e. *Development of Sediment Quality Objectives. Phase I Options.* Ontario Ministry of the Environment, Toronto, Ontario, Canada.

Anon., 1988f. *Development of Sediment Quality Guidelines. Phase II - Guideline Development*. Ontario Ministry of the Environment, Toronto, Ontario, Canada.

Anon., 1988g. *Guidelines for the Management of Dredged Material*. Ontario Ministry of Environment, Toronto, Ontario, Canada.

Archibald, P.A. (Ed.), 1982. *Environmental Biology State-of-the-Art Seminar*. USEPA 600/9-82-007. U.S. Environmental Protection Agency, Washington, D.C.

Athey, L.A., J.M. Thomas, J.R. Skalski & W.E. Miller, 1987. *Role of Acute Toxicity Bioassays in the Remedial Action Process at Hazardous Waste Sites*. EPA/ 600/8-87/044. U.S. Environmental Protection Agency, ERL, Corvallis, OR: 106 pp.

Athey, L.A., J.M. Thomas, W.E. Miller & J.Q. Word, 1989. Evaluation of bioassays for designing sediment cleanup strategies at a wood treatment site. *Environ. Toxicol. Chem.* 8: 223-230.

Atkinson, D.S., N.M. Ram & M.S. Switzenbaum, 1985. *Evaluation of the Microtox® Analyzer for Assessment of Sediment Toxicity*. Env. Eng. Report No. 86-85-3. University of Massachusetts, Amherst, MA: 63 pp.

Attar, E.N. & E.J. Maly, 1982. Acute toxicity of cadmium, zinc and cadmium-zinc mixtures to *Daphnia magna*. *Arch. Environ. Contam. Toxicol.* 11: 291-296.

Babich, H. & G. Stotsky, 1977. Reductions in the toxicity of cadmium to microorganisms by clay minerals. *Appl. Environ. Microbiol.* 33: 696-705.

Babich, H. & G. Stotsky, 1985. Heavy metal toxicity to microbe-mediated ecologic processes: A review and potential application to regulatory policies. *Environ. Res.* 36: 111-137.

Bahnick, D.A., W.A. Swenson, T.P. Markee, D.J. Call, C.A. Anderson, & R.T. Morris, 1980. *Development of Bioassay Procedures for Defining Pollution of Harbor Sediments*. Final Report of the U.S. EPA Project No. R-804918-01. Center for Lake Superior Environmental Studies, University of Wisconsin, Superior, WI.

Baker, J.H. & R.P. Griffiths, 1984. Effects of oil on bacterial activity in marine freshwater sediments. In *Proc. 3rd Int. Symp. Current Perspectives in Microbial Ecology*. American Society for Microbiology, Washington, D.C.: 546-551.

Batac-Catalan, Z. & D.S. White, 1982. Effect of chromium on larval Chironomidae as determined by the optical fiber light-interruption biomonitoring system. *Entomol. News* 93: 54-58.

Batley, G.E. & M.S. Giles, 1980. A solvent displacement technique for the separation of sediment interstitial waters. In R.A. Baker (Ed.), *Contaminants and Sediments*, Vol. 2. Ann Arbor Science, Ann Arbor, MI: 101-117.

Baudouin, M.F. & P. Scoppa, 1974. Acute toxicity of various metals to freshwater zooplankton. *Bull. Environ. Contam. Toxicol.* 12: 745-751.

Beckmann, J., P. Tazik & R. Gorden, 1984. Effects of two herbicides on selected aquatic bacteria. *Bull. Environ. Contam. Toxicol.* 32: 243-250.

Bedford, K., A. Carlson, A. El Shaarawi, J. Giesy, A. Mudroch, M. Munawar, T. Reynolds, K. Simpson & R. Thomas, 1987. Recommended methods for quantitative assessment of sediments in areas of concern. In *Guidance on Characterization of Toxic Substances Problems in Areas of Concern in the Great Lakes Basin*. Report to the Great Lakes Water Quality Board. International Joint Commission, Windsor, Ontario, Canada: 125-156.

Bellar, T.A., J.L. Lichtenberg & S.C. Conneman, 1980. Recovery of organic compounds from environmentally contaminated bottom materials. In R.A. Baker (Ed.), *Contaminants and Sediments*, Vol. 2. Ann Arbor Science, Ann Arbor, MI: 57-100.

Berglund, D.L. & S. Eversman, 1988. Flow cytometric measurement of pollutant stresses on algal cells. *Cytometry* 9: 150-155.

Berglund, D.L., S. Strobel, F. Sugawara & G.A. Strobel, 1988. Flow cytometry as a method for assaying the biological activity of phytotoxins. *Plant Sci.* 56:183-188.

Bertram, P.E. & B.A. Hart, 1979. Longevity and reproduction of *Daphnia pulex* (de Geer) exposed to cadmium contaminated water. *Environ. Pollut.* 19: 295-305.

Blaise, C., R. Legault, N. Bermingham, R. Van Coillie & P. Vasseur, 1986. A simple microplate algal assay technique for aquatic toxicity assessment. *Tox. Assess.* 1: 261-281.

Blanck, H., 1984. Species dependent variation among aquatic organisms in their sensitivity to chemicals. *Ecol. Bull.* (*Stockholm*) 36: 107–119.

Braginsky, L.P. & E.P. Scherban, 1978. Acute toxicity of heavy metals to aquatic invertebrates at different temperatures. *Hydrobiol. J.* 14: 78–82.

Bringmann, G. & R. Kuhn, 1980. Comparison of the toxicity thresholds of water pollutants to bacteria, algae and protozoa in the cell multiplication inhibition test. *Water Res.* 14: 231–241.

Brooks, L.T., D.J. Call, D.L. Geiger & C.E. Northcott (Eds.), 1984. *Acute Toxicities of Organic Chemicals to Fathead Minnows (Pimephales promelas)*. Center for Lake Superior Environmental Studies, University of Wisconsin-Superior: 414 pp.

Buikema, A.L., Jr. & E.F. Benfield, 1979. Use of macroinvertebrate life history information in toxicity tests. *J. Fish. Res. Bd. Can.* 36: 321–328.

Buikema, A.L., Jr., J.G. Geiger & D.R. Lee, 1980. *Daphnia* toxicity tests. In A.L. Buikema & J. Cairns, Jr. (Eds.), *Aquatic Invertebrate Bioassays*. ASTM 715. American Society for Testing and Materials, Philadelphia, PA: 48–69.

Buikema, A.L., Jr., B.R. Niederlehner & J. Cairns, Jr., 1982. Biological monitoring. IV. Toxicity testing. *Water Res.* 16: 239–262.

Bulich, A.A., 1984. Microtox® – A bacterial toxicity test with several environmental applications. In D. Liu & B.J. Dutka (Eds.), *Toxicity Screening Procedures Using Bacterial Systems*. Marcel Dekker, New York: 55–64.

Bulich, A.A., M.W. Greene & D.L. Isenberg, 1981. Reliability of the bacterial luminescence assay for determination of the toxicity of pure compounds and complex effluents. In J.G. Pearson, R.B. Foster & W.E. Bishop (Eds.), *Aquatic Toxicology and Hazard Assessment*. ASTM STP 737. American Society for Testing and Materials, Philadelphia, PA: 338–347.

Burkhard, L.P. & G.T. Ankley, 1989. Effluent toxicity limits: a historical perspective and overview of the EPA's toxicity-based approach to identifying toxicants in effluents. *Environ. Sci. Technol.* 23(12): 1438–1443.

Burton, G.A. & G.R. Lanza, 1987. Aquatic microbial activity and macrofaunal profiles of an Oklahoma stream. *Water Res.* 21: 1173–1182.

Burton, G.A., D. Nimmo, D. Murphy & F. Payne, 1987. Stream profile determinations using microbial activity assays and *Ceriodaphnia*. *Environ. Toxicol. Chem.* 6: 505–513.

Burton, G.A. & B.L. Stemmer, 1988. Evaluation of surrogate tests in toxicant impact assessments. *Tox. Assess.* 3: 255–269.

Cairns, J., Jr., 1979. A strategy for use of protozoans in the evaluation of hazardous substances. In S. James (Ed.), *Biological Indicators of Water Quality*. University of Newcastle upon Tyne, U.K.: 6–1 to 6–17.

Cairns, M.A., A.V. Nebeker, J.H. Gakstatter & W. Griffis, 1984. Toxicity of copper-spiked sediments to freshwater invertebrates. *Environ. Toxicol. Chem.* 3: 435–446.

Canton, J.H. & D.M.M. Adema, 1978. Reproducibility of short-term and reproduction toxicity experiments with *Daphnia magna* and comparison of the sensitivity of *D. magna* with *D. pulex* and *D. encullata* in short term experiments. *Hydrobiologia* 59: 135.

Chapman, P.M., 1986. Sediment quality criteria from the sediment quality triad: An example. *Environ. Toxicol. Chem.* 5: 957-964.

Chapman, P.M., 1987. Marine sediment toxicity tests. In *Symposium on Chemical and Biological Characterization of Sludges, Sediments, Dredge Spoils and Drilling Muds*. ASTM STP 976. American Society for Testing and Materials, Philadelphia, PA: 391-402.

Chapman, P.M. & R.O. Brinkhurst, 1984. Lethal and sublethal tolerances of aquatic oligochaetes with reference to their use as a biotic index of pollution. *Hydrobiologia* 115: 139-144.

Chapman, P.M. & R. Fink, 1984. Effects of Puget Sound sediments and their elutriates on the life cycle of *Capitella capitata*. *Bull. Environ. Contamin. Toxicol.* 33: 451-459.

Chapman, P.M., G.A. Vigers, M.A. Farrell, R.N. Dexter, E.A. Quinlan, R.M. Kocan & M. Landolt, 1982a. *Survey of Biological Effects of Toxicants upon Puget Sound Biota. I. Broad-Scale Toxicity Survey*. NOAA Technical Memorandum OMPA-25. U.S. Department of Commerce, National Oceanographic and Atmospheric Administration, Washington, D.C.

Chapman, P.M., R.N. Dexter & E.R. Long, 1987. Synoptic measures of sediment contamination, toxicity and infaunal community composition (the sediment quality triad) in San Francisco Bay. *Mar. Ecol. Prog. Ser.* 37: 75-96.

Ciborowski, J.J.H. & L.D. Corkum, 1988. Organic contaminants in adult aquatic insects of the St. Clair and Detroit Rivers, Ontario, Canada. *J. Great Lakes Res.* 14: 148-156.

Coleman, R.N. & A.A. Qureshi, 1985. Microtox and *Spirillum volutans* tests for assessing toxicity of environmental samples. *Bull. Environ. Contam. Toxicol.* 35: 443-451.

Connell, D.W., M. Bowman & D.W. Hawker, 1988. Bioconcentration of chlorinated hydrocarbons from sediment by oligochaetes. *Ecotoxicol. Environ. Safety* 16(3): 293-302.

Cooley, W.W. & P.R. Lohnes, 1971. *Multivariate Data Analysis*. John Wiley and Sons, Inc., New York: 364 pp.

Cowgill, U.M., 1987. Critical analysis of factors affecting the sensitivity of zooplankton and the reproducibility of toxicity test results. *Water Res.* 21(12): 1453-1462.

Cowgill, U.M., I.T. Takahashi & S.L. Applegath, 1984. A comparison of the effect of four benchmark chemicals on *Daphnia magna* and *Ceriodaphnia dubia-affinis* tested at two different temperatures. *Environ. Toxicol. Chem.* 4: 415-422.

Cummins, K.W., 1973. Trophic relations of aquatic insects. *Annu. Rev. Entomol.* 18: 183-206.

Curtis, C., A. Lima, S.J. Lozano & G.D. Veith, 1982. Evaluation of a bacterial bioluminescence bioassay as a method for predicting acute toxicity of organic chemicals to fish. In J.G. Pearson, R.B. Foster & W.E. Bishop (Eds.), *Aquatic Toxicology and Hazard Assessment*. ASTM-STP 766. American Society for Testing and Materials, Philadelphia, PA: 170-178.

Daube, D.D., D.S. Daly & C.S. Abernathy, 1985. Factors affecting the growth and survival of the Asiatic clam, *Corbicula* sp. under controlled laboratory conditions. In R.D. Cardwell, R. Purdy & R.C. Bahner (Eds.), *Aquatic Toxicology and Hazard Assessment*. ASTM STP 854. American Society for Testing and Materials, Philadelphia, PA: 134–144.

Davis, J.C., 1977. Standardization and protocols of bioassays — their role and significance for monitoring research and regulatory usage. In W.R. Parker, E. Pessah, P.G. Wells & G.F. Westlake (Eds.), *Proceedings of the 3rd Aquatic Toxicity Workshop*. EPS-5-AR-77-1. Environmental Protection Service Report, Halifax, Nova Scotia, Canada: 1–14.

Davis, W.S., T.J. Denbow & J.E. Lathrup, 1987. Aquatic Sediments. *J. Water Pollut. Control. Fed.* 59: 586–597.

Dawson, D.A., E.F. Stebler, S.L. Burks & J.A. Bantle, 1988. Evaluation of the developmental toxicity of metal-contaminated sediments using short-term fathead minnow and frog embryo-larval assays. *Environ. Toxicol. Chem.* 7: 27–34.

Depinto, J.V., T.L. Theis, T.C. Young, D. Vanetti, M. Waltman & S. Leach, 1987. *Exposure and Biological Effects of In-Place Pollutants: Sediment Exposure Potential and Particle Contaminant Interactions*. Interim report to U.S. EPA. U.S. EPA, Office of Research and Development, ERL, Duluth, MN and Large Lakes Research Station, Grosse Ile, MI: 75 pp.

Devillers, J., A. Elmousaffek, D. Zakarya, & M. Chastrette, 1988. Comparison of ecotoxicological data by means of an approach combining cluster and correspondence factor analysis. *Chemosphere* 17(4): 633–646.

DeZwart, D. & W. Slooff, 1983. The Microtox as an alternative assay in the acute toxicity assessment of water pollutants. *Aquat. Toxicol.* 4: 129–138.

Dickson, K.L. & J.H. Rodgers, Jr., 1985. Assessing the hazards of effluents in the aquatic environment. In H.L. Bergman, R.A. Kimerle & A.W. Maki (Eds.,) *Environmental Hazard Assessment of Effluents*. Pergamon Press, Elmsford, NY: 209–215.

Dickson, K.L., A.W. Maki & W.A. Brungs (Eds.), 1987. *Fate and Effects of Sediment-bound Chemicals in Aquatic Systems*. Pergamon Press, New York: 445 pp.

DiToro, D.M., 1989. *A Review of the Data Supporting the Equilibrium Partitioning Approach to Establishing Sediment Quality Criteria*. Report to the National Research Council.

Dutka, B.J. & K.K. Kwan, 1988. Battery of screening tests approach applied to sediment extracts. *Tox. Assess.* 3: 303–314.

Dutka, B.J., K. Jones, K.K. Kwan, H. Bailey & R. McInnis, 1988. Use of microbial and toxicant screening tests for priority site selection of degraded areas in water bodies. *Water Res.* 22(4): 503–510.

Effler, S.W. & C.T. Driscoll, 1985. Calcium chemistry and deposition in ionically enriched Onondaga Lake, New York. *Environ. Sci. Technol.* 19: 716–720.

Elnabarawy, M.T., A.N. Wetter & R.R. Robideau, 1986. Relative sensitivity of three daphnid species to selected organic and inorganic chemicals. *Environ. Toxicol. Chem.* 4: 393–398.

Elnabarawy, M.T., R.R. Robideau & S.A. Beach, 1988. Comparison of three rapid toxicity test procedures: Microtox®, Polytox® and activated sludge respiration inhibition. *Tox. Assess.* 3: 361–370.

Fallon, M.E. & F.J. Horvath, 1985. Preliminary assessment of contaminants in soft sediments of the Detroit River. *J. Great Lakes Res.* 11: 373-387.

Finney, D.J., 1971. *Probit Analysis*, 3rd ed. Cambridge University Press, Cambridge, England.

Finney, D.J., 1978. *Statistical Method in Biological Assay*, 3rd ed. Charles Griffin and Company, Ltd., London.

Fisher, N.S., J.L. Teyssie, S. Krishnaswami & M. Baskaran, 1987. Accumulation of Th, Pb, U, and Ra in marine phytoplankton and its geochemical significance. *Limnol. Oceanogr.* 32: 131-141.

Fremling, C.R. & W.L. Mauck, 1980. Methods for using nymphs of burrowing mayflies (*Ephemeroptera, Hexagenia*) as toxicity test organisms. In A.L. Buikema, Jr. and John Cairns, Jr. (Eds.), *Aquatic Invertebrate Bioassays.* ASTM STP 715. American Society for Testing and Materials, Philadelphia, PA: 81-97.

Geckler, J.R., W.B. Horning, T.M. Neiheisel, Q.H. Pickering & E.L. Robinson, 1976. *Validity of Laboratory Tests for Predicting Copper Toxicity in Streams.* EPA-600/3-76/116. National Technical Information Service, Springfield, VA.

Geiger, D.L., C.E. Northcott, D.J. Call & L.T. Brooke (Eds.), 1985. *Acute Toxicities of Organic Chemicals to Fathead Minnows (Pimephales promelas)*, Vol. 2. Center for Lake Superior Environmental Studies, University of Wisconsin, Superior: 326 pp.

Gerould, S., P. Landrum & J.P. Giesy, 1983. Anthracene bioconcentration and biotransformation in chironomids: Effects of temperature and concentration. *Environ. Pollut.* 30A: 175-188.

Gersich, F.M., 1984. Evaluation of a static renewal chronic toxicity test method for *Daphnia magna* (Straus) using boric acid. *Environ. Toxicol. Chem.* 3: 89-94.

Gersich, F.M., F.A. Blanchard, S.L. Applegath & C.N. Park, 1986. The precision of daphnid (*Daphnia magna* Straus, 1820) static acute toxicity tests. *Arch. Environ. Contam. Toxicol.* 15: 741-749.

Giesy, J.P. & P.M. Allred, 1985. Replicability of aquatic multispecies test systems. In J. Cairns, Jr. (Ed.), *Multispecies Toxicity Testing.* Pergamon Press, New York: chap. 12.

Giesy, J.P. & R.A. Hoke, 1988. *Toxicity of Sediment from Western Lake Erie and the Maumee River at Toledo, Ohio, 1987.* Completion report to U.S. Army Corps of Engineers, Buffalo District Office, Buffalo, NY: 81 pp.

Giesy, J.P. & R.A. Hoke, 1989. Freshwater sediment toxicity bioassessment: Rationale for species selection and test design. *J. Great Lakes Res.* 15: 539-569.

Giesy, J.P., R.L. Graney, J.L. Newsted, C.J. Rosiu, A. Benda, R.G. Kreis, Jr. & F.J. Horvath, 1988a. Comparison of three sediment bioassay methods using Detroit River sediments. *Environ. Toxicol. Chem.* 7(6): 483-498.

Giesy, J.P., C.J. Rosiu, R.L. Graney, J.L. Newsted, A. Benda, R.G. Kreis, Jr. & F.J Horvath, 1988b. Toxicity of Detroit River sediment interstitial water to the bacterium, *Photobacterium phosphoreum. J. Great Lakes Res.* 14: 502-513.

Giesy, J.P., C.J. Rosiu, R.L. Graney & M.G. Henry, 1990. Benthic invertebrate bioassays with toxic sediment and pore water. *Environ. Toxicol. Chem.* 9: 233-248.

Gillespie, P.A. & R.F. Vaccaro, 1978. A bacterial bioassay for measuring the copper-chelation capacity of sea water. *Limnol. Oceanogr.* 23: 543–548.

Goatcher, L.J., A.A. Qureshi & I.D. Goudet, 1984. Evaluation and refinement of the *Spirillum volutans* test for use in toxicity screening. In D. Liu and B. J. Dutka (Eds.), *Toxicity Screening Procedures Using Bacterial Systems*. Marcel Dekker, Inc., New York: 89–107.

Goulden, C.E., R.M. Comotto, J.A. Hendrickson, L.L. Horning & K.L. Johnson, 1982. Procedures and recommendations for the culture and use of *Daphnia* in bioassay studies. In J.G. Pearson, R.B. Foster & W.E. Bishop (Eds.), *Aquatic Toxicology and Hazard Assessment*. American Society for Testing and Materials, Philadelphia, PA: 139–160.

Greene, J.C., W.E. Miller, M.K. Debacon, M.A. Long & C.L. Bartels, 1985. A comparison of three microbial assay procedures for measuring toxicity to chemical residues. *Arch. Environ. Contam. Toxicol.* 14: 659–667.

Hall, W.S., K.L. Dickson, F.Y. Saleh, H.H. Rodgers, Jr., D. Wilcox & A. Entazami, 1986. Effects of suspended solids on the acute toxicity of zinc to *Daphnia magna* and *Pimephales promelas*. *Water Res. Bull.* 22: 913–920.

Hart, D.R., J. Fitchko & P.M. McKee, 1988. *Development of Sediment Quality Guidelines. Phase II - Guideline Development*. Beak Consultants, Ltd., Brampton, Ontario, Canada.

Hatakeyama, S., 1987. Chronic effects of Cd on reproduction of *Polypedilum nubifer* (Chironomidae) through water and food. *Environ. Pollut.* 48: 249–261.

Hermanns, J., H. Canton, N. Steyger & R. Wegman, 1984. Joint effects of a mixture of 14 chemicals on mortality and inhibition of reproduction of *Daphnia magna*. *Aquat. Toxicol.* 5: 315–322.

Hermanns, J., F. Busser, P. Leevwangh & A. Musch, 1985. Quantitative structure-activity relationships and mixture toxicity of organic chemicals in *Photobacterium phosphoreum*: The Microtox® test. *Ecotoxicol. Environ. Safety* 9: 17–25.

Hinwood, A.L. & M.J. McCormick, 1987. The effect of ionic strength of solutes on EC-50 values measured using the Microtox test. *Tox. Assess.* 2: 449–461.

Hodson, R.E., F. Azam & R.F. Lee, 1977. Effects of four oils on marine bacterial populations: Controlled ecosystem pollution experiment. *Bull. Mar. Sci.* 27: 119–126.

Hoke, R.A. & B.L. Prater, 1980. Relationship of percent mortality of four species of aquatic biota from 96-hour sediment bioassays of five Lake Michigan harbors and elutriate chemistry of the sediments. *Bull. Environ. Contam. Toxicol.* 25: 394–399.

Hoke, R.A., J.P. Giesy & J.R. Adams, 1990a. The use of linear orthogonal contrasts in the analysis of environmental data. *Environ. Toxicol. Chem.* (in press).

Hoke, R.A., J.P. Giesy, G.T. Ankley, J.L. Newsted & J. Adams, 1990b. Toxicity of sediments from western Lake Erie and Maumee River at Toledo, Ohio, 1987: Implications for current dredged material disposal practices. *J. Great Lakes Res.* (in press).

Hoke, R. A., J. P. Giesy & C. J. Rosiu, 1990c. Microtox evaluation of ammonia toxicity in sediment porewaters from the Lower Fox River and Green Bay, Wisconsin. *Environ. Toxicol. Chem.* (manuscript submitted).

Horning, C.E., 1980. *Use of Aquatic Oligochaete, Lumbriculus variegatus, for Effluent Biomonitoring.* EPA-600/D-80-005. Environ. Res. Brief, U.S. EPA.

Hunter, R.S., F.D. Culver, J.R. Hill & A. Fitzgerald, 1987. *QSAR System User Manual: A Structure-Activity Based Chemical Modeling System.* Institute for Biological and Chemical Process Analysis, Montana State University, Bozeman, MT.

Hutchinson, G.E., 1957. *A Treatise on Limnology.* John Wiley and Sons, New York: 1015 pp.

Indorato, A.M., K.B. Snyder & P.J. Usinowicz, 1984. Toxicity screening using Microtox analyzer. In D. Liu & B.J. Dutka (Eds.), *Toxicity Screening Procedures Using Bacterial Systems.* Marcel Dekker, New York: 37–54.

Iwan, G.R. & G.E. Cella, 1981. On-site critical life-stage bioassay with the fathead minnow, *Pimephales promelas*, on effluent from the wastewater treatment facilities of Austin, Minnesota. In D.R. Branson & K.L. Dickson (Eds.), *Aquatic Toxicology and Hazard Assessment.* ASTM STP 737. American Society for Testing and Materials, Philadelphia, PA: 312–323.

Jenne, E.A., V.C. Kennedy, J.M. Burchard & J.W. Ball, 1980. Sediment extraction and processing for selective extraction and for total trace element analyses. In R.A. Baker (Ed.), *Contaminants and Sediments*, Vol. 2. Ann Arbor Science, Ann Arbor, MI: 169–190.

Johnson, B.T. & V.I. Romanenko, 1984. Xenobiotic perturbation of microbial growth as measured by CO_2 uptake in aquatic heterotrophic bacteria. *J. Great Lakes Res.* 10: 245–250.

Johnson, W.W. & M.T. Finley, 1980. *Handbook of Acute Toxicity of Chemicals to Fish and Aquatic Invertebrates.* Publication No. 137. U.S. Department of Interior, Fish and Wildlife Service, Washington, D.C.

Jones, R.A. & G.F. Lee, 1978. *Evaluation of the Elutriate Test as a Method of Predicting Contaminant Release During Open Water Disposal of Dredged Sediment and Environmental Impact of Open Water Dredged Material Disposal.* Vol. 1: *Discussion.* Technical Report D78-45. U.S. Army Corps of Engineers, Waterways Experiment Station, Vicksburg, MS.

Jones, R.B., C.C. Gilmore, D.L. Stoner, M.M. Weir & J.H. Tuttle, 1984. Comparison of methods to measure acute metal and organometal toxicity to natural aquatic microbial communities. *Appl. Environ. Microbiol.* 47: 1005–1011.

Kangarot, B.S. & P.K. Ray, 1987. Correlation between heavy metal acute toxicity values in *Daphnia magna* and fish. *Bull. Environ. Contam. Toxicol.* 38: 722–726.

Karickhoff, S.W., 1980. Sorption kinetics of hydrophobic pollutants in natural sediments. In R.A. Baker (Ed.), *Contaminants and Sediments*, Vol. 2. Ann Arbor Science, Ann Arbor, MI: 193–205.

Keilty, T.J., D.S. White & P.F. Landrum, 1988a. Short-term lethality and sediment avoidance assays with endrin-contaminated sediment and two oligochaetes from Lake Michigan. *Arch. Environ. Contam. Toxicol.* 17: 95–101.

Keilty, T.J., D.S. White & P.F. Landrum, 1988b. Sublethal responses to endrin in sediment by *Limnodrilus hoffmeisteri* (Tubificidae), and in mixed-culture with *Stylodrilus heringianus* (Lumbriculidae). *Aquat. Toxicol.* 13(3): 227–250.

Khan, M.A., R.A. Gupta & M.P. Mohamed, 1986. Toxicity of zinc and mercury to chironomid larvae. *Indian J. Environ. Health* 28: 34–38.

Knezovich, J.P. & F.L. Harrison, 1988. The bioavailability of sediment-sorbed chlorobenzenes to larvae of the midge, *Chironomus decorus. Ecotoxicol. Environ. Safety* 15: 226–241.

Knezovich, J.P., F.L. Harrison & R.G. Wilhelm, 1987. The bioavailability of sediment-sorbed organic chemicals: A review. *Water Air Soil Pollut.* 32: 233–245.

Kosalwat, P. & A.W. Knight, 1987a. Acute toxicity of aqueous and substrate-bound copper to the midge, *Chironomus decorus. Arch. Environ. Contam. Toxicol.* 16: 275–282.

Kosalwat, P. & A.W. Knight, 1987b. Chronic toxicity of copper to a partial life cycle of the midge, *Chironomus decorus. Arch. Environ. Contam. Toxicol.* 16: 283–290.

Kreis, R.G. (Ed.), 1988. *Integrated Study of Exposure and Biological Effects of In-place Sediment Pollutants in the Upper Connecting Channel: Interim Results, Final Report to the Upper Great Lakes Connecting Channel Activities Work Groups.* U.S. Environmental Protection Agency, Office of Research and Development, ERL, Duluth, MN and Large Lakes Research Station, Grosse Ile, MI: 1200 pp.

LeBlanc, G.A., 1980. Acute toxicity of priority pollutants to water flea (*Daphnia magna*). *Bull. Environ. Contam. Toxicol.* 21: 684–691.

Lebsack, M.E., A.D. Anderson, G.M. DeGrave & H.L. Bergman, 1981. Comparison of bacterial luminescence and fish bioassay results for fossil-fuel process waters and phenolic constituents. In J.G. Pearson, R.B. Foster & W.E. Bishop (Eds.), *Aquatic Toxicology and Hazard Assessment*. ASTM STP 737. American Society for Testing and Materials, Philadelphia, PA: 348–356.

Lee, G.F. & R.H. Plumb, 1974. *Literature Review on Research Study for the Development of Dredged Material Disposal Criteria*. Report No. D-74-1. U.S. Army Corps of Engineers, Waterways Experiment Station, Vicksburg, MS: 145 pp.

Lee, G.F. & R.A. Jones, 1982. Discussion of dredged material evaluations: Correlations between chemical and biological procedures. *J. Water Pollut. Control Fed.* 54: 406–407.

Leeuivangh, P., 1978. Toxicity tests with daphnids: Its application in the management of water quality. *Hydrobiologia* 59: 145–148.

Leonard, S.L., 1979. Effects on survival, growth and reproduction of *Daphnia magna*. In E. Scherer (Ed.), Toxicity Tests for Freshwater Organisms. *Can. J. Fish. Aquat. Sci.*, Special Publication No. 44: 91–103.

Lewis, M.A. & R.L. Perry, 1981. Acute toxicities of equimolar and equitoxic surfactant mixtures to *Daphnia magna* and *Lepomis macrochirus*. In D.R. Branson & K.L. Dickson (Eds.), *Aquatic Toxicology and Hazard Assessment*. ASTM STP 737. American Society for Testing and Materials, Philadelphia, PA: 402–418.

Lewis, P.A. & C.I. Weber, 1985. A study of the reliability of *Daphnia* acute toxicity tests. In R.D. Cardwell, R. Purdy & R.C. Bahner (Eds.), *Aquatic Toxicology and Hazard Assessment*. ASTM STP 854. American Society for Testing and Materials, Philadelphia, PA: 73–86.

Litchfield, J.T. & F. Wilcoxon, 1949. A simplified method of evaluating dose/effect experiments. *J. Pharmacol. Exp. Ther.* 96: 99–113.

Liu, D. & B.J. Dutka (Eds.), 1984. *Toxicity Screening Procedures Using Bacterial Systems.* Marcel Dekker, Inc., New York: 476 pp.

Macek, K.J. & B.H. Sleight, 1977. Utility of toxicity tests with embryos and fry of fish in evaluating hazards associated with the chronic toxicity of chemicals to fishes. In F.L. Mayer & J.L. Hamelink (Eds.), *Aquatic Toxicology and Hazard Evaluation.* ASTM STP 634. American Society for Testing and Materials, Philadelphia, PA: 137–146.

Maki, A.W., 1979. Correlations between *Daphnia magna* and fathead minnow (*Pimephales promelas*) chronic toxicity values for several classes of test substances. *J. Fish. Res. Bd. Can.* 36: 411–421.

Malueg, K.W., G.S. Schuytema, J.H. Gakstatter & D.F. Krawczyk, 1983. Effect of *Hexagenia* on *Daphnia* response in sediment toxicity tests. *Environ. Toxicol. Chem.* 2: 73–82.

Malueg, K.W., G.S. Schuytema, D.F. Krawczyk & J.H. Gakstatter, 1984a. Laboratory sediment toxicity tests, sediment chemistry and distribution of benthic invertebrates in sediments from Keweenaw Waterway, Michigan. *Environ. Toxicol. Chem.* 3: 233–242.

Malueg, K.W., G.S. Schuytema, J.H. Gakstatter & D.F. Krawczyk, 1984b. Toxicity of sediments from three metal-contaminated areas. *Environ. Toxicol. Chem.* 3: 279–291.

Martell, F.L., R.T. Motekaitis & R.M. Smith, 1988. Structure-stability relationships of metal complexes and metal speciation in environmental aqueous solutions. *Environ. Toxicol. Chem.* 7: 417–434.

Mayer, F.L., Jr. & M.R. Ellersieck, 1986. *Manual of Acute Toxicity: Interpretation and Data Base for 410 Chemicals and 66 Species of Freshwater Animals.* USFWS Publication No. 160, U.S. Fisheries and Wildlife Service, Washington, D.C.

McIntyre, A.D., 1984. What happened to biological effects monitoring? *Mar. Pollut. Bull.* 15: 391–392.

McKim, J.M., 1977. Evaluation of tests with early life stages of fish for predicting long-term toxicity. *J. Fish. Res. Bd. Can.* 34: 1148–1154.

Morrison, D.F., 1976. *Multivariate Statistical Analysis.* McGraw-Hill, New York.

Mosher, R.B. & W.J. Adams, 1982. *Method for Conducting Acute Toxicity Tests with the Midge Chironomus tentans.* Environmental Sciences Report No. EAS-82-AOP-44. Monsanto, St. Louis, MO.

Mosher, R.B., R.A. Kimerle & W.J. Adams, 1982. *MIC Environmental Assessment Method for Conducting 14-Day Partial Life Cycle Flow-Through and Static Sediment Exposure Toxicity Tests with the Midge Chironomus tentans.* Environmental Sciences Report ES-82-M-10. Monsanto, St. Louis, MO.

Mount, D.I., 1988. *Methods for Aquatic Toxicity Identification Evaluations: Phase III Toxicity Confirmation Procedures.* EPA/600-3-88/036. U.S. Environmental Protection Agency, Duluth, MN.

Mount, D.I. & T.J. Norberg, 1984. A seven-day life-cycle cladoceran toxicity test. *Environ. Toxicol. Chem.* 3: 425–434.

Mount, D.I. & L. Anderson-Carnahan, 1988a. *Methods for Aquatic Toxicity Iden-*

tification Evaluations: Phase I Toxicity Characterization Procedures. EPA/ 600-3-88/034. U.S. Environmental Protection Agency, Duluth, MN.

Mount, D.I. & L. Anderson-Carnahan, 1988b. Methods for Aquatic Toxicity Identification Evaluations: Phase II Toxicity Characterization Procedures. EPA/ 600-3-88/034. U.S. Environmental Protection Agency, Duluth, MN.

Mudroch, A., L. Sarazin, A. Leaney-East, T. Lomas & C. deBarros, 1986. Report on the Progress of the Revision of the MOE Guidelines for Dredged Material Open Water Disposal, 1984/1985. Environment Canada, Inland Waters Directorate, Environmental Contaminants Division: 15 pp.

Munawar, M., 1982. Toxicity Studies on Natural Phytoplankton Assemblages by Means of Fractionation Bioassays. Technical Report of Fisheries and Aquatic Sciences, No. 1152. Canadian Department of Fisheries and Oceans, Great Lakes Research Branch.

Munawar, M. & I.F. Munawar, 1987. Phytoplankton bioassays for evaluating toxicity of in situ sediment contaminants. Hydrobiologia 149: 87-105.

Munawar, M., A. Mudroch, I.F. Munawar & R.L. Thomas, 1983. The impact of sediment-associated contaminants from the Niagara River mouth on various size assemblages of phytoplankton. J. Great Lakes Res. 9: 303-313.

Munawar, M., R.L. Thomas, H. Shear, P. McKee & A. Mudroch, 1984. An Overview of Sediment-Associated Contaminants and Their Bioassessment. Technical Report of Fisheries and Aquatic Sciences, No. 1253. Canadian Department of Fisheries and Oceans: 136 pp.

Munawar, M., R.L. Thomas, W. Norwood & A. Mudroch, 1985. Toxicity of Detroit River sediment-bound contaminants to ultraplankton. J. Great Lakes Res.11: 264-274.

Nacci, D., E. Jackim & R. Walsh, 1986. Comparative evaluation of three rapid marine toxicity tests: Sea urchin early embryo growth test, sea urchin sperm cell toxicity test and Microtox®. Environ. Toxicol. Chem. 5: 521-525.

Nebeker, A.V. & C.E. Miller, 1988. Use of the amphipod crustacean Hyalella azteca for freshwater and estuarine sediment toxicity tests. Environ. Toxicol. Chem. 7(12): 1027-1033.

Nebeker, A.V., J.K. McCrady, R.M. Shar & C.K. McAuliffe, 1983. Relative sensitivity of Daphnia magna, rainbow trout and fathead minnows to endosulfan. Environ. Toxicol. Chem. 2: 69-72.

Nebeker, A.V., M.A. Cairns & C.M. Wise, 1984a. Relative sensitivity of Chironomus tentans life stages to copper. Environ. Toxicol. Chem. 3: 151-158.

Nebeker, A.V., M.A. Cairns, J.H. Gakstatter, K.W. Malueg, G.S. Schuytema & D.F. Krawczyk, 1984b. Biological methods for determining toxicity of contaminated freshwater sediments to invertebrates. Environ. Toxicol. Chem. 3: 617-630.

Nebeker, A.V., S.T. Onjukka, M.A. Cairns & D.F. Krawczyk, 1986. Survival of Daphnia magna and Hyalella azteca in cadmium spiked water and sediment. Environ. Toxicol. Chem. 5: 933-938.

Neff, J.M., J.Q. Word & T.C. Gulbransen, 1985. Recalculation of Screening Level Concentrations for Non-Polar Organic Contaminants in Marine Sediments. U.S.

Environmental Protection Agency, Office of Water Regulations and Standards, Criteria and Standards Division, Washington, D.C. (unpublished report).

Norberg, T.J. & D.E. Mount, 1985. A new fathead minnow (*Pimephales promelas*) subchronic toxicity test. *Environ. Toxicol. Chem.* 4: 711–718.

Obst, U., 1985. Test instruction for measuring the microbial metabolic activity in sewage samples. *Fresenius Z. Anal. Chem.* 321: 166–168.

Oliver, B.A., 1985. Desorption of chlorinated hydrocarbons from spiked and anthropogenically contaminated sediments. *Chemosphere* 14: 1087–1106.

Olson, R.J., S.L. Frankel, S.W. Chisholm & H.M. Shapiro, 1983. An inexpensive flow cytometer for the analysis of florescence signals in phytoplankton: Chlorophyll and DNA distributions. *J. Exp. Mar. Biol. Ecol.* 68: 129–144.

Patrick, W.H., Jr., R.P. Gambrell & R.A. Khalid, 1977. Physicochemical factors regulating solubility and bioavailability of toxic heavy metals in contaminated dredged sediment. *J. Environ. Sci. Health* A12: 475–492.

Persaud, D., R. Jaagumagi & A. Hayton, 1989. *Development of Provincial Sediment Quality Guidelines*. Ontario Ministry of the Environment, Water Resources Branch, Aquatic Biology Section, Toronto, Ontario, Canada: 19 pp.

Powlesland, C. & J. George, 1986. Acute and chronic toxicity of nickel to larvae of *Chironomus riparis* (Meigen). *Environ. Pollut. Ser. A* 42: 47–64.

Pranckevicius, P.E., 1986. *1982 Detroit Michigan Area Survey*. EPA Report No. 905-4-86-002. U.S. Environmental Protection Agency, Chicago, IL.

Prater, B.L. & M. Anderson, 1977. A 96-hr sediment bioassay of Duluth and Superior harbor basins using *Hexagenia limbata, Asellus communis, Daphnia magna* and *Pimephales promelas* as test organisms. *Bull. Environ. Contam. Toxicol.* 13: 159–169.

Qureshi, A.A., K.W. Flood, S. M. Thompson, C.S. Inniss & R.A. Rokosh, 1982. Comparison of a luminescent bacterial test with other bioassays for determining toxicity of pure compounds and complex effluents. In J.G. Pearson, R.B. Foster & W.E. Bishop (Eds)., *Aquatic Toxicology and Hazard Assessment*. ASTM STP 766. American Society for Testing and Materials, Philadelphia, PA: 179–195.

Qureshi, A.A., R.N. Coleman & J.H. Paran, 1984. Evaluation and refinement of the Microtox® test for use in toxicity screening. In D. Liu & B.J. Dutka (Eds.), *Toxicity Screening Procedures Using Bacterial Systems*. Marcel Dekker Publishers, New York: 1–22.

Rawson, D.M., A.J. Willmer & M.F. Cardosi, 1987. The development of whole cell biosensors for on-line screening of herbicide pollution of surface waters. *Tox. Assess.* 2(3): 325–340.

Rosiu, C.J., J.P. Giesy & R.G. Kreis, Jr., 1989. Toxicity of sediments in the Trenton Channel, Detroit River, Michigan to *Chironomus tentans* (Insecta: Chironomida). *J. Great Lakes Res.* 15(4): 570–580.

Salomons, W., N.M. de Rooij, H. Kerdijk & J. Bril, 1987. Sediments as a source for contaminants. *Hydrobiologia* 149: 13–30.

Samoiloff, M.R., J. Bell, D.A. Birkholz, G.R.B. Webster, E.G. Arott, R. Pulak & A. Madrid, 1983. Combined bioassay–chemical fractionation scheme for the determination and ranking of toxic chemicals in sediments. *Environ. Sci. Technol.* 17: 329–334.

Sasa, M. & M. Yasuno, 1982. Chironomids as biological indicators of environmental pollution. Man and the biosphere program in Japan. UNESCO, 5-7:78–88.

Schiewe, M.H., E.G. Hawk, D.I. Actor & M.M Krahn, 1985. Use of a bacterial bioluminescence assay to assess toxicity of contaminated marine sediments. *Can. J. Fish. Aquat. Sci.* 42: 1244–1248.

Schuytema, G.S., P.O. Nelson, K.W. Malueg, A.V. Nebeker, D.F. Krawczyk, A.K. Ratcliff & J.H. Gakstatter, 1984. Toxicity of cadmium in water and sediment to *Daphnia magna. Environ. Toxicol. Chem.* 3: 293–308.

Seelye, J.G. & M.J. Mac, 1982. *Bioaccumulation of Toxic Substances During Dredging.* Report AD-14-F-1-529-0 of interagency agreement to U.S. EPA. Great Lakes National Program Office, Chicago, IL.

Shea, D., 1988. Developing national sediment quality criteria. *Environ. Sci. Technol.* 22: 1256–1261.

Shaner, S.W. & A.W. Knight, 1985. The role of alkalinity in the mortality of *Daphnia magna* in bioassays of sediment-bound copper. *Comp. Biochem. Physiol.* 82C: 273–277.

Sinex, S.A., A.Y. Cantillo & G.R. Helz, 1980. Accuracy of acid extraction methods of trace metals in sediments. *Anal. Chem.* 52: 2341–2346.

Slooff, W., 1985. The role of multispecies testing in aquatic toxicology. In J. Cairns, Jr. (Ed.), *Multispecies Toxicity Testing.* Pergamon Press, New York: 45–60.

Sly, P.G., 1988. Interstitial water quality of lake trout spawning habitat. *J. Great Lakes Res.* 14: 301–315.

Sokal, R.R. & F.J. Rohlf, 1981. *Biometry*, 2nd ed. W.H. Freeman and Co., New York: 859 pp.

Steele, R.G.D. & J.H. Torrie, 1980. *Principles and Procedures of Statistics*, 2nd ed. McGraw-Hill Book Co., New York: 633 pp.

Stephan, C.E., 1977. Methods for calculating an LC-50. In F.L. Mayer & J.C. Hamelink (Eds.), *Aquatic Toxicology and Hazard Evaluation.* ASTM STP 634. American Society for Testing and Materials. Philadelphia, PA: 65–84.

Sullivan, J., J. Ball, E. Brick, S. Hausmann, G. Pilarski & D. Sopcich, 1985. *Report of the Technical Subcommittee on Determination of Dredge Material Suitability for In-Water Diposal.* Wisconsin Department of Natural Resources, Madison, Wisconsin: 44 pp.

Sunda, W.G. & P.A. Gillespie, 1979. The response of a marine bacterium to cupric ion and its use to estimate cupric ion activity in sea water. *J. Mar. Res.* 37: 761–777.

Swartz, R.C., G.R. Ditsworth, D.W. Schults & J.O. Lamberson, 1985. Sediment toxicity to a marine infaunal amphipod: Cadmium and its interaction with sewage sludge. *Mar. Environ. Res.* 18: 133–153.

Swartz, R.C., P.F. Kemp, D.W. Schults & J.O. Lamberson, 1988. Effects of mixtures of sediment contaminants on the marine infaunal amphipod, *Rhepoxynius abronius. Environ. Toxicol. Chem.* 7: 1013–1020.

Swindoll, C.M. & F.M. Applehans, 1987. Factors influencing the accumulation of sediment-sorbed hexachlorobiphenyl by midge larvae. *Bull. Environ. Contam. Toxicol.* 39: 1055–1062.

Takahashi, I.T., U.M. Cowgill & P.C. Murphy, 1987. Comparison of ethanol toxicity to *Daphnia magna* and *Ceriodaphnia dubia* tested at two different temperatures: Static acute toxicity test results. *Bull. Environ. Contam. Toxicol.* 39: 229–236.

Tarkpea, M., M. Hansson & B. Samuelson, 1986. Comparison of the Microtox® test with the 96-hr. LC-50 test for the harpacticoid *Nitocra spinipes*. *Ecotoxicol. Environ. Safety* 11: 127–143.

Tessier, A. & P.G.C. Campbell, 1987. Partitioning of trace metals in sediments: Relationships with bioavailability. *Hydrobiologia* 149: 43–52.

Thomas, J.M., J.R. Skalski, J.F. Cline, M.C. McShane, J.C. Simpson, W.E. Miller, S.A. Peterson, C.A. Callahan & J.C. Greene, 1986. Characterization of chemical waste site contamination and determination of its extent using bioassays. *Environ. Toxicol. Chem.* 5: 487–501.

Van Coillie, R., P. Couture & S.A. Visser, 1983. Use of algae in aquatic ecotoxicology. In J.O. Nriagu (Ed.), *Aquatic Toxicology*. John Wiley and Sons, New York: 487–512.

van de Guchte, C. & J.L. Maas-Diepeveen, 1988. Screening sediments for toxicity: A water-concentration related problem. In *Proc. 14th Annual Aquatic Toxicity Workshop*. November 1–4, 1987, Toronto, Ontario, Canada. *Can. Tech. Rep. Fish. Aquat. Sci.* No. 1607: 81–91.

Van Leeuwen, C.J., W.J. Luttmer & P.S. Griffioen, 1985. The use of cohorts and populations in chronic toxicity studies with *Daphnia magna*: A cadmium example. *Ecotoxicol. Environ. Safety.* 9: 26–39.

Vitkus, T., P.E. Gaffney & E.P. Lewis, 1985. Bioassay system for industrial chemical effects on the water treatment process: PCB interactions. *J. Water Pollut. Control. Fed.* 57: 935–941.

Ward, T.J. & P.C. Young, 1984. Effects of metals and sediment particle size on the species composition of the epifauna of *Pinna bicolor* near a lead smelter, Spencer Gulf, South Australia. *Estuar. Coastal Shelf Sci.* 18: 79–95.

Wentsel, R., A. McIntosh & G. Atchison, G., 1977. Sublethal effects of heavy metals contaminated sediment on midge larvae (*Chironomus tentans*). *Hydrobiologia* 56: 153–156.

Wentsel, R., A. McIntosh & W.P. McCafferty, 1978. Emergence of the midge *Chironomus tentans* when exposed to heavy metal contaminated sediment. *Hydrobiologia* 57: 195–196.

White, D. S., 1988. Persistent toxic substances and zoobenthos in the Great Lakes. In M.S. Evans (Ed.), *Toxic Contaminants and Ecosystem Health; A Great Lakes Focus*. John Wiley and Sons, New York: 537–548.

Williams, L.G., P.M. Chapman & T.C. Ginn, 1986. A comparative evaluation of marine sediment toxicity using bacterial luminescence, oyster embryo and amphipod sediment bioassays. *Mar. Environ. Res.* 19: 225–249.

Williams, B.K., 1987. The use of analysis of variance procedures in biological studies. *Appl. Stoich. Models Data Anal.* 3: 207–226.

Winner, R.W., 1988. Evaluation of the relative sensitivities of four day *Daphnia magna* and *Ceriodaphnia dubia* toxicity tests for cadmium and sodium penta-chlorophenate. *Environ. Toxicol. Chem.* 2: 153–160.

Zepp, R.G. & P.F. Schlotzhauer, 1983. Influence of algae on photolysis rates of chemicals in water. *Environ. Sci. Technol.* 17: 462–468.

Ziegenfuss, P.S. & W.J. Adams, 1985. *A Method for Assessing the Acute Toxicity of Contaminated Sediments and Soils with Daphnia magna and Chironomus tentans.* Environmental Sciences Report ESC-EAG-M-85-01. Monsanto, St. Louis, MO: 11 pp.

Ziegenfuss, P.S., W.J. Renoudette & W.J. Adams, 1986. Methodology for assessing the acute toxicity of chemicals sorbed to sediments: testing the equilibrium partitioning theory. In *Aquatic Toxicology and Environmental Fate*. ASTM STP 921. American Society for Testing and Materials, Philadelphia, PA: 479–493.

Zimmerman, M.C., T.E. Wissing & R.P. Rutter, 1975. Bioenergetics of the burrowing mayfly, *Hexagenia limbata*, in a pond ecosystem. *Verh. Int. Verein. Limnol.* 19: 3039–3049.

CHAPTER 10

Pollution and Recovery of Lake Orta (Northern Italy)

Rosario Mosello and Alcide Calderoni

1.0 INTRODUCTION

Lake Orta is one of the most important water bodies in the main Italian lacustrine area which is formed by the great marginal lakes Maggiore, Lugano, Como, Iseo, and Garda, south of the Alps. It occupies the south-western part of the Lake Maggiore drainage basin. Lake Orta has for decades suffered dramatic industrial pollution. The cumulative effects of large ammonium loads, heavy metal discharges, and the consequent in-lake transformation of the pollutants drastically changed the chemistry and biology of the water body, which was originally very similar in its features to the other southern Alpine lakes. The limnological history of the lake is well documented (Pavesi 1879, Giaj Levra 1926, Monti 1930, Baldi 1949, Moretti 1954, Corbella *et al.* 1958, Tonolli 1961, Tonolli & Vollenweider 1961a, 1961b, Vollenweider 1963, Chiaudani 1969, Bonacina 1970, Barbanti *et al.* 1972, Bonacina *et al.* 1973, Bonacina & Bonomi 1974, Bonacina & Bonomi 1984, Mosello *et al.* 1986a, Mosello *et al.* 1986b, Bonacina *et al.* 1988a, Bonacina *et al.* 1988b, Mosello *et al.* 1990, Camusso *et al.* 1990, van Dam & Martens 1990). Since 1981 the ammonium load has been drastically reduced by a treatment plant. Also, two other treatment plants for urban sewage and, in part, for wastes containing heavy metals have been in operation since 1982. The marked reduction in the pollution load has resulted in an improvement of the water quality, but the lake has remained acidic because of the oxidation of the ammonium to nitrate. This chapter will describe the pollution and recovery of Lake Orta and illustrate the results of the mass budget of the main ions, used as a tool to predict the stages of the recovery and to plan a liming treatment which has been designed to buffer the acidity of the water and hasten the recovery of Lake Orta.

2.0 THE LAKE AND ITS WATERSHED

Lake Orta, located in Northern Italy, has an area of 18 km², a volume of 1.3×10^9 m³, a maximum and mean depth of 143 and 71 m, respectively, and a drainage basin of 116 km² (Figure 1). The outlet, which leaves the lake at its northern end, is a tributary of the larger Lake Maggiore. Both Lake Orta and its drainage basin have a long, narrow shape extending in a north-south direction. The distribution of the areas in the various altimetric bands of the watershed is shown by the hypsographic curve (Figure 2), which shows a minimum elevation (lake surface) of 290 m above sea level and a median and maximum elevation of 595 and 1643 m, respectively. The main subbasins are located in the western part of the watershed, except for the basin of the River Pescone (18 km², Figure 1). The lake basin was probably formed by ice erosion on a preexisting river valley of a quaternary glacial

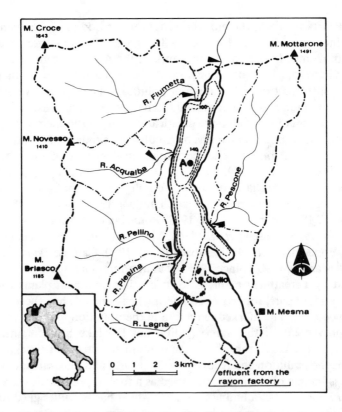

Figure 1. Watershed of Lake Orta and subbasins of the studied rivers. Sampling sites for chemical analyses: (A) lake water sample, (▼) rivers, (■) atmospheric deposition.

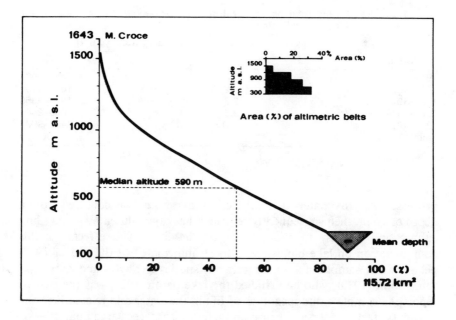

Figure 2. Hypsographic curve of Lake Orta drainage basin and distribution of the altimetric belts. (Reprinted from Mosello *et al.* 1986b. With permission.)

mass which had its origin as a branch of the massive glacier of the Ossola Valley. There are extensive outcrops of morainic deposits in the southwestern and southern part of the basin, while in the rest of the area there are in general smaller morainic deposits. The prequaternary rocks are mostly formed of gneiss, micaschist, and granites of the Strona-Ceneri Zone (Boriani & Sacchi 1974), while there are porphyry outcrops in the southeastern portion of the region.

Mean annual rainfall is 1900 mm year^{-1}, but the large volume of the lake makes the water renewal very slow. Based on the mean discharge of 4.93 m^3 sec^{-1}, the theoretical water renewal time has been estimated to be 8.5 years. The lake is classified as warm monomictic, with a minimum winter temperature of 5°C.

3.0 HISTORY OF CONTAMINATION

The pollution of Lake Orta started at the end of 1926, when a rayon factory (Bemberg) was set up at the southern end of the lake, which is opposite to the end of the lake where the outflow is located (Figure 1). The factory, which is still operating, needs a large quantity of water for its

Table 1. Copper (Cu) and Total Inorganic Nitrogen (TIN) Loadings Discharged into the Lake from the Rayon Factory

Period	Loading (tons year^{-1})	
	Cu	TIN
1926–1947	ca. 30	ca. 1000
1948–1957	32–80	1000–1800
1958–1974	3.7–9.5	1950–3350
1975–1979	2.8–3.1	2000–2400
1980–1981	0.8	1400–900
From 1982 on	0.3	30

processes. Approximately 0.15 m^3 sec^{-1} of lake water was collected from a depth of 70 m, then subsequently returned, heavily polluted by ammonium sulfate and copper, to the surface of the lake. The effects of this contamination on lake biota were almost immediate. At the end of 1927, pelagic water samples did not contain any species of phyto- and zooplankton. Monti (1930), who had studied the lake before 1926, was the first to report the complete disappearance of life in the water and related it to the biocidal effect of copper. The rayon factory discharged large quantities of ammonium and copper to Lake Orta (Table 1). The greatest discharges of ammonium, which were between 2000 and 3000 tons year^{-1}, were reached in the 1960s and 1970s, while the copper loads, which had been extremely high in the period 1927–1958, were reduced in the subsequent years to between 3 and 5 tons year^{-1}. Pollution from the Bemberg factory was further reduced in 1981, when an efficient plant for the recovery of ammonium sulfate went into operation; this reduced the load of ammonium to values similar to those derived from all other sources in the watershed and greatly reduced the quantities of copper discharged from the plant (Table 1).

Since the mid-1950s, Lake Orta has been affected by a further source of industrial pollution on the southwestern shore, in the watershed of the River Lagna (Figure 1), where there has been a proliferation of small factories which manufacture bathroom accessories. These factories discharge the heavy metals Cu, Zn, Ni, and Cr, anionic detergents, and cyanides into Lake Orta. The Cu and Cr loading from the area of greatest industrial activity was between 5 and 14 tons year^{-1}, respectively, in 1983 but had been reduced to 1.2 and 8 tons year^{-1}, respectively, in 1985 (Bonacina et al. 1986). These wastes are treated at a plant which at the moment is only partially operational, so that most of the metals reach the lake.

A third source of pollution for Lake Orta is the urban sewage of the approximately 30,000 inhabitants of the watershed. A load of about 40 tons year^{-1} of total phosphorus is discharged to the lake via the tributaries or the drains along the shoreline. Since 1988, phosphorus loads have been reduced by partially collecting and treating the sewage outside the drainage basin (Bonacina *et al.* 1986).

4.0 EFFECTS ON THE LAKE

After the first acute effects of the Cu on the biota of Lake Orta, other slower but important changes in the water chemistry occurred. Because of the geology of the watershed, originally the water of Lake Orta was poorly buffered, with total alkalinity of 0.3 meq L^{-1} (analysis done in 1926, reported by Baldi 1949). The in-lake oxidation of the ammonium to nitrate started to produce acidity, which reduced the alkalinity of the water. Thus, although the lake was receiving bicarbonate from the tributaries, the pH of the water slowly decreased to a minimum of 3.8 over the whole water column in February 1985 (Figure 3). In the 1950s all of the ammonium discharged into the lake was oxidized, which produced an increase of the nitrate concentration (Figure 3). In the 1960s and 1970s the ammonium discharged exceeded the oxidation capacity of the lake, and thus the concentration of ammonium started to increase. Concentrations of both ammonium and nitrate reached maximum values of between 0.3 and 0.5 meq L^{-1} in the 1970s. In the first years after the adoption of the plant to reduce the ammonium load (1981) a considerable reduction of the ammonium concentration was observed, while since 1985 the rate of decrease in ammonium concentration has been less (Figure 3). The decrease of the concentration of nitrate has not been as evident as that of ammonium because of the continuous production from the ammonium still stored in the water mass.

The conductivity of Lake Orta water has increased from 87 μS cm^{-1} in January 1956 to 153 μS cm^{-1} in February 1985 (Figure 3). This increase was due both to the increase in ionic concentrations (1.36 and 2.04 meq L^{-1}, respectively) and to the greater concentration of hydrogen ion, which was 5 and 126 μeq L^{-1} in 1956 and 1985, respectively. The mean concentrations at the overturn of the other chemical compounds measured since 1960, when the first complete and reliable ionic balance of the lake water was calculated by Vollenweider (1963), are shown in Table 2. In addition to the variations of conductivity, ammonium, nitrate, and pH already discussed, a comparison of the concentrations during the 1960s and the period between 1972 and 1981 shows marked increases in concentrations of sulfate, sodium, and chloride, due to the industrial discharge, and a decrease in copper. The main changes observed in the 1980s, after the reduction of the pollution from the rayon factory, have been the increase of calcium, used in the factory since 1981 to buffer the water discharged, and the reduction of concentrations of copper and chromium. A more complete review of the changes in the concentrations of these two metals is given in Figure 4, which summarizes the

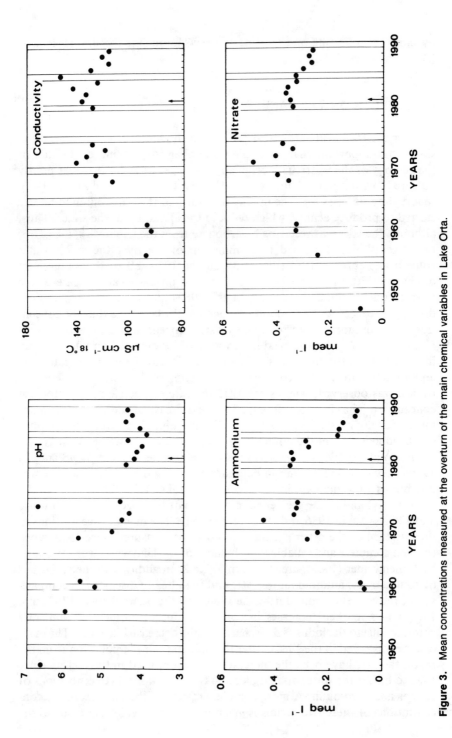

Figure 3. Mean concentrations measured at the overturn of the main chemical variables in Lake Orta.

Table 2. Chemical Characteristics of the Water of Lake Orta Before and After the Adoption of the Ammonium Recovery Plant (1982)[a]

		1960	1972	1981	1985	1989
Conductivity	$\mu S\ cm^{-1}$ (18°C)	85	133	136	153	115
pH		5.5	4.32	4.20	3.90	4.37
Ca^{2+}	$\mu eq\ L^{-1}$	271	279	264	329	453
Mg^{2+}	$\mu eq\ L^{-1}$	143	126	123	123	129
Na^+	$\mu eq\ L^{-1}$	160	251	248	239	217
K^+	$\mu eq\ L^{-1}$	20	28	29	28	27
$N\text{-}NH_4^+$	$\mu eq\ L^{-1}$	66	327	332	164	90
SO_4^{2-}	$\mu eq\ L^{-1}$	390	650	657	653	639
$N\text{-}NO_3^-$	$\mu eq\ L^{-1}$	274	370	343	321	256
Cl^-	$\mu eq\ L^{-1}$	28	73	74	66	68
RP	$\mu g\ L^{-1}$	4	2	2	1	3
TP	$\mu g\ L^{-1}$	–	–	5	3	5
RSi	$mg\ Si\ L^{-1}$	3.4	3.7	4.4	4.0	4.6
Cu	$\mu g\ L^{-1}$	86	58	40	35	35
Zn	$\mu g\ L^{-1}$	–	–	–	60	60
Cr	$\mu g\ L^{-1}$	–	8	6	5	2
Al	$\mu g\ L^{-1}$	–	–	–	83	108
Ni	$\mu g\ L^{-1}$	–	–	–	22	–
Fe	$\mu g\ L^{-1}$	–	–	–	80	77
Mn	$\mu g\ L^{-1}$	–	–	–	100	110

[a]Mean volume-weighted concentrations at the overturn (February).

results of different papers. As regards the period before 1950, Picotti (1958) pointed to measurements carried out in 1930, which showed a copper concentration of 80 μg Cu/L. Fewer data are available for zinc; however, a concentration of about 85 μg Zn/L was observed in 1972 by Barbanti et al. (1972), while Gerletti & Provini (1978) reported a concentration of 105 μg Zn/L in 1975. Present concentrations (1989) of seven different metals, compared with the 1985 values, demonstrate that the concentrations at overturn are still very great and sufficient to be toxic to fish and other living organisms (Table 2).

During thermal stratification the epilimnetic waters reach pH values which are greater than those at the overturn. The difference is in part due to the effect of the waters of the tributaries and to the uptake of carbon dioxide due to primary production. The epilimnetic pH increase during summer varies with the amount of precipitation during spring and summer. The greatest values of 6.0–6.5 were observed in September 1984 (Mosello et al. 1986b). A corresponding decrease in the concentrations of trace metals has been observed as a result of the precipitation of Cu and Cr due to incorporation into the less soluble oxy-hydroxides of aluminum and iron (Mosello et al. 1986a). The same changes in concentrations have been observed in laboratory experiments performed by adding calcium carbonate to lake water, which have been conducted to determine the effects of liming on the lake (Mosello et al. 1989).

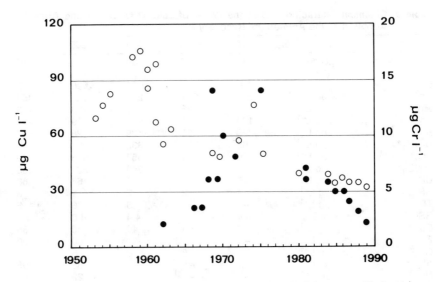

Figure 4.   Trend of copper and chromium concentrations in lake water. (Updated from Mosello *et al.* 1986a. With permission.)

The peculiar chemical conditions of the lake have determined the composition of the biological community, which for years was characterized by just one copepod species (*Cyclops abyssorum*) and only a few rotifer species, with occasional blooms; open-water cladocerans and diaptomids are absent (Bonacina *et al.* 1986). A population of *Daphnia obtusa* has been present since the spring of 1986 (Bonacina *et al.* 1988). The abundant phytoplankton is structurally monotonous: a few chlorophycean species (*Coccomyxa minor* and *Scenedesmus* sp.) and a blue-green alga (*Oscillatoria limnetica*). The fish present are pike (*Esox lucius*), eel (*Anguilla anguilla*), pumpkin seed sunfish (*Lepomis gibbosus*), spined loach (*Cobitis taenia*), chub (*Squalius cephalus cabeda*), and bleak (*Alburnus alburnus alborella*). All of the fish, except for the bleak, are littoral species. Pelagic fish (white fish, shad, trout) are absent from Lake Orta (Bonacina *et al.* 1986). An increase in perch was observed for the first time in 1987.

5.0 CHEMICAL BUDGET

In the period from 1984 until 1987 monthly samples were collected from six tributaries and from the outlet of Lake Orta. In 1984 and 1985 only, the effluent from the rayon factory was also sampled (Figure 1).

Because of the small ratio between the watershed and the lake surface (5.4), atmospheric deposition directly falling on the lake surface contributes

Table 3. pH, Conductivity, and Mean Concentrations of Atmospheric Depositions (A) and Inflowing Waters (B) for the period 1984–1987, Compared with the Lake Chemistry (C) at the Overturn (1989)

Parameter		A	B	C
pH		4.42	7.4	4.37
Conductivity	μS cm^{-1}(18°C)	30	67	115
H$^+$	μeq L^{-1}	38	0	43
NH$_4$$^+$	μeq L^{-1}	62	27	90
Ca^{2+}	μeq L^{-1}	38	298	453
Mg^{2+}	μeq L^{-1}	9	119	129
NA$^+$	μeq L^{-1}	13	213	217
K$^+$	μeq L^{-1}	4	25	27
HCO$_3$$^-$	μeq L^{-1}	0	293	0
SO$_4$H^{2-}	μeq L^{-1}	91	218	639
NO$_3$H$^-$	μeq L^{-1}	53	96	256
Cl$^-$	μeq L^{-1}	16	75	68
Σ cations	μeq L^{-1}	164	682	959
Σ anions	μeq L^{-1}	160	682	963

significantly to the total load. For this reason, a sampling station for bulk deposition was established at Mount Mesma (Figure 1), which is a site undisturbed by local domestic heating or industry.

Mean composition of atmospheric deposition, expressed as mean volume-weighted concentration, compared with the mean concentration of the six tributaries, weighted on the surfaces of their watershed, and with the lake concentration, is given in Table 3. Though the atmospheric deposition in the watershed of Lake Orta is decidedly acidic (Table 3) and constitutes a hydrogen ion input on the lake surface, the water of the tributaries is buffered. The mean total alkalinity of the tributaries is about 0.3 meq L^{-1} and pH is generally above 7.0. Even during heavy rains, which are quite frequent in this area, no episode of acid input from the tributaries has been detected. Thus, the acidification of the lake is due to in-lake processes and is primarily due to the oxidation of ammonium to nitrate.

Weathering of rocks and soils in the watershed is the most important source of inorganic ions, with the exception of calcium, sulfate, and ammonium, a substantial amount of which comes from the rayon factory (Table 4). Calcium hydroxide is used to adjust the pH of the lake water, which is between 4.2 and 4.4, before being used in the industrial processes.

Atmospheric deposition contributes significantly to the hydrogen and ammonium loads to the lake surface, while the weathering of the drainage basin, which is mainly formed of gneiss, micaschist, and granites, results in a slight contribution of alkalinity to the lake. Comparison of the masses entering and leaving the lake (Table 4) demonstrates a considerable imbalance in the case of ammonium and nitrate, whose output is greater than their input because of the huge store of these substances in the lake.

Table 4. Chemical Budget of Lake Orta for the Period from January 1984 Until December 1987.[a]

	H$^+$	NH$_4$$^+$	Ca^{2+}	Mg^{2+}	Na$^+$	K$^+$	HCO$_3$$^-$	SO$_4$$^{2-}$	Cl$^-$	NO$_3$$^-$
Drainage basin	0	2.0	34	14	25	3.0	33	24	8.1	11
Atmospheric deposition	1.1	1.9	1.1	0.3	0.4	0.1	0	2.5	0.5	1.5
Rayon factory sewage	0	2.1	56	1.0	5.5	0	1.7	60	1.0	0.4
Total input	1.1	6.0	91	15	31	3.1	35	87	9.6	13
Outflow	0.5	21	54	17	32	3.6	1.0	90	9.3	38
Difference	0.6	−15	37	n.s.	n.s.	−0.5	34	n.s.	n.s.	−25

[a]Units in 10^6 eq; n.s. = not significant.

Bicarbonate is almost entirely retained and consumed by the in-lake production of acidity. The small output is due to the summer outflow of poorly buffered epilimnetic water. On the other hand, the output of hydrogen ion occurs during winter and spring, when pH is low in the whole lake water mass.

In the case of calcium the input is greater than the output because of the amounts discharged from the rayon factory, which has resulted in a net increase in the Ca concentration (Table 2) in Lake Orta since 1981. If the 4-year period from January 1984 to December 1987 is considered, a mass budget of the whole lake which takes into account the loads entering (I) and leaving (O) the lake and the variations of the in-lake masses of ions at the beginning (M_0) and end (M_1) of the study period (Table 5) is possible. The difference $(O + M_1) - (I + M_0)$ is zero if further sources or sinks in the lake are absent. Differences which are smaller than 10% of $(M_0 + M_1)/2$ are considered not significant because of the errors associated with the values calculated.

The differences are not significant for any of the ions except ammonium and bicarbonate, both of which are consumed in the lake. The loss of

Table 5. Mass Budget of Lake Orta[a]

	I	M$_0$	O	M$_1$	(O + M$_1$) − (I + M$_0$)
H$^+$	4	59	2	60	n.s.
NH$_4$$^+$	28	350	84	140	−150
Ca^{2+}	360	390	220	550	n.s.
Mg^{2+}	60	160	68	160	n.s.
Na$^+$	120	300	130	280	n.s.
K$^+$	12	34	14	32	n.s.
HCO$_3$$^-$	140	0	4	0	−140
SO$_4$$^{2-}$	350	800	360	790	n.s.
Cl$^-$	38	91	37	85	n.s.
NO$_3$$^-$	52	430	150	350	n.s.

[a]M_0 and M_1: in-lake masses at January 1984 and December 1987, respectively; I and O: input and output during the 4-year period. Units in 10^6 eq; n.s. = not significant.

ammonium may be explained by the oxidation processes, while there is no evidence of a direct assimilation of ammonium by phytoplankton due to photosynthesis. Though to different extents, both of these processes produce acidity (Schuurkes & Mosello 1988), which is responsible for the consumption of the bicarbonate input from the watershed. Production of nitrate, which might be expected from the complete oxidation of ammonium, is not observed (Table 5). This may be at least partially explained by the photosynthetic uptake of nitrate and its subsequent loss through sedimentation or via the outflow. On the other hand, the great concentration of nitrate in Lake Orta indicates that this ion must have been produced to a great degree, at least at some period during the long history of the pollution of the lake.

6.0 CHEMICAL EVOLUTION OF LAKE ORTA

The substantial reduction in ammonium input since 1981 (Table 1) has led to a gradual recovery of the water quality of the lake. However, the remaining strong acidity and the metal concentrations in the lake water are still the main obstacles to the reestablishment of a normal biological community in the lake.

In order to forecast the time necessary for Lake Orta water to reach its original alkalinity, a simple input-output model, which takes account the acidity/alkalinity variation with time, was used. The model considers the annual mean chemical characteristics measured at overturn (end of February) at the station of maximum depth. A starting condition of negative alkalinity (total acidity) is quantified from hydrogen ion concentration (actual acidity), calculated from pH and from the ammonium still present (potential acidity), which will be oxidized in the next few years. On the basis of previous studies (Vollenweider 1963, Mosello et al. 1986b), a factor of 1.5 to express the hydrogen ion production deriving from the oxidation of one equivalent of ammonium was chosen. This figure should be taken as an empirical value which includes the effects of other oxidation-reduction processes going on in the lake, such as production and demolition of organic matter, iron and manganese speciation, and the effect of sedimentation or other loss for both alkalinity and ammonium, which were not considered in the model. A relationship developed by Vollenweider (1975) was used to calculate total alkalinity (Equation 1).

$$[m_w]_t = [m_w]_{to}\, e^{-r\,(t\,-\,to)} + Im\,[1 - e^{r\,(t\,-\,to)}]/r_w \qquad (1)$$

where $[m_w]_t$ = concentration of total alkalinity, calculated as the sum of hydrogen ion and 1.5 times the ammonium concentration, at time t

$[m_w]_{to}$ = −580 μeq L^{-1} is the total alkalinity value measured at the overturn of 1982, the first year after the end of ammonium pollution

r_w = 0.124 year^{-1} is the theoretical flushing coefficient, based on the historical mean discharge of the outlet in a 30-year period

I_m = 19 μeq L^{-1} year^{-1} is the volumnar loading of net alkalinity, taking into account the bicarbonate and ammonium load, relative to the 4-period

The predicted variations of alkalinity as a function of time are compared with the experimental values (full line) (Figure 5). The function is also plotted considering the extreme annual values measured in the 4-year period (dashed and dotted lines, respectively) as an indication of the variability of the estimates (Equations 2 and 3):

$$\text{maximum } I_m = 25 \ \mu\text{eq L}^{-1} \text{ year}^{-1}; \ r_w = 0.150 \text{ year}^{-1} \qquad (2)$$

$$\text{minimum } I_m = 17 \ \mu\text{eq L}^{-1} \text{ year}^{-1}; \ r_w = 0.080 \text{ year}^{-1} \qquad (3)$$

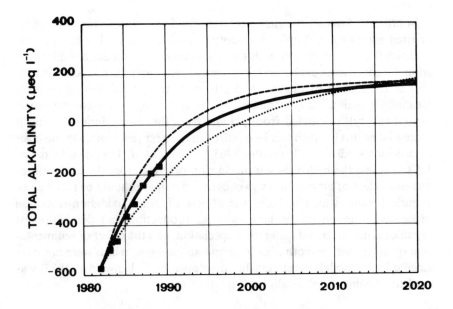

Figure 5. Variation of total alkalinity in time following the input-output model, compared with the experimental measurements (squares). For details see the text.

The three curves indicate that, in the absence of any treatment, the condition of alkalinity = 0 will be reached in about 5 years, while it will be about 10–15 years before alkalinity is between 100 and 200 μeq L^{-1}, a value more suitable for biological activity.

7.0 THE RECOVERY OF THE LAKE

On the basis of these considerations, the Istituto Italiano di Idrobiologia of the National Research Council made the proposal in 1986 to lime Lake Orta (Bonacina *et al.* 1988a,b). The proposal was approved by the Provincial and Regional Administrations and sponsored by the Ministry of the Environment. The project involves spreading on the lake surface a slurry of calcium carbonate to a total of 12,000 tons of pure $CaCO_3$ (dry weight), using finely powdered natural limestone with a very low content of trace metals. On the basis of the model used, this amount should result in an alkalinity of about 50 μeq L^{-1}, even after the complete oxidation of ammonium. The liming technique was devised in the light of experiences in Northern Europe, adapted for the peculiar characteristics of Lake Orta. Further details on the chemistry of the lime used and of the response to liming of the oxidation processes and trace metal concentrations have been provided by Mosello *et al.* (1989).

The liming has been in progress since May of 1989. The project has involved the construction of a boat capable of spraying about 120 tons of calcium carbonate per day; at present, about 4700 tons have been spread. The portion of the lake being treated is the southernmost part, which is farthest from the outflow. The different limnological aspects of the liming are being studied in an extensive research program, coordinated by this Institute and carried out in collaboration with two other Institutes (C.N.R. Istituto Ricerche sulle Acque, Brugherio, and J.R.C. Istituto dell'Ambiente, Ispra). The first results, after the spreading of about 4000 tons of calcium carbonate, show positive effects on the lake chemistry. Today it is reasonable to predict a complete recovery of the lake in the nineties, with stable chemical and biological equilibria, suitable for the creation of conditions adequate for the re-establishment of a normal biological community.

ACKNOWLEDGMENTS

We wish to thank A. Carollo and V. Libera, of the Istituto Italiano di Idrobiologia, for the unpublished data of water inflow used for the calculation of chemical budget.

REFERENCES

Baldi, E., 1949. Il Lago d'Orta, suo declino e condizioni attuali. *Mem. Ist. Ital. Idrobiol.* 5: 145–188.

Barbanti, L., C. Bonacina, G. Bonomi & D. Ruggiu, 1972. Lago d'Orta: situazione attuale e previsioni sulla sua evoluzione in base ad alcune ipotesi di intervento. *Ed. Ist. Ital. Idrobiol.* Pallanza: 113 pp.

Bonacina, C., 1970. Il Lago d'Orta: ulteriore evoluzione della situazione chimica e della struttura della biocenosi planctonica. *Mem. Ist. Ital. Idrobiol.* 26: 141–204.

Bonacina, C., G. Bonomi & D. Ruggiu, 1973. Reduction of the industrial pollution of Lake Orta (N. Italy): An attempt to evaluate its consequences. *Mem. Ist. Ital. Idrobiol.* 30: 149–168.

Bonacina, C. & G. Bonomi, 1974. La conoscenza dell'origine ed evoluzione dell'inquinamento del Lago d'Orta come base per la formulazione di strumenti previsionali indispensabili ad una politica di intervento. In *Atti I Convegno Internazionale sull'Ambiente e sulla Crisi dell'Energia*, Vol. 4. Torino, May 8–12, 1974: 25 pp.

Bonacina, C. & G. Bonomi, 1984. I grandiosi effetti ambientali determinati dalle prime fasi del disinquinamento del Lago d'Orta. *Doc. Ist. Ital. Idrobiol.* 2: 1–24.

Bonacina, C., G. Bonomi & R. Mosello, 1986. Notes on the present recovery of Lake Orta: An acid, industrially polluted, deep lake in North Italy. *Mem. Ist. Ital. Idrobiol.* 44: 97–115.

Bonacina, C., G. Bonomi, L. Barbanti, R. Mosello, D. Ruggiu & G. Tartari, 1988a. Lake Orta (N. Italy): Recovery after the adoption of restoration plans. In N.W. Schmidtke (Ed.), *Toxic Contamination in Large Lakes*, Volume 2. *Impact of Toxic Contaminants on Fisheries Management*. Lewis Publishers, Ann Arbor, MI: 101–130.

Bonacina, C., G. Bonomi, L. Barbanti, R. Mosello & D. Ruggiu, 1988b. Recovery of an industrially acidified, ammonia and heavy metal polluted lake (Lake Orta, N. Italy), due to the adoption of treatment plants. *Verh. Int. Verein. Limnol.* 23: 535–544.

Boriani, A. & R. Sacchi, 1974. The "Insubric" and other tectonic lines in the southern Alps (NW Italy). *Mem. Soc. Geol. It.* 13: 1–11.

Camusso, M., G. Tartari, L. Previtali & A. Zirino, 1990. Regulatory mechanism and behaviour of copper in an industrially-polluted acidic lake in Northern Italy. *Int. Ver. Verein. Limnol.* (in press).

Chiaudani, G., 1969. Contenuti normali ed accumuli di rame in *Phragmites communis* L. come risposta a quelli nei sedimenti in sei laghi italiani. *Mem. Ist. Ital. Idrobiol.* 25: 81–95.

Corbella, C., V. Tonolli & L. Tonolli, 1958. I sedimenti del Lago d'Orta, testimoni di una disastrosa polluzione cupro-ammoniacale. *Mem. Ist. Ital. Idrobiol.* 10: 9–52.

Gerletti, M. & A. Provini, 1978. Effect of nitrification in Orta Lake. *Prog. Water Technol.* 10: 839–851.

Giaj Levra, P., 1926. Diatomee del Lago d'Orta. *Atti Soc. Ligustica Sci. Lett.* 5: 66–82.

Monti, R., 1930. La graduale entinzione della vita nel Lago d'Orta. *Rend. R. Ist. Lomb. Sci. Lett.* 63: 3–22.

Moretti, G.P., 1954. Il limnobio neritico dei Tricotteri a testimonianza dell'attuale situazione biologica del Lago d'Orta. *Boll. Soc. Eustachiana* 47: 59–117.

Mosello, R., R. Baudo & G.A. Tartari, 1986a. Metal concentrations in a highly acidic lake: L. Orta (Northern Italy). *Mem. Ist. Ital. Idrobiol.* 44: 73–96.

Mosello, R., C. Bonacina, A. Carollo, V. Libera & G.A. Tartari, 1986b. Acidification due to in-lake ammonia oxidation: An attempt to quantify the proton production in a highly polluted subalpine Italian lake (Lake Orta). *Mem. Ist. Ital. Idrobiol.* 44: 47–71.

Mosello, R., A. Calderoni & G.A. Tartari, 1989. pH related variations of trace metal concentrations in L. Orta. *Sci. Total Environ.* 87/88: 255–268.

Mosello, R., A. Calderoni & R. de Bernardi. 1990. Mass budget as a tool for predicting the response to liming of the acidified, ammonium polluted L. Orta. *Int. Ver. Verein. Limnol.* (in press).

Pavesi, P., 1879. Nuova serie di ricerche sulla fauna pelagica nei laghi italiani. *Rend. Ist. Lomb. Sci. Lett.*: 474–483.

Picotti, M., 1958. Ricerche nel Lago d'Orta. *Boll. Pesca Piscic. Idrobiol.* 12: 126–158.

Schuurkes, J.A.A.R. & R. Mosello, 1988. The role of external ammonium inputs in freshwater acidification. *Schweiz. Z. Hydrol.* 50: 71–86.

Tonolli, L., 1961. La polluzione cuprica del Lago d'Orta: comportamento di alcune popolazioni di Diatomee. *Verh. Int. Ver. Limnol.* 14: 900–904.

Tonolli, V. & R.A. Vollenweider, 1961a. *Rapporto sulle Ricerche Eseguite sul Lago d'Orta nel Periodo dal 1959 al 1961.* Pallanza, October 1961: 1–18.

Tonolli, V. & R.A. Vollenweider, 1961b. Le vicende del Lago d'Orta inquinato da scarichi cupro-ammoniacali. In *Atti "Convegno Acque di Scarico Industriali"*, Milano, April 4–7, 1960: 99–109.

van Dam, H. & A. Mertens. A comparison of recent epilithic diatom assemblages from the industrially acidified and copper polluted Lake Orta (Northern Italy) with old literature data. *Diatom Res.* (in press).

Vollenweider, R.A., 1963. Studi sulla situazione attuale del regime chimico e biologico del Lago d'Orta. *Mem. Ist. Ital. Idrobiol.* 16: 21–125.

Vollenweider, R.A., 1975. Input-output models. *Schweiz. Z. Hydrol.* 37: 53–84.

CHAPTER 11

An Operative System for Environmental Consequence Analysis for Aquatic Ecosystems

Lars Håkanson

1.0 INTRODUCTION AND AIM

Within environmental research there is a completely fundamental question: What is the ecological effect caused by a given discharge, and why? Naturally, in order to be able to answer this question, it is necessary to make it more concrete: What type of discharge is it? Where does the discharge take place? What is meant by ecological effect? A discharge can consist of anything from thousands of different substances to single substances in numerous different chemical forms. If the discharge takes place in a lake, the lake can be characterized by numerous biological, chemical, or physical parameters which may also influence the character of the discharge. Consequently, the analysis of environmental consequences is extremely complex, approaching the chaotic. This is seen as a challenge by many scientists throughout the world, not only for purely scientific reasons but also because the environmental questions have today attained such importance that they concern our survival and quality of life on Earth.

The intention of this chapter is to present one type of environmental consequence analysis (ECA) for aquatic systems. There are other, supplementary systems for ECA which focus on approaches ranging from tests at the cell level to computer models of what might happen in thousands of years to come (Mackey & Peterson 1982, O'Neill et al. 1982, Cairns & Pratt 1987). The present system is intended for natural lakes, rivers, coastal areas, and seas. At the outset, it should be noted that this operative system is meant to use many well-known concepts, results, and ideas put together in a new, practical, "Linnean" manner.

The operative system can be described if an analogy is made with the world of music: In order to create music, a symphony or a ballad, not only a system of notes is required, so that the musicians know how to play together, but also a set of musical instruments. This article concerns a "note system" for ECA, i.e., the scientific framework for assessing environmental consequences (Håkanson 1984). The "instruments", i.e., the tools which concern the practical handling of the scientific concepts, are not complete today but are being developed in the flexible and user-friendly programming language called HyperCard. However, this chapter does not deal with the programming aspects. It may also be said that the operative system is based on a number of fundamental concepts (effect, dose, sensitivity, area and time compatibility, target indicators, etc.) which have approximately the same relationship to ECA as letters have to language: Using a few letters (26 in the English alphabet) not only tens of thousands of English words can be formed but also tens of thousands of Swedish and Gaelic words. In the same way, it is hoped that the immensely complex relationships between discharge, environmental character, and environmental effect will be described by a limited number of general basic concepts which, in principle, will apply to all types of substances, in all types of aquatic systems, and for all types of environmental effects.

The system is intended to be:

- synthesizing, such that old and new knowledge may be collected, processed, and included in the system in a structured manner
- surveyable; i.e., the information will be stored in such a manner that it will be easy, rapid, and simple to find
- flexible and extendible when new knowledge becomes available

All this is easy to say, but how can it be achieved? Before looking further at that question, it must be emphasized that a lot of work still remains to be done with the operative system and that even in a highly developed form the system cannot be used for everything; its task is simply to fill a niche. This chapter only presents the fundamental definitions and illustrates the possibilities. The system can be used to assess the implications of in-place pollutants, and the models included in the system can also be used in a forecasting mode.

2.0 EFFECT, DOSE, AND SENSITIVITY

Let us consider an official at an environmental authority who has the task of examining an industrial discharge. The basic question he must ask himself is: What ecological/biological effects can this discharge cause? The operative system should be an aid in answering the question. The whole process thus starts with this goal in mind, and the intention is that the user can find his way through a system where a definitive answer to the question can be found at the end. It must also be assumed that the information is available in

HyperCard as a number of linked bundles, where each bundle consists of a number of cards with structured information. The user must first click the so-called mouse in order to get a so-called Basecard (Figure 1).

The Basecard consists of a number of buttons for different environments, rivers, lakes, and marine. If the discharge takes place in a lake, then the "lake" button is pressed and one proceeds to the next level in the system (Figure 2). Here, there are four possibilities: Is it a question of discharge of nutrients, which may lead to eutrophication, acidifying substances, toxic substances, or complex wastewater?

Here, the system will be followed from the start to the end product, the load diagram. Today, this can only be done for a few substances. To do so one presses the button "toxic substances" and three different alternatives are displayed: Chlorinated organic materials, metals, and radioisotopes. It is assumed here that the discharge mainly contains mercury, and therefore the "Hg" button is pressed. Naturally, one must also consider buttons for all metals in the periodical system at the level where the Hg button is found. When one presses the Hg button a new button is displayed with the question: Is the effect term known?

The *effect term* is a key concept in this system. If one wants an explanation of this term the button "explain" is pressed. It is of vital importance that the user realizes what is meant and not meant by the effect term. If the user is not aware of the environmental effects caused by a given substance, in this case Hg, there is a system whereby effects can be determined. The "no" button is pressed and the *"mesocosm test"* module is entered (Figure 3). A mesocosm is a reproduction of the environment (e.g., a given lake type) that is as close to reality as possible in a reasonably large "laboratory scale" (Landner 1989). The mesocosm should contain the fundamental

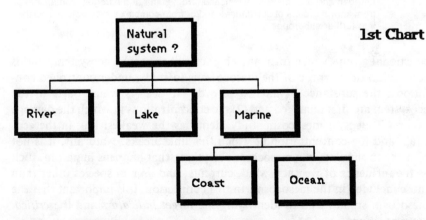

Figure 1. The first chart (HyperCard) in the operative system for environmental consequence analysis.

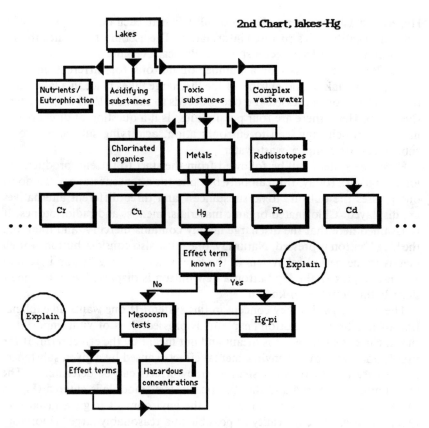

Figure 2. The second chart in the operative system. Any threat to a lake system could first be categorized into effect categories such as eutrophication from nutrients nitrogen and phosphorus, acidification from sulfur and nitrogen, contamination from known groups of substances and/or from complex waste waters, and then into further subgroups.

functional groups which form and characterize the actual ecosystem. In this connection the purpose of the mesocosm is to study, under controlled conditions, the substance of interest in order to see (1) which parts of the ecosystem are first damaged and (2) the concentrations at which the damage occurs. That is, a mesocosm study identifies the weakest link in the eco-chain and the concentration at which this link breaks. Naturally, it is not possible to simulate in a mesocosm everything that happens in nature, such as the influence of weather, wind, currents, and animal species other than those included in the mesocosm. In this connection, it is important that the mesocosm studies lead to identifying the *target indicators* and the *critical concentrations*.

The importance of identifying target indicators is illustrated in Figure 4.

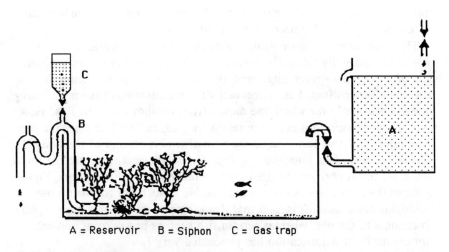

Figure 3. An example of a mesocosm: Model of the shallow-water ecosystem of the Baltic in an outdoor basin (8 m³).

This figure schematically illustrates, on the left, that effects of the metal zinc can be obtained in a very wide range of concentrations in short-term and long-term laboratory experiments with various types of biological indicator organisms. The most sensitive indicator organisms, those in focus here, react to Zn concentrations in water of about 100 ppm, whereas the

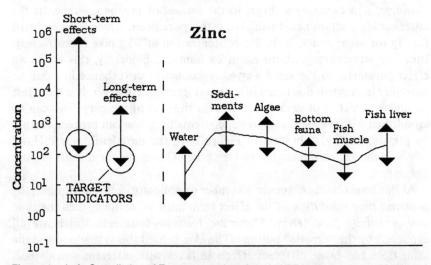

Figure 4. Left: Compilation of Zn concentrations (ppm) yielding short-term and long-term toxic effects on various biological indicator organisms in different laboratory tests and illustration of target indicators. Right: Concentrations of Zn in natural waters, sediments, algae, bottom fauna, and fish (ppm, dry weight).

most zinc-resistant indicator organisms do not react until the concentration of Zn in water is 3-4 powers of ten greater.

The right side of Figure 4 illustrates that Zn concentrations in natural waters are generally relatively low, that the Zn concentrations in sediments are often comparatively high, and that zinc does not bioaccumulate or become biomagnified, but in contrast Zn concentrations (like most heavy metals) become lower when one moves from bottom fauna to fish. However, the concentrations are often relatively great in fish liver since the liver and kidneys of fish function as the detoxification centers of the body. In this connection, it is important to find out which organisms are the most sensitive. In order to emphasize the importance of target indicators, Figure 5 shows the animal species which are suitable as such not for metals but for acidifying substances. It is clear that, for example, crustaceans react rapidly to changes in the pH, whereas certain fish such as brook trout and eels do not die until the acidification has proceeded very far.

There are about 50,000 publications dealing with mercury as an environmental pollutant. However, contrary to common belief, Hg does not appear to be a major threat to aquatic life. The fact that it is difficult to find true environmental effects of mercury and of other heavy metals in natural waters depends, for example, on the affinity of metals to become bound to different types of carrier particles (e.g., clay minerals, humus, algae), whereupon the biological uptake of the metals is more difficult. Thus, mercury does not constitute any known threat to the aquatic ecosystem. However, it is certainly a threat to the fetuses of pregnant women if the latter eat Hg-contaminated fish. The effect parameter, which has been used for Hg for many years, is the Hg concentration in 1-kg pike (*Esox lucius*). Pike is a stationary predator eaten by humans. Evidently, this is not an effect parameter in the same way as reproductive disturbance in roach or mortality in crayfish due to acidification (Figure 5). However, it is the effect parameter which is of greatest interest for the case of mercury. This can be established without mesocosm tests; consequently, one can proceed further by pressing the "yes" button (in Figure 2) to the next level, where Hg-pi refers to the Hg concentration in 1-kg pike. This is the third chart (Figure 6).

At this level one must answer a number of obligatory questions. First, the *area and time resolution* of the effect term must be defined, then the *dose and sensitivity parameters*. These are four key concepts which are all explained by the "explain" buttons. The idea behind this is that one and the same dose can cause different effects in lakes with different sensitivities. The sensitivity may be said to regulate the route between dose and effect. However, in order to be able to compare effect, dose, and sensitivity, one must first demonstrate that comparable data are available. Therefore, one

pH (dose) vs Ecological/biological effects

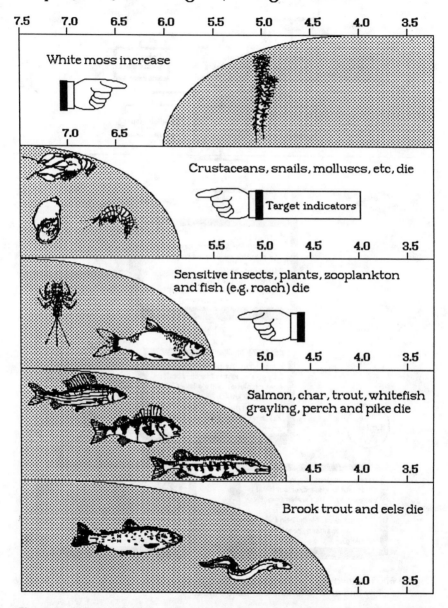

Figure 5. Lake acidification versus biological/ecological effects and illustration of target indicators.

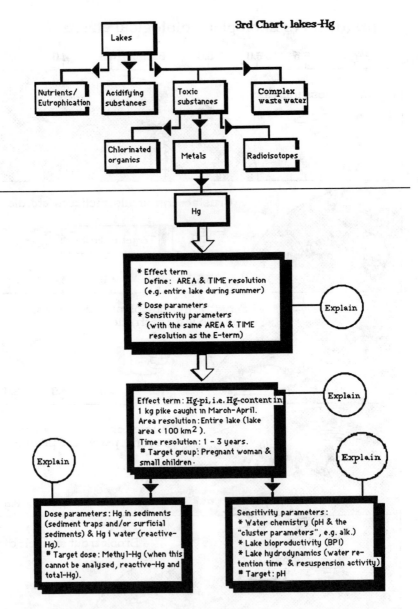

Figure 6.   The third chart in the operative system. Definitions of effect, dose, and sensitivity and area- and time-compatible data.

must first explain in greater detail what is meant by time- and area-compatible data.

3.0 TIME AND AREA RESOLUTION

As an example, the time resolution of the effect parameter Hg-pi will be examined (Figure 7). The Hg-pi is the Hg content in the pike caught during the spawning period in March/April. The value depends on how the pike has lived and what it has eaten during a fairly long period before being caught. The Hg concentration in pike is an integrated value for the entire

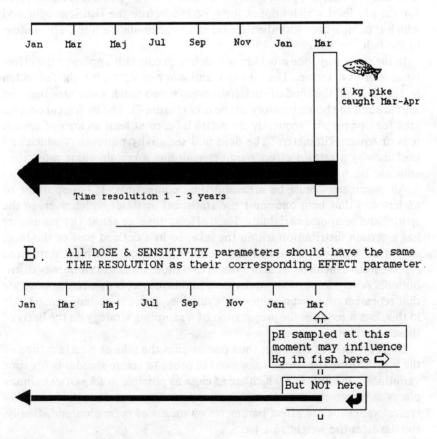

Figure 7. Illustration of time-compatible data.

environment of the pike and its prey. It may be said that the Hg content in pike has a certain half-life; i.e., if an Hg-contaminated pike was placed in a "Hg-free" environment, it would take 1–3 years before the Hg content had decreased to half its original level. That is the time needed to build up a stable Hg content in pike. Thus, the Hg content in pike depends on the Hg dose supplied to the entire environment and not only to the clump of reeds where the pike was caught, as well as to the biological, chemical, and physical conditions in this environment over a long period, since these conditions influence the distribution of the Hg dose on different carrier particles and the bioavailability of the Hg dose. The pH of the water is important for the binding of mercury to different types of carrier particles and for how Hg is distributed among different Hg forms, such as Hg°, Hg^{2+}, and methyl-Hg. It is not, however, the pH of the water when the fish is caught that is of interest but the pH of the water during a long period previously. Thus, one must look for a mean value or a corresponding value for the pH level which applies for a period before the fish is caught and which is comparable with the time taken to accumulate a stable Hg content in the fish.

In the same way, one must know which area constant applies to the effect parameter in question. The pike is a stationary predator, but the fish eaten by the pike and the food of such fish come from a much wider area than the area around the home territory of the pike (Figure 8). The biological contact area for one pike is frequently the entire lake, or at least an area of several tens of square kilometers. The dose and sensitivity parameters which can explain why a certain effect parameter attains a certain value must then emanate from all this area.

All parameters must be area and time compatible. It is important to understand that here one must use statistical methods which average the spatial and temporal variability. Each effect, dose, or sensitivity parameter has a certain distribution within the lake, or in a defined part of the lake, during the defined period of time; in this connection, one must start from representative mean or median values from these statistical frequency distributions. A very important requirement in this respect is then to demonstrate that representative mean values are available. There is nothing remarkable in this, but it requires the preparation of a sampling strategy on the basis of these conditions.

Consider the Hg example; then one catches the pike at a certain time of the year (during the spring spawning) in order to create standards for time variations. One uses fish which are as close as possible to 1 kg from as many places as possible within the lake in question. Then one determines an area-typical value for the effect parameter by means of regression and, finally, standardizes the weight to 1 kg.

Mercury concentrations in water are generally low and expensive to deter-

Area resolution

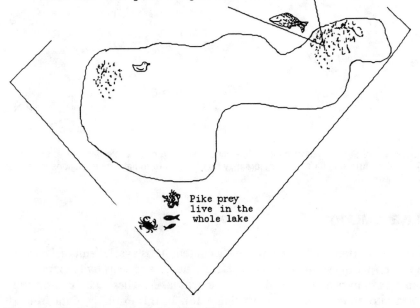

Figure 8. Illustration of area-compatible data.

mine. In order to establish the Hg dose one should, in the ideal situation, place a number of sediment traps at different sites in the lake for 1–3 years and analyze the material collected in the traps for Hg (Figure 9). This may be measured as total Hg or different fractions of Hg, such as methyl-Hg. Naturally, very little of the mercury found in the sediment traps will enter the fish. The mean Hg content from the sediment traps will provide an indirect measure of the Hg load to the lake for the registration period.

In the same way, area- and time-compatible data can be determined for assessory chemical parameters such as pH, alkalinity, hardness, and total P content, which are known to be of importance for how a certain Hg dose is spread in the lake and accumulated by fish (Håkanson *et al.* 1988). Thus, these sensitivity parameters are based on statistical distributions of empirical data. Most of the water-chemical parameters show some degree of intercorrelation. Alkalinity, hardness, and conductivity are, for instance, very closely related. They represent, however, different lake characteristics and are determined using different methods.

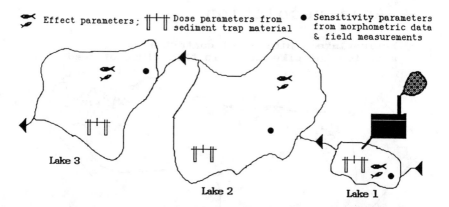

Figure 9. Schematic illustration of how time- and area-compatible data for effect, dose, and sensitivity parameters may be obtained from field investigations in natural lakes.

4.0 ECOMETRY

It is important to know why the Hg content in fish varies among lakes. To answer this question, the actual effect, dose, and sensitivity parameters must be empirically determined for several lakes. These data can be compiled into an *ecometric matrix* (Figure 10). In this example, data from 18 different lakes and/or lake types have been used (Håkanson 1980). The effect parameter is Hg content in 1-kg pike (Hg-pi), the dose parameter is the median Hg content in surface sediments from a number of sampling places in each lake (Hg50), and the sensitivity parameters are lake area, mean lake pH, and the degree of bioproduction in the lakes (BPI). The question now is: How much of the variation among lakes in the effect parameter (from 0.45 to 1.5 in Figure 10) can be explained by the variation in the dose and the sensitivity parameters?

This can be tested with stepwise multivariate regression analysis (Figure 11). The degree of explanation (r^2 = coefficient of determination) is about 37% for pH alone, which, thus, is the most important parameter in this connection (Figure 11A). This is interesting and demonstrates that the two large environmental problems, acidification and contamination, are associated. In acidified lakes with low pH the fish accumulate a greater proportion of a given Hg dose. There are also causal explanations of this statistical relationship, but these will not be discussed here (Björnberg *et al.* 1988). In this ecometric situation, the point is that there are many factors which could potentially have influenced the Hg content in fish. By means of these calculations, one obtains a scientifically relevant possibility of ranking all these factors. This is useful from a practical viewpoint of environmental manage-

A	B	C	D	E	F
A small ecometric matrix					
Lake	Effect	Dose	Sensitivity, W1	Sensitivity, W2	Sensitivity, W3
	Hg-pi, mg/kg ww	Hg50, ng/g ds	Lake area, km2	pH	BPI
Vättern	0.45	80	1856	7.6	2.9
Stora Aspen	0.45	470	5.9	7.2	4.9
Blacken	0.5	260	91	7.3	5
11 Värml.sjöar	0.5	160	1.14	6	3.4
Bysjön	0.6	340	5.1	6.8	4.2
Dalbosjön	0.7	160	2066	7.1	3.1
Leran	0.7	650	28	6.9	5.2
S. Barken	0.7	470	20.5	7	5.1
15 Värml.sjöar	0.7	220	9.38	5.9	3.8
19 Smål.sjöar	0.7	250	0.42	5.7	3.9
7 Värml.sjöar	0.9	220	1.36	5.9	3.5
9 Smål.sjöar	0.9	260	1.07	5.3	3.1
3 Smål.sjaör	1	290	0.23	5.9	2.5
Värml.sjön	1.1	730	3582	7.2	2.9
9 Värml.sjöar	1.1	280	0.38	5.5	3.8
Övre Hillen	1.2	2040	4.4	7.1	4.9
8 Värml.sjöar	1.5	300	1.35	5.1	3.4
5 Smål.sjöar	1.5	300	0.31	4.8	3.3

Effect parameters E1 E2 E3	Dose parameters D1 D2 D3	Sensitivity parameters W1 W2 W3 .. ..
0.1 . .	12 . .	299 . .
0.3 . .	17 . .	281 . .
0.5 . .	21 . .	270 . .
. . .	. . .	. . .
. . .	. . .	. . .
2.6 . .	199 . .	98 . .

The ecometric matrix

Figure 10. The ecometric matrix in two fashions, a small ecometric matrix for mercury in 18 Swedish lakes/lake types and the general setup of an ecometric matrix.

ment but also for research, since it is possible in this way to look for true causal explanations to the statistical relationships in a structured manner.

While pH in this example explains the greatest proportion of the variation in the effect term, other factors are also important. The next most important factor, according to the stepwise regression, is the Hg dose to the lake (Hg50). It is interesting to note that a sensitivity factor can, in fact, be more important than a dose factor in explaining the variation in an effect term in nature. If pH and Hg50 are pooled, 63% of the variation in Hg-pi among lakes can be explained statistically. If one further sensitivity factor is added, the degree of bioproduction (BPI), then the degree of explanation increases to 74%. This is as far as one can proceed with this set of data. The latter link is also interesting, since it demonstrates that the Hg concentration in fish is not only

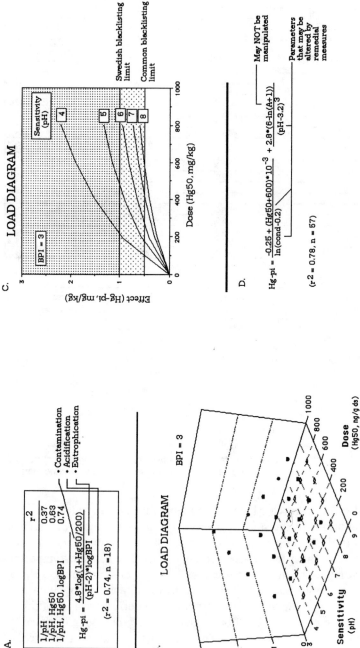

Figure 11. (A) Results from stepwise multiple regression analysis for mercury and illustration of how the three major threats to lake ecosystems interact in one effect-dose-sensitivity model. (B) A load diagram for mercury. (C) Another setup of the same load diagram for mercury. (D) Another empirical effect-dose-sensitivity model for mercury.

increased by acidification (synergism) but also with the third major environmental problem in aquatic systems, namely eutrophication. The greater the BPI the greater the production of algae, plankton, and fish and the greater the amount of biomass in the lake. This implies that a given Hg dose is spread over a larger biomass whereupon the Hg content in the biomass becomes less (antagonism). This is one explanation why the BPI value is such an important factor. Another explanation concerns the internal correlation: If the bioproduction in a lake is increased, then the entire character of the ecosystem is changed, which influences practically all other water chemical and biological factors with which a lake ecosystem can be described.

5.0 LOAD DIAGRAM

5.1 Mercury

The equation that gave the best adaptation to the actual data set (Figure 10) is given in Figure 11A. The formula is illustrated graphically in a 3-D diagram (Figure 11B), which is one example of a *load diagram*. It shows, in this case for a constant BPI value, how the Hg content in fish (effect) is greater when the Hg dose is greater and the pH (sensitivity) is less. On the basis of this relationship one may determine which measures could be manipulated in order to decrease the Hg content in fish. The dose could be decreased, the pH could be increased, and/or the bioproduction could be increased. Naturally, the bioproduction must not be increased indiscriminately in order to reduce the Hg content in fish. This would lead to a eutrophication problem! However, this is not any particular problem since the high Hg contents in fish mainly occur in low-productive (oligotrophic) lakes. In such lakes it may be practical, economical, and ecologically relevant to, for example, lime the lakes in order to increase the pH. This is presently being tested in a major Swedish project. Adding nutrients to the lakes by, for example, starting a fish cage farm where the feed waste from the farm is utilized to increase lake bioproduction is also being tested in Sweden. Naturally, the best way in the long run to reduce Hg concentrations in fish is to stop inputs of Hg.

Since the load diagram is the objective in this operative system, another variant of the load diagram for the given Hg equation is shown in Figure 11C. In this diagram, there are also two practical guidelines for the effect parameter. In Sweden, lakes are "blacklisted" if the Hg concentration in 1-kg pike exceeds 1 mg Hg/kg. Fish from such lakes may not be marketed or even given away. In many countries a lower guideline of 0.5 mg/kg wet weight is used.

It is important to emphasize that this equation is based on a given empirical data set. This data set must be accurately reported, and the relationship only applies under the conditions thereby defined. The actual equation in Figure 11A is based on a manipulated data set (Håkanson 1980) and is only included here to demonstrate the central principles in this operative system. There are other possible equations for predicting the Hg content in fish (Håkanson et al. 1988). One such empirical relationship is given in Figure 11D. There are no BPI values for these lakes. This equation is based on the same two primary parameters: Hg dose (Hg50) and the sensitivity parameter pH. In addition, the conductivity of the lakes and the lake area are included. The greater the conductivity the smaller the Hg concentration in fish. The greater the surface area the smaller the Hg concentration in fish. This relationship has been included here to demonstrate that different kinds of equations can be obtained when different types of empirical data are used as well as to demonstrate that the parameters included may be of different character with regard to the remedial measures to be introduced: The area of the lakes cannot, for example, for all practical purposes be manipulated by humans, whereas pH, conductivity, and total P can be influenced by liming and fertilizing. The arguments for liming lakes, not only to reduce acidification damage (Figure 5) but also to reduce the Hg content in fish, are strengthened by the results summarized in the formula in Figure 11D since a liming will directly influence not only pH but also conductivity.

Thus, in this example, we have proceeded along the entire pathway in the operative system, from pressing the "lake" button to the load diagram for Hg. Naturally, there are many comments to be made in connection with the different steps and the details in each step, but these comments must wait until another occasion. The aim of the present paper is primarily to describe the operative system and its vital parts. Is it possible to apply the same reasoning used for Hg to other substances?

5.2 Radioactive Cesium

The Chernobyl accident occurred in April 1986. As a result of unfavorable wind/weather conditions, Sweden received a large fallout of radiation. How is the radioactive cesium transported in soil, water, and biological material? Which lakes are particularly sensitive? Which measures should be introduced to speed up the recovery? Which effect parameters should be used? What does a load diagram look like for radioactive cesium in lakes? These are typical questions which should be, and have been, asked in regard to radioactive cesium from Chernobyl. Several of these questions can now also be answered using this type of environmental consequence analysis.

The analysis is started by pressing the "lake" button in the Basecard (Figure 1) and arriving at Figure 12. By pressing the button again for toxic substances and the button for radioactive substances one arrives at the button for Cs-137. If one presses this button, the same types of basic questions concerning the choice of effect, dose, and sensitivity parameters and their time and area compatibility are displayed. For Cs-137 in lakes, as for Hg, there are no generally accepted ecological/biological effect parame-

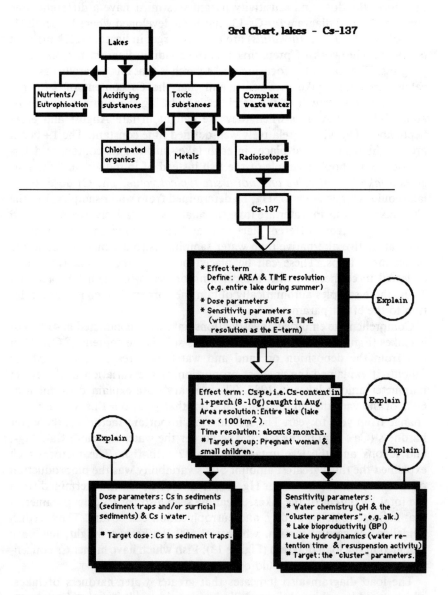

Figure 12. The third chart in the operative system for radioactive cesium.

ters. The threat does not appear to be directed against life in the lakes but is mainly directed at humans, primarily at the fetuses of pregnant women and at small children. Adult men and women are thus not the primary target group. This means that also for Cs-137 one should focus on fish as effect parameters. Radioactive cesium has a shorter half-life than mercury in fish; therefore, the dose and sensitivity parameters must have a different time constant. A load diagram for Cs-137 has been developed where 1+ perch (1-year-old perch, *Perca fluviatilis*) are used as target indicators (Håkanson *et al.* 1989). The work of preparing a corresponding diagram for pike is in progress. The reason for focusing on 1+ perch is that it is generally easier to catch perch than pike in Sweden and that the Cs content of 1+ perch depends on the growth of the perch mainly during the summer months of June, July, and August when they are caught in late August and early September. This gives a relatively well-defined time constant. The 1+ perch are caught from several places in each lake, and the Cs content is determined from a pooled sample of the fish instead of in individual fish. This gives a *lake-typical value and not a site-typical value*. The Cs dose to the lake could, as was done for Hg, be determined from water samples, but the Cs concentrations in water are low, the analysis is relatively expensive, and many samples from different places and at different times must be taken. A more attractive alternative than water sampling is to use material collected in sediment traps. These can be placed out after the spring flood and collected in connection with the catching of perch in August/September. Thus, these samples automatically obtain the correct time constant in relation to the effect parameter.

Comprehensive empirical investigations have been conducted in 41 Swedish lakes (Figure 13). The primary dose (Cs-soil) is the content of Cs-137 in soil from the deposition on land and water as a result of atmospheric fallout. It explained the greatest proportion of the variation in the effect parameter among lakes in 1987. The primary dose explained as much as 69% of the variation in the material from that year, but this will certainly change from year to year. The second most important factor was the water hardness (Ca + Mg) in the lakes: The harder the water the more Ca-, Mg-, and K-ions and the less uptake of Cs-137 into fish. The factor which explained the third greatest amount of Cs variability was the bioproduction of the lakes, just as it was for Hg. In the case of Cs it was determined from the total P content of the lakes. It can be seen also for Cs how parameters that indicate toxicity, acidity, and eutrophication can covary. The Swedish guideline for cesium in fish, which is 1.500 Bq/kg wet weight, has been included in the load diagram (Figure 13). Fish which have higher Cs concentrations than this value should only be eaten restrictively.

The load diagram also indicates that greater water hardness of lakes, which could be achieved through liming, would result in lower Cs concen-

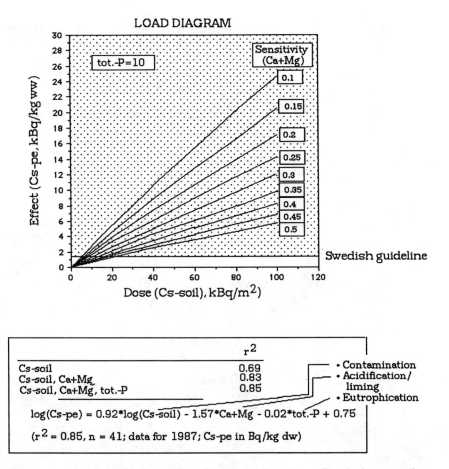

Figure 13. Upper: A load diagram for radioactive cesium. Lower: Results from stepwise multiple regression analysis for cesium and illustration of one effect-dose-sensitivity model.

trations in fish. In this case, liming and fertilizing would probably be effective remedial actions for excessive concentrations of Cs in fish. If one had a load diagram, the remediation could be adapted to the specific requirements of the lake.

5.3 Phosphorus

Eutrophication, acidification, and contamination are the three major problem areas within aquatic environmental management. The eutrophication debate was intensive in the late 1960s and then Vollenweider (1968) developed a load diagram for phosphorus in lakes. This diagram is based on a combination of mass balance calculations and empirics/statistics (Figure

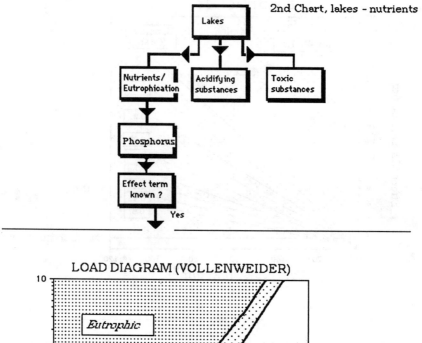

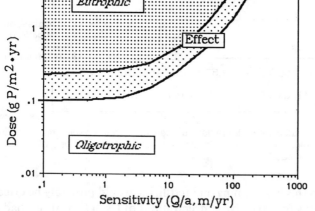

Figure 14. The second chart in the operative system for phosphorus in lakes and the load diagram for lake eutrophication.

14). In the operative system, one reaches this load diagram by first pressing the "lake" button (Figure 1), then the "nutrient/eutrophication" button, and finally the "phosphorus" button. When one reaches the question about effect terms, the answer is known in this case and does not need to be found by means of mesocosm tests. The relationship between the nutrient dose of phosphorus and eutrophication effect with regard to increased trophic

Table 1. Characteristic Features in Lakes of Various Trophic Level

Trophic level	Primary prod. (g C/m^2·year)	Secchi d. transp. (m)	Chloro-phyll a[a] (mg/m^3)	Algal volume[a] (g/m^3)	Total P[b] (mg/m^3)	Total N[b] (mg/m^3)	Dominant fish[c]
Oligotrophic	<30	>5	<2	<0.8	<5	<300	Trout, WF
Mesotrophic	25–60	3–6	2–8	0.5–1.9	5–20	300–500	WF, Perch
Eutrophic	40–200	1–4	6–35	1.2–2.5	20–100	350–600	Perch, roach
Hypertrophic	130–600	0–2	30–400	2.1–20	>100	>1000	Roach, bream

[a]Mean value for the growing period (May–Oct.).
[b]Mean value for the spring circulation.
[c]WF = white fish.

level, increased primary production, decreased water transparency, and characteristic fish population is given in Table 1.

From the load diagram one can calculate how much the phosphorus dose to the lake must be reduced for the conditions to change from eutrophic (nutrient-rich, overfertilized) to oligotrophic (nutrient-poor). This depends on the sensitivity of the lake to nutrients, which depends, among other things, on the water turnover in the lake. In general the Vollenweider diagram gives a good environmental description in relatively low-productive lakes but a poorer description in more productive systems. This is probably due to the fact that the load diagram in this form does not take the internal nutrient load into consideration. The flow of nutrients from the sediments can be considerable, particularly in shallow lakes with a high resuspension activity. Such lakes are generally eutrophic and this internal loading can be important (see Chapter 5 in this volume).

5.4 Chlorinated Organic Compounds

Discharge of chlorinated organic materials from forest industries has attracted a great deal of attention during recent years, especially in Sweden. This is a complex group of substances which includes some of the most well-known and discussed environmental toxins such as dioxins and probably a considerable number of substances about which little is known today as regards their properties and environmental effects. Environmental consequence analysis for natural aquatic ecosystems has not been done for chlorinated organic materials. Lacking such an analysis there is ample space for comments and speculations about which ecological effects occur as a result of given discharges and given substances and what can be linked to other causes.

If one returns to Figure 1 and presses the button for "marine systems", then Figure 15 comes up. It gives one example of how to structure the very complex group of chlorinated organic material. There are several levels here, where TOCl stands for total chlorinated organic material, EOCl for extractable organically bound chlorine, EPOCl for extractable persistent

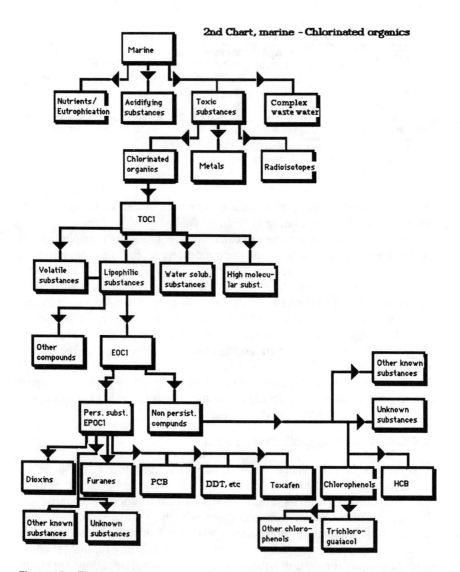

Figure 15. The second chart in the operative system for chlorinated organic compounds.

organically bound chlorine, etc. Collective parameters such as TOCl and EOCl can be characterized by, for instance, their ability to bioaccumulate, their degree of lipophilicity, their particle affinity, and their molecular weight. Such general methods may also be used to characterize different types of wastewater under that button.

The steps in the environmental consequence analysis start from the chosen effect parameters and what they represent. All dose and sensitivity parameters must be adjusted to the time and area resolution valid for the effect parameters. If, for example, one chooses perch as the biological indicator organism (Figure 16) and then, for example, uses the EOCl content in perch muscle, the liver somatic index (LSI), or physiological parameters such as cytochrome P-54 activity in liver as effect parameters, it is important to know the time and area resolution of these effect parameters for perch.

It is also important to note that one must be able to determine all parameters of interest in a scientifically relevant and, preferably, simple manner.

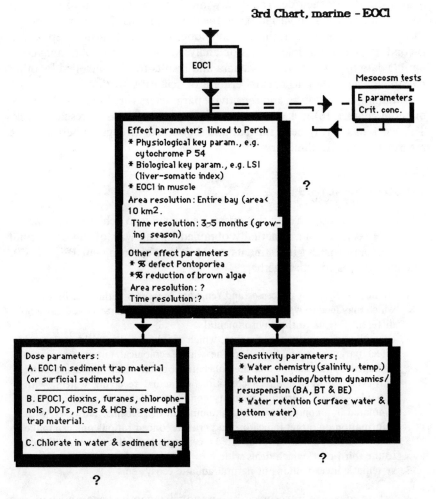

Figure 16. The third chart in the operative system for EOCl and examples of possible effect, dose, and sensitivity parameters.

We know by experience that many of these sensitivity parameters, such as water temperature, may vary considerably within a coastal area and with time. One single value then has little representativity and gives a low degree of explanation if it is linked to an effect parameter. This implies that field measurements should be done during different wind/weather conditions in each area.

When all field measurements have been carried out, and it has been demonstrated that area- and time-compatible effect, dose, and sensitivity parameters are available, these values can be entered into the ecometric matrix (Figure 10). Using known statistical methods of multiple regression analysis, one can then determine how much of the variation in the effect parameters depends on variations in the dose and sensitivity parameters. It should be emphasized that the difficulty does *not* lie in the final step, which is conducting the multiple regression analysis, but rather in obtaining compatible data in a way that allows data and results to be controlled by other researchers according to normal criteria of scientific work.

Obtaining effect, dose, and sensitivity parameters for chlorinated organic material (Figure 16) is a very important and comprehensive research task. This work has hardly been started, and the information given here must be regarded as very preliminary.

6.0 COST ANALYSIS

An important objective in this operative system is to achieve a *cost analysis*, to determine what is actually obtained in the way of environmental improvements for the investments made in the environment (Figure 17). The intention is to link together:

1. studies of different processes and remedial measures within the industry, which may lead to wastewaters of different chemical character and causing different threats to the environmental
2. controlled mesocosm experiments, which are designed to yield ecological effect parameters and to determine which ecological target groups are damaged by discharges and the concentrations at which damage occurs. However, the very complex conditions in actual receiving waters, with variable weather, currents, visibility conditions, light, etc., can never be simulated in a completely relevant manner in mesocosms. However, with information on target indicators and critical concentrations one can make field studies in a number of selected receiving waters. This, then, is linked to the third part, which deals with
3. systematic investigations of natural aquatic ecosystems

A certain measure introduced by an industry can be accomplished at a certain cost (x1 $). This reduces the discharge by y1 ($/year). The environ-

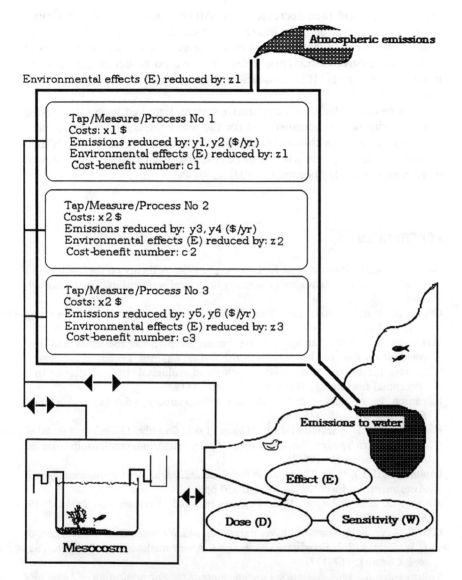

Figure 17. Schematic illustration of the operative system in practical use where various remedial measures/processes in an industry are tested in a strict scientific manner to optimize the economic investments and minimize the ecological threats. The most accurate target indicators are determined from mesocosm tests and used in the natural systems together with dose and sensitivity parameters so that real, ecological effects in natural systems may be linked to the various measures in the industry.

mental effects will then decrease by $z1$. All this gives a cost-benefit number of $c1$. It would be most interesting to conduct analyses of this kind, since it would imply that inputs of money were invested where they would achieve the best environmental effect. This would be beneficial for both industry and society. It is remarkable that so little has been done in this connection.

It can be concluded that the operative system described herein is intended not primarily for the scientists but for the water managers to create more order and structure within a very complicated sector. It is hoped that a final product will be a system which can be continuously extended and renewed when new knowledge becomes available.

REFERENCES

Björnberg, A., L. Håkanson, & K. Lundbergh, 1988. A theory on the mechanisms regulating the bioavailability of mercury in natural waters. *Environ. Pollut.* 49: 53–61.

Cairns, Jr., J. & J.R. Pratt, 1987. Ecotoxicological effect indices: A rapidly evolving system. *Water Sci. Technol.* 19: 1–12.

Håkanson, L., 1980. The quantitative impact of pH, bioproduction and Hg-contamination on the Hg-content in fish (pike). *Environ. Pollut.* 1: 285–304.

Håkanson, L., 1984. Aquatic contamination and ecological risk. An attempt to a conceptual framework. *Water Res.* 18: 1107–1118.

Håkanson, L., Å. Nilsson, & T. Andersson, 1988. Mercury in fish in Swedish lakes. *Environ. Pollut.* 48: 145–162.

Håkanson, L., T. Andersson, & Å. Nilsson, 1989. Caesium-137 in perch in lakes from northern Sweden after Chernobyl—Present situation, relationships, trends. *Environ. Pollut.* 58: 195–212.

Landner, L. (Ed.), 1989. *Chemicals in the Aquatic Environment. Advanced Hazard Assessment.* Springer-Verlag, Berlin: 350 pp.

Mackey, D. & S. Paterson, 1982. Fugacity revisited. *Environ. Sci. Technol.* 16: 654–660.

O'Neill, R.V., R.H. Gardner, L.W. Barnthouse, G.W. Suter, S.G. Hildebrand, & C.W. Gehrs, 1982. Ecosystem risk analysis: A new methodology. *Environ. Toxicol. Chem.* 1: 167–177.

Vollenweider, R., 1968. *Scientific Fundamentals of the Eutrophication of Lakes and Flowing Waters, with Particular Reference to Nitrogen and Phosphorus as Factors in Eutrophication.* Rep. OECD/DAS/SCI/68.27. Organization for Economic Cooperation and Development, Paris: 192 pp.

List of Authors

Donald D. Adams, Center for Earth and Environmental Science, State University of New York, Plattsburgh, New York 12901, U.S.A.

Renato Baudo, C.N.R. Istituto Italiano di Idrobiologia, Largo Tonolli 50/52, I-28048 – Pallanza (NO), ITALY

Alcide Calderoni, C.N.R. Istituto Italiano di Idrobiologia, Largo Tonolli 50/52, I-28048 – Pallanza (NO), ITALY

Richard G. Carlton, W.K. Kellogg Biological Station, Michigan State University, 3700 East Gull Lake Road, Hickory Corners, Michigan 49060, U.S.A.

Frank M. D'Itri, Institute of Water Research and Department of Fisheries and Wildlife, Michigan State University, East Lansing, Michigan 48824–1222, U.S.A.

Nicholas J. Fendinger, U.S. Department of Agriculture – ARS, Environmental Chemistry Lab, Beltsville, Maryland 20705, U.S.A.

Ulrich Förstner, Technische Universität Hamburg-Harburg, Eissendorferstr. 40, D-2100 Hamburg 90, WEST GERMANY

René Gächter, Lake Research Laboratory, EAWAG/ETH, CH-6047 Kastanienbaum, SWITZERLAND

John P. Giesy, Department of Fisheries and Wildlife, Pesticide Research Center, Center for Environmental Toxicology, Michigan State University, East Lansing, Michigan 48824–1222, U.S.A.

Dwight E. Glotfelty, U.S. Department of Agriculture – ARS, Environmental Chemistry Lab, Beltsville, Maryland 20705, U.S.A.

Lars Håkanson, Department of Hydrology, University of Uppsala, V. Ågatan 24, S-75220 Uppsala, SWEDEN

Robert A. Hoke, Department of Fisheries and Wildlife, Center for Environmental Toxicology, Michigan State University, East Lansing, Michigan 48824–1222, U.S.A.

Michael J. Klug, W.K. Kellogg Biological Station, Michigan State University, 3700 East Gull Lake Road, Hickory Corners, Michigan 49060, U.S.A.

Peter F. Landrum, Great Lakes Environmental Research Laboratory, NOAA, 2205 Commonwealth Blvd., Ann Arbor, Michigan 48103, U.S.A.

Rosario Mosello, C.N.R. Istituto Italiano di Idrobiologia, Largo Tonolli 50/52, I-28048 — Pallanza (NO), ITALY

Herbert Muntau, CCR Ispra, Environmental Institute, I-21020 Ispra, ITALY

Joseph S. Meyer, Lake Research Laboratory, EAWAG/ETH, CH-6047 Kastanienbaum, SWITZERLAND

John A. Robbins, Great Lakes Environmental Research Laboratory/ NOAA, 2205 Commonwealth Blvd., Ann Arbor, Michigan 48103, U.S.A.

Index